FIRST PEOPLES IN A NEW WORLD

FIRST PEOPLES
IN A NEW WORLD

Colonizing Ice Age America

David J. Meltzer

UNIVERSITY OF CALIFORNIA PRESS
Berkeley   Los Angeles   London

THE PUBLISHER GRATEFULLY ACKNOWLEDGES THE GENEROUS SUPPORT
OF THE GENERAL ENDOWMENT FUND OF THE UNIVERSITY OF
CALIFORNIA PRESS FOUNDATION

University of California Press, one of the most distinguished university
presses in the United States, enriches lives around the world by advancing
scholarship in the humanities, social sciences, and natural sciences. Its
activities are supported by the UC Press Foundation and by philanthropic
contributions from individuals and institutions. For more information, visit
www.ucpress.edu.

University of California Press
Berkeley and Los Angeles, California

University of California Press, Ltd.
London, England

Library of Congress Cataloging-in-Publication Data

Meltzer, David J.
First peoples in a new world : colonizing ice age America / David J. Meltzer.
        p. cm.
    Includes bibliographical references and index.
    ISBN 978-0-520-25052-9 (cloth : alk. paper).
    1. Paleo-Indians—North America. 2. Glacial epoch—North America.
    3. North America—Antiquities. I. Title.
    377.9.M45   2009
    970.01—dc22   2008035901

Manufactured in the United States of America
16 15 14 13 12 11 10
10 9 8 7 6 5 4 3 2

The paper used in this publication meets the minimum requirements of
ANSI/NISO Z39.48-1992 (R 1997) (*Permanence of Paper*).

Cover illustration: Above: A biface from the deGraffenreid Clovis cache,
Texas. (Photograph courtesy of J. David Kilby, used with permission of
Mark Mullins.) Below: The Meander glacier flowing out of the Borchgrevink
Mountains toward the Ross Sea, northern Victoria Land, Antarctica.
(Photograph courtesy of Steve Emslie.)

To Florence Garner Meltzer,
who arranged for her fifteen-year-old son to
spend the summer excavating at a Paleoindian site,
and thereby started him on a career

CONTENTS

Plates appear following page 238

PREFACE

An *Albuquerque Journal* reporter was on the phone. "Have you heard of the recent discoveries at Pendejo Cave here in New Mexico?" he asked and then added, laughing, "Do you know what 'Pendejo' means in Spanish? Our readers sure do!" I had heard. I did know. And what I said next—foolishly, in retrospect—nearly got me pummeled one night in a hotel bar in Brazil by Scotty MacNeish: excavator of Pendejo Cave, grand old man of archaeology . . . and former Golden Gloves boxing champion.

The fight was about a discovery as profound—or trivial—as fingerprints. Not just any fingerprints: MacNeish came out of Pendejo Cave and announced he'd found *human* fingerprints that were upwards of 37,000 years old, instantly tripling the then oldest accepted antiquity for the arrival of humans in the New World (the Clovis archaeological presence, dated to nearly 11,500 years ago). When the reporter asked what I thought of MacNeish's claim, I replied, "You're not going to convince me until you've fingerprinted the crew."

Granted, it was a flip response. But I thought the point reasonable. To persuade an extremely skeptical archaeological community to accept this unparalleled discovery, MacNeish would have to demonstrate those fingerprints were just as old as advertised, and not odd clay globs his excavators had inadvertently imprinted and later mistook for archaeological specimens. Extraordinary claims require extraordinary proof. I thought I was being helpful. MacNeish thought otherwise. It surely didn't help that my response led to his Pendejo Cave claim being named one of that year's *Albuquerque Journal* Cowchip Award winners (I don't think I need to explain why the Cowchip is not a coveted award).

When we bumped into each other in that bar a year later—ironically, we'd both been invited to Brazil to examine another purportedly ancient site—MacNeish swore furiously at me for accusing his crew of faking evidence. Faking evidence!? Only after fifteen minutes of very fast talking, spent just beyond the distance I guessed he could still throw a punch at age seventy-five, was I able to convince him that wasn't my point at all. I am not sure he ever believed me. I know he never forgave me. But he did send me a reprint of the article he published on the Pendejo Cave fingerprints, and even auto-graphed it: "Finally got it published," he scrawled across the top, "*in spite of you.*"

Fair enough.

I framed the reprint, and it's prominently displayed in my office—perfect witness to the heat that's generated in the search for America's first peoples.

Not that this is anything new. Questions about the origins, antiquity, and adapta-tions of the first Americans, although easily asked, have proven extraordinarily difficult to answer, and have been contentious since first posed in modern form in the 1860s. Those questions are still the focus of research today, albeit using vastly different theo-retical, analytical, and archaeological tools; involving a far wider range of contributing disciplines; and producing a stream of publications that in the last several decades has become a raging academic torrent. The intervening century and a half has witnessed multiple site discoveries, conceptual breakthroughs, pivotal moments that have pro-pelled and guided research, and cycles of bitter controversy and grudging, short-lived periods of peace. We've learned a great deal.

Still, in just the last dozen years much of what we knew—or thought we knew— about the peopling of the Americas has been turned on its head by new discoveries, new analyses, and new controversies, all of which cut across multiple disciplinary lines. The biggest difference? Before, we spoke of the possibility of a pre-Clovis presence in the New World in hypothetical terms; now it is a reality, and it's a whole new archaeological world as a result. In the scramble to right ourselves, many ideas—some controversial, others outlandish—are being tried on for size. It's the natural course of affairs in scien-tific change, and no cause for alarm. Yet.

So much has changed that my previous book on the topic, *Search for the First Americans,* published in 1993, is now woefully out of date (more embarrassing: used copies are now sell-ing for one dollar on the Web, and the press that published it has folded; these two facts, I choose to believe, are unrelated). At the time I wrote *Search for the First Americans,* geneticists were only just beginning to peer into corners of the human genome to use DNA to trace our collective ancestry; the excavation and analysis of the Monte Verde site in Chile—then one of several candidates for great antiquity in the New World, and not the first among all—was just being wrapped up; Pleistocene geologists had only glimpsed the complexity of the causes as well as the climatic and ecological consequences of the Ice Age (Pleistocene), especially the frenetic changes at its end, which were occurring just as the first Americans were radiating out across the continent; and the now infamous Kennewick skeleton still lay buried in the banks of the Columbia River, yet to make its *60 Minutes* debut with Leslie Stahl or become the centerpiece of a costly lawsuit that exposed deep rifts within the archaeological community, and especially between those who study the past and those—Native Americans—who are its living descendants. These and a gaggle of other developments have wrought a sea change in our approach to and understanding of the first Americans. It's time for a fresh look.

This book was originally intended as a second edition of *Search for the First Americans*, but my attempt to gently insert new material, delete stale parts, and patch up original but still-serviceable bits proved impossible, and I soon gave up the effort. I had underestimated just how much had changed—including my own thinking on many of these matters. So I instead tore down each of the original chapters to their foundation timbers, discarded unwanted parts (and even one unwanted chapter), added several new chapters, and then rebuilt the whole from the ground up to reflect all the changes in evidence, emphasis, and thinking. The basic framework remains, and some of the load-bearing elements of a chapter, if judged sufficiently robust, were allowed to stay, along with stories that were just too good not to retell. Much of the new material is based on articles I've published in recent years in a variety of scholarly journals and books, and road tested in my classes, so it has been through the wringer of peer review from colleagues and— perhaps a more stringent test—has had to pass muster with my students, undergraduate and graduate alike. The result is the book before you: more than twice as long, far wider in range, more detailed in coverage than *Search for the First Americans*, and because it supersedes and replaces the earlier volume, deserving of a new title: *First Peoples in a New World: Colonizing Ice Age America*.

WHAT THIS VOLUME IS . . .

First Peoples in a New World is my effort to explain the twists and turns of the search for the first Americans, the controversy that has long enveloped it, and what we've learned of who they were, when and from where they came, and how they colonized what was then, truly, a New World. Although I am an archaeologist, I am by nature eclectic in my approach to scientific problems, and have spent a fair amount of time on search and seizure missions behind interdisciplinary lines looking for help from geneticists, geologists, linguists, and physical anthropologists in answering stubborn archaeological questions. And there are few questions more stubborn or that lend themselves so readily to an interdisciplinary solution as the peopling of the Americas. *First Peoples in a New World* is thus not just a synthesis of the intellectual history and current state of the archaeological understanding of the peopling of the Americas, it's also a close look at the evidence being brought to bear by non-archaeologists to this problem—and an effort to see whether we can all get along.

In fact, this book centers around two interlocking themes. The first is *what* we know of the first Americans—about who they were, where they came from, when we think they arrived, how many early migratory pulses there may have been, and by which route(s) they came to the Americas. It's also about the climatic and ecological conditions of the

Ice Age terrain they traveled and the diverse landscapes they encountered, their adaptive responses to the challenges of colonizing an uninhabited and unfamiliar world, the speed with which these pedestrian hunter-gatherers moved across their new world, their effect on the native animals of the Americas—and whether they had a hand in the extinction of some thirty-five genera of Pleistocene mammals. Finally, it is about the evolutionary processes and pathways they blazed, and the long-term consequences of their prehistory.

The other theme is about *how* we know what we know about the first Americans. It is about the methods archaeologists, geologists, linguists, physical anthropologists, and geneticists are bringing to bear on the problem of origins, antiquity, and adaptations (my non-archaeological colleagues will, I hope, forgive my trespassing). Because these approaches yield very different kinds of evidence, they are not easily reconciled, nor do the practitioners in each field sing in harmony. Hence, it is important to understand how they (we) arrive at our conclusions, and just how reliable those conclusions might be.

Admittedly, talking about how we know what we know is not nearly as satisfying as talking about what we know, but it's vital all the same, especially in light of how this topic is often portrayed in the popular media. Our contentiousness encourages journalists, science writers, and filmmakers to pitch a story of the peopling of the Americas around colorful characters, raging controversy, and outrageous theories—we don't lack for any of these—especially if it involves that hackneyed theme of an iconoclastic scholarly David fighting the establishment Goliath to prove the revolutionary idea that (fill in the blank) proves everything science has ever thought was wrong. The headlines fairly leap from page and screen: "American Indians were not the first ones here!" "Siberian hunter claims extinct Ice Age bears still alive!" And, for conspiracy buffs, "The suppressed story of the people who discovered the New World." One doesn't have to make these up: the last is the subtitle of a just-released book.

Those of us in the business are not without sin. We feed the beast, holding press conferences to announce the discovery of the (latest) oldest site in the Americas, make claims on camera that would never pass muster in the professional journals, or give flip comments to reporters about our colleagues' discoveries (like, say, "Fingerprint the crew") that stoke the fires of controversy. Indeed, as I was writing this book, a group of geologists and archaeologists launched a press campaign proclaiming that a comet blasted the earth in the late Pleistocene, an unwelcome ET that wreaked havoc on global climates, destroyed North America's megafauna, and devastated Paleoindian populations. They might be right, but it's customary in science to build the case and publish the evidence *before* issuing the press release about the conclusions.

However entertaining the often-gossipy popular accounts of this controversy—especially when it's not your ox being gored—they rarely provide accurate or complete details of the science behind it all, or its results. Having long been a participant in the pre-Clovis controversy, and particularly its tipping point at Monte Verde, I can easily see that commentary on it by individuals who view it from the outside bears only a passing resemblance to what I saw actually happening. In fact, not only do these err on what happened publicly, but they also naturally miss much of what went on behind the scenes, and those who put dialogue in our mouths to recreate events are usually completely wrong.

Moreover, there is the inescapable fact that beneath all the tabloid talk, there are legitimate scientific and substantive reasons why we disagree about issues, why the same archaeological (or linguistic, or genetic, or skeletal) evidence can and often is viewed very differently by different investigators, and why there is ambiguity and disputed interpretation. Challenging though it may be at times, to truly understand the peopling of the Americas requires probing deeply into how this knowledge is created, shaped, and put to use. Only with that understanding is it possible to appreciate, despite evidence converging from so many diverse fields, why questions about the first Americans are among the most contentious in anthropology, and may remain so.

Finally, and though it may go without saying, I confess I am not without sin. My voice has long contributed to the din over the origins and antiquity and adaptations of the first Americans, and I have been directly involved in disputes over key pre-Clovis sites, in contesting the claim the Americas were colonized from Ice Age Europe, in debating the role of Paleoindians in the extinction of the Pleistocene megafauna, and in seeking to understand how hunter-gatherers met the challenges of moving across and adapting to the vast, unknown, diverse, and changing landscapes of Pleistocene North America.

I have had my own ox gored.

I will nonetheless do my best to present the different sides of a disputed issue, but the reader is forewarned. *Caveat lector.*

. . . AND WHAT THIS VOLUME IS NOT

This book is not about the Ice Age peopling or Paleoindian archaeology of the entire New World: it's mostly about *North* America. This is so for several reasons, not least that Pleistocene glaciation, climates, and environments play out in very different ways in the Northern and Southern hemispheres, and so, too, the archaeological records are dissimilar. Even using the term pre-Clovis in South America is a misnomer since Clovis fluted points, strictly speaking, only reach as far south as Panama. Covering the entire hemisphere would double the size of an already large book, and is unnecessary in any case since there are several volumes that ably cover the South American ground, leaving me free to concentrate on North America, which is the region of my own archaeological field research and expertise.

That said, the South American record is not ignored. I examine hemisphere-wide evidence from language, teeth, genes, and crania relevant to questions of the peopling of North America, as well as the South American sites that figure prominently in the pre-Clovis debate. The latter are archaeologically relevant since the ancestors of the first South Americans must have come via North America; we haven't a shred of evidence to indicate South America was peopled directly by ocean crossing. If the oldest accepted sites in the New World are in the Southern Hemisphere—as is the case at the moment—then there must be ones older still in the Northern Hemisphere. Only, we've not found them yet—or at least not agreed we've found them.

I wrote this book for the general reader and not my archaeological colleagues, who've perhaps heard quite enough from me on this subject already. The difference is largely a matter of style rather than substance, but also of coverage. The constraints of space and the demands of the narrative forbade me from mentioning every important site, researcher, argument, or claim (sometimes, I confess, I was glad of it). Accordingly, rather than provide encyclopedic, site-by-site lists of what was found where—the sort of thing only an archaeologist could love—I instead highlight finds that help illustrate broad archaeological patterns and adaptive processes. I provide details as needed, but there's too good a story to be told here to become bogged down in archaeological minutiae. To further ease the narrative for my intended readers, I have gone against all my scholarly instincts and omitted citations from the text. But I cannot fully shed my obligation to give credit (or blame) for ideas and discoveries. Thus, I have embedded endnotes throughout the book that provide citations to source material, along with occasional follow-up comments.[1]

Many voices will be heard here, save for an obvious one: those of the *descendants* of the first peoples, American Indians. I do not omit discussion of their traditional origin narratives out of either disinterest or disdain, or because I think American Indians are unrelated to the first peoples in America. I don't. Rather, it is because my expertise lies elsewhere. Even so, I am acutely aware that questions of the Pleistocene peopling of the Americas bear on contemporary issues of Native American identity and ancestry, and of "ownership" of the past and present. It's not a rhetorical question to ask, as Vine Deloria has, if American Indians had "barely unpacked before Columbus came knocking on the door," will people doubt their claims to the land and its resources? And I am sympathetic to the anger provoked among Native groups by speculations by some archaeologists and physical anthropologists that the Americas were originally peopled from Europe, or not by ancestral Native Americans. Such claims cannot be made lightly nor without unimpeachable evidence, though as will be seen, they have been.

I also recognize that Native American views of their origins are not always consonant with those of archaeology. In some cases—as, for example, Deloria's piercing *Red Earth, White Lies*—they furiously condemn it. There are archaeologists who agree: we need to downplay "solid archaeological dogma such as the Bering Land Bridge migration route to the Americas," they say.[2] Here's my view: the past is large enough to accommodate many different uses (as Robert McGhee put it), and I am content to co-exist.

But more important, I won't be shy about casting a critical scientific eye on what archaeologists and anthropologists know and don't know about the peopling of the Americas. After all, it's only dogma if it's left unexamined. That won't happen here.

Finally, a comment on my use of the terms *colonization* and *New World*. I well understand the baggage that comes with both: colonization conjures painful images of the displacement and destruction of indigenous peoples and culture in America after 1492. The word itself is rooted even deeper, in the settlements established in territories conquered by the legions of the Roman Empire (from the Latin *coloniae*[3]). However, in the 2,000 years since the Romans, and in the centuries since European global expansion, "colonize" has acquired a much broader and more neutral meaning in the sciences to refer to the dispersal of a population or species, and its settlement in a different place. It is in this unencumbered ecological sense that the word is used here, and often interchangeably with *peopling* or *migration*.

The term *New World* was, of course, one applied by Europeans to the Americas.[4] At the tail end of the fifteenth century, the American continent was new. To them. It was hardly new to the Native Americans who were here to greet them, for they were the descendants of peoples who had been living here for millennia. Yet, to speak of the Ice Age colonization of the New World is unquestionably appropriate in this context, for when the first people reached America more than 12,500 years ago, this truly *was* a New World. In fact, as we shall see, Ice Age America was new in more ways than just a world uninhabited.

ACKNOWLEDGMENTS

Blake Edgar, acquisitions editor for the University of California Press, has my thanks for asking—just as my previous volume with the University of California Press (*Folsom*, 2006) was rolling off the presses—whether I had another book in the offing. Here 'tis.

Blake, as well as Wendy Ashmore, Tom Dillehay, Donald Grayson, Torrey Rick, and David Hurst Thomas read the entire manuscript, and provided detailed and helpful comments. Daniel Mann and Stephen Zegura, geologist and geneticist, respectively, provided constructive, detailed critiques of chapters 2 (Mann), and 5 and 6 (Zegura), thereby saving me possible embarrassment in venturing into disciplines not my own. I am extremely grateful for the help of all, and hereby absolve them of blame for any lingering errors.

As with all my work, I've been able to rely on friends and colleagues who've kindly answered questions (or asked pointed ones), given advice, supplied references or unpublished manuscripts, or otherwise rendered aid along the way. For that I'd like to thank Michael Adler, Tony Baker, Douglas Bamforth, Lewis Binford, Deborah Bolnick, Michael Cannon, Tom Dillehay, Arthur Dyke, Sunday Eiselt, Michael Gramly, Donald Grayson, Michael Hammer, Vance Holliday, Karl Hutchings, Brian Kemp, Jeffrey Long, Dan Mann, Daniel Moermann, Connie Mulligan, James O'Connell, Nick Patterson, Torrey Rick, Theodore Schurr, Vin Steponaitis, Noreen Tuross, Richard Waitt, Cathy Whitlock, and Don Yeager.

For providing photographs or help with other illustrations, I am grateful to Jim Adovasio, Tony Baker, Alex Barker, Charlotte Beck, Michael Collins, Deborah Confer,

Judith Cooper, Joseph Dent, Tom Dillehay, Boyce Driskell, Steve Emslie, Michael Gramly, Eugene Hattori, Louis Jacobs, George Jones, David Kilby, Jason LaBelle, Anne Meltzer, Ethan Meltzer, Jeff Rasic, Richard Reanier, Torrey Rick, Richard Rose, Mark Stiger, Lawrence Straus, Renee Walker, Fred Wendorf, David Willers, David Wilson, Michael Wilson, Tom Wolff, and David Yesner. I am particularly indebted to Katherine Monigal, who put her considerable artistic and computer talents to work on many of the figures in the book, and thereby saved me from having to get over my loathing of Adobe Photoshop (Adobe Illustrator and I still remain on quite good terms).

Several years ago I was fortunate to participate in a series of fascinating National Science Foundation–sponsored conferences organized by John Moore and Bill Durham that brought together archaeologists, geneticists, linguists, and social anthropologists to explore issues related to the colonization of new landscapes—and not just the Americas, but also Australia and the islands of Oceania. These sessions and my fellow participants contributed immeasurably to my thinking about these matters.

This book was written in the spring of 2007 while I was on a Research Fellowship Leave from Southern Methodist University, for which I am most grateful. My archaeological fieldwork and historical research touched on here have been supported by grants from the National Science Foundation, the Potts and Sibley Foundation, and especially (over the last decade) by the Quest Archaeological Research Fund, generously established at SMU by Joseph and Ruth Cramer.

In the fall of 1992, when I was writing *Search for the First Americans*, my children were four and six years old. And so I wrote at my university office to avoid the wonderful and welcome distractions of home. The children have since flown the nest for their own universities far away. And so I wrote *First Peoples* in my son's room—now converted to an office. But I miss those wonderful distractions that used to be home. Shadow the dog tried his best to fill the void, but it was just not the same.

David J. Meltzer
Dallas, Texas, January 2008

1

OVERTURE

It was the final act in the prehistoric settlement of the earth. As we envision it, sometime before 12,500 years ago, a band of hardy Stone Age hunter-gatherers headed east across the vast steppe of northern Asia and Siberia, into the region of what is now the Bering Sea but was then grassy plain. Without realizing they were leaving one hemisphere for another, they slipped across the unmarked border separating the Old World from the New. From there they moved south, skirting past vast glaciers, and one day found themselves in a warmer, greener, and infinitely trackless land no human had ever seen before. It was a world rich in plants and animals that became ever more exotic as they moved south. It was a world where great beasts lumbered past on their way to extinction, where climates were frigidly cold and extraordinarily mild. In this New World, massive ice sheets extended to the far horizons, the Bering Sea was dry land, the Great Lakes had not yet been born, and the ancestral Great Salt Lake was about to die.

They made prehistory, those latter-day Asians who, by jumping continents, became the first Americans. Theirs was a colonization the likes and scale of which was virtually unique in the lifetime of our species, and one that would never be repeated. But they were surely unaware of what they had achieved, at least initially: Alaska looked little different from their Siberian homeland, and there were hardly any barriers separating the two. Even so, that relatively unassuming event, the move eastward from Siberia into Alaska and the turn south that followed, was one of the colonizing triumphs of modern humans, and became one of the great questions and enduring controversies of American archaeology. Those first Americans could little imagine our intense interest

Beringia

Old Crow

Cordilleran ice sheet

Laurentide ice sheet

Kennewick

Calico
Pendejo
Folsom
Clovis

Meadowcroft

Approximate position of
coastline at 18,000 BP

Valsequillo

Taima-Taima

Pedra
Furada

Monte Verde

Cerro Sota & Pali Aike

in their accomplishment thousands of years later, and would almost certainly be puzzled—if not bemused—at how seemingly inconsequential details of their coming sparked a wide-ranging, bitter, and long-playing controversy, ranking among the greatest in anthropology and entangling many other sciences.

Here are the bare and (mostly) noncontroversial facts of the case. The first Americans came during the Pleistocene or Ice Age, a time when the earth appeared vastly different than it does today. Tilts and wobbles in the earth's spin, axis, and orbit had altered the amount of incoming solar radiation, cooling Northern Hemisphere climates and triggering cycles of worldwide glacial growth. Two immense ice sheets up to three kilometers high, the Laurentide and Cordilleran, expanded to blanket Canada and reach into the northern United States (while smaller glaciers capped the high mountains of western North America).

As the vast ice sheets rose, global sea levels fell approximately 120 meters, since much of the rain and snow that came down over the land froze into glacial ice and failed to return to the oceans. Rivers cut deep to meet seas that were then hundreds of kilometers beyond modern shorelines (Figure 1). Lower ocean levels exposed shallow continental shelf, including that beneath the Bering Sea, thereby forming a land bridge—*Beringia*—that connected Asia and America (which are today separated by at least ninety kilometers of cold and rough Arctic waters). When Beringia existed, it was possible to walk from Siberia to Alaska. Of course, once people made it to Alaska, those same glaciers presented a formidable barrier to movement further south—depending, that is, on precisely when they arrived in this far corner of the continent.

These ice sheets changed North America's topography, climate, and environment in still more profound ways. It was colder, of course, during the Ice Age, but paradoxically winters across much of the land were warmer. And the jet stream, displaced southward by the continental ice sheets, brought rainfall and freshwater lakes to what is now western desert and plains, while today's Great Lakes were then mere soft spots in bedrock beneath millions of tons of glacial ice grinding slowly overhead.

A whole zoo of giant mammals (*megafauna*, we call them) soon to become extinct roamed this land. There were multi-ton American elephants—several species of mammoth and the mastodon—ground sloths taller than giraffes and weighing nearly three tons, camels, horses, and two dozen more herbivores including the glyptodont, a slow-moving mammal encased in a turtle-like shell and bearing an uncanny resemblance to a 1966 Volkswagen Beetle—or at least a submersible one with an armored tail. Feeding on these herbivores was a gang of formidable predators: huge lions, saber-toothed cats,

---

FIGURE 1.

Map of the Western Hemisphere, showing the extent of glacial ice at the Last Glacial Maximum (LGM) 18,000 years ago, the approximate position of the coastline at the time, and some of the key early sites, archaeological and otherwise, hemisphere-wide.

and giant bears. All of these mammals were part of richly mixed animal communities of Arctic species that browsed and grazed alongside animals of the forests and plains.

But this was no fixed stage. From 18,000 years ago, at the frigid depths of the most recent glacial episode—the Last Glacial Maximum (LGM) it's called—until 10,000 years ago when the Pleistocene came to an end (and the earth entered the Holocene or Recent geological period), the climate, environment, landscapes, and surrounding seascapes of North America were changing. Many changes happened so slowly as to be imperceptible on a human scale; others possibly were not. Certainly, however, the world of the first Americans was unlike anything experienced by any human being on this continent since.

Once they got to America, these colonists and their descendants lived in utter isolation from their distant kin scattered across the planet. Over the next dozen or so millennia, in both the Old World and the New, agriculture was invented, human populations grew to the millions, cities and empires rose and fell, and yet no humans on either side of the Atlantic or Pacific oceans was aware of the others' existence, let alone knew of their doings.

It would not be until Europeans started venturing west across the Atlantic that humanity's global encircling was finally complete. Peoples of the Old World and the New first encountered one another in a remote corner of northeast Canada around AD 1000.[1] But that initial contact between Norse and American Indians was brief, often violent, and mostly served to thwart the Vikings' colonizing dreams and drive them back to Greenland and Iceland. It had none of the profound, long-term consequences that followed Columbus's splashing ashore on a Caribbean island that October day of 1492.

Europeans, of course, were profoundly puzzled by what they soon realized was far more than a series of islands, but instead a continent and peoples about whom the Bible—then the primary historical source for earth and human history—said absolutely nothing. We can presume Native Americans were just as perplexed by these strange-looking men, but their initial reactions went largely unrecorded by them or contemporary Europeans. Over the next several centuries, Europeans sought to answer questions about who the American Indians were, where they had come from, when they had arrived in the Americas, and by what route. The idea that they must be related to some historically known group—say, the Lost Tribes of Israel—held sway until the mid-nineteenth century, when it became clear that wherever their origins, they had arrived well before any historically recorded moment. The answer would have to be found in the ground in the artifacts, bones, and sites left behind from a far more ancient time.

But *how* ancient would prove a matter of much dispute. In 1927, and after centuries of speculation and more than fifty years of intense archaeological debate, a discovery at the Folsom site in New Mexico finally demonstrated the first Americans had arrived at least by Ice Age times. The smoking gun?—a distinctive, *fluted* spear point found embedded between the ribs of an extinct Pleistocene bison. A hunter had killed that Ice Age beast (see Plate 1).

A half-dozen years later, outside the town of Clovis (also in New Mexico), larger, less finely made, and apparently still older fluted spear points than those at Folsom were found—this time alongside the skeletal remains of mammoth. As best matters could then be determined, these were the traces of the most distant ancestors of Native Americans. *Paleoindians,* they were named, to recognize their great antiquity and their ancestry to American Indians.

But were these the very *first* Americans, and if so, just when had they arrived? A more precise measure of their antiquity would have to wait on chemist Willard Libby's Nobel Prize–winning development of radiocarbon dating in the 1950s. By the early 1960s, that technique showed that the Folsom occupation was at least 10,800 years old, while Clovis dated to almost 11,500 radiocarbon years before the present (BP).[2] This was relatively new by Old World standards—humans had lived there for millions of years—but it was certainly old by New World standards.

Better still, the Clovis radiocarbon ages apparently affirmed the suspicion this archaeological culture represented the first Americans, for the dates coincided beautifully with the retreat of North America's vast continental glaciers that, it was widely believed, had long obstructed travel to the south and forced any would-be first Americans to cool their heels in Alaska.

As those glaciers retreated, an "ice-free" corridor opened between them (around 12,000 years ago) along the eastern flanks of the Rocky Mountains, forming a passageway for travel into unglaciated, lower-latitude North America. Emerging from the southern end of the corridor onto the northern plains fast on the heels of its opening, the first Americans radiated across the length and breadth of North America with apparently breathtaking speed, spreading Clovis and Clovis-like artifacts across North America within a matter of centuries. Nor did they stop at the border: their descendants evidently continued racing south, arriving in Tierra del Fuego within 1,000 years of leaving Alaska (having developed en route artifacts that were no longer recognizably Clovis). It's an astonishing act of colonization, especially given it took our species more than 100,000 years just to reach the western edge of Beringia.

Indeed, the possibility that Clovis groups traversed North America in what may have been barely 500 years is all the more striking given that North America was then in the midst of geologically rapid climatic and environmental change. Yet, Clovis groups seemingly handled the challenge of adapting to this unfamiliar, ecologically diverse, and changing landscape with ease. Their toolkit, including its signature fluted points, is remarkably uniform across the continent. That lack of variability is taken as testimony to the rapidity of their dispersal (that is, it happened so quickly there was hardly time for new point styles to emerge).

That some of those points were found embedded in the skeletons of mammoth and bison suggested an answer to the question of how Paleoindians had moved so quickly and effortlessly: they were apparently big-game hunters, whose pursuit of now-extinct animals pulled them across the continent. Some took the argument a step further: it was their relentless slaughter that drove the Pleistocene megafauna to extinction.

ON DATES AND DATING

Throughout this book, time is denoted in years before present, abbreviated simply as BP. In regard to deep geological time, as with the onset of glaciation 2.5 million years ago, little need be said by way of qualification. Such ages are, at best, well-rounded estimates derived by a variety of geochemical dating methods, and are certainly accurate at the scale of hundreds of thousands of years, which is sufficient for our purposes. However, when attention turns to the last 50,000 years, the period of particular interest here, we seek more precise chronological control.

For that span, radiocarbon dating is the method of choice. It works off a straightforward decay principle (illustrated in Figure 2): when cosmic ray neutrons

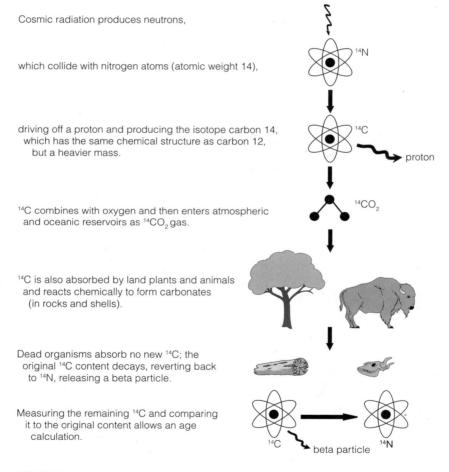

Cosmic radiation produces neutrons,

which collide with nitrogen atoms (atomic weight 14),

^{14}N

driving off a proton and producing the isotope carbon 14, which has the same chemical structure as carbon 12, but a heavier mass.

^{14}C

proton

^{14}C combines with oxygen and then enters atmospheric and oceanic reservoirs as $^{14}CO_2$ gas.

$^{14}CO_2$

^{14}C is also absorbed by land plants and animals and reacts chemically to form carbonates (in rocks and shells).

Dead organisms absorb no new ^{14}C; the original ^{14}C content decays, reverting back to ^{14}N, releasing a beta particle.

Measuring the remaining ^{14}C and comparing it to the original content allows an age calculation.

^{14}C beta particle ^{14}N

FIGURE 2.

The radiocarbon process in schematic form (see text for a fuller explanation).

bombard the earth's upper atmosphere, they react with nitrogen (^{14}N) to drive off a proton to form radioactive carbon or radiocarbon (^{14}C), one of several isotopes (*isotope* = same element, different mass) of carbon. Radiocarbon has the same chemical structure as elemental carbon (^{12}C), but a heavier mass (maintaining nitrogen's atomic mass of 14). And like ^{12}C, radiocarbon combines with oxygen to form carbon dioxide (CO_2), which is then absorbed by plants via photosynthesis, and which moves up the food chain into the animals that feed on those plants.

When a plant or animal dies, its supply of ^{14}C is no longer being replenished, and the resident ^{14}C slowly begins to revert back to ^{14}N, and in this decay process releases a radioactive emission (beta particle). Immediately after death, ^{14}C decay produces roughly 15 beta emissions/gram/minute. After 5,730 years, half of the ^{14}C is gone, and the decay process yields roughly 7.5 beta emissions/gram/minute. That lapsed period is called a *half-life*. After another 5,730 years have passed (that is, 11,460 years after the organism died), another half of the original ^{14}C is now gone (we are down to 25% remaining), and the decay process yields roughly 3.75 beta emissions/gm/minute. And so on.

Thus, by measuring the amount of radiocarbon still present in a sample, one can determine the approximate date that the organism died. By consensus, all radiocarbon ages are expressed as years before present, present being arbitrarily set at 1950, the year the first successful dates were reported by Willard Libby, the chemist who invented the technique (for which he received a Nobel Prize). We set all our radiocarbon clocks to years before 1950 to avoid the confusion that would follow when comparing the ages of different samples whose radioactivity was measured at different times (e.g., 1950 vs. 2000).

Radioactivity is a statistically random process. When it's measured, the result is an estimate of the average amount of ^{14}C in the sample, with an accompanying standard deviation to show the estimated error (the true value should fall within one standard deviation 68% of the time). A date of 10,130 ± 60 BP means that the estimated age of the sample based on the mean of the emissions was 10,130 years, and the chances are two out of three that the true age lies between 10,070 and 10,190 BP.

Theoretically, radiocarbon decay takes place until all the ^{14}C is gone from a sample—and that takes about ten half-lives. In principle we should be able to date material that old, but problems of preservation, the difficulty of detecting the tiniest amounts of ^{14}C, and the potential for contamination of ancient samples, put the present reliable upper limit of radiocarbon dating at about 50,000 years.

In terms of detection, measuring the amount of ^{14}C in a sample can be done in one of two ways: the conventional decay-counting method is to prepare a sample as a liquid or a gas, put it in a radioactive counter, and wait for beta emissions to happen. Older samples with less ^{14}C obviously have fewer and more widely spaced beta emissions, and obtaining a statistically reliable count of them can take days, weeks, and sometimes months.

The alternative technique, Accelerator Mass Spectrometry (AMS) dating, uses particle accelerators to count ^{14}C atoms directly by sending a sample at high speeds around a circular or oval particle accelerator. The lighter ^{12}C atoms can take the tight turns; the heavier ^{14}C atoms can't and fly off the molecular racetrack and crash into a strategically placed mass spectrometer, which counts the number of atoms. AMS dating takes only minutes or hours, not days or weeks, and standard errors are often less than fifty years. Best of all, because atoms are counted directly, large samples are no longer necessary. Prior to the advent of AMS dating, approximately 5 grams of carbon were required; now, it is on the order of 1 *milligram*. That's the difference between needing the entire limb bone of a bison, as opposed to the single tooth of a rodent.

Since AMS dating became available in the 1980s, it has greatly expanded our ability to date sites. But radiocarbon dating is not without complications, especially because the amount of radiocarbon in the atmosphere and ocean has varied over time. In effect, we cannot assume that all plants and animals over time started with the same amount. That variation is driven by how much radiocarbon is produced in the upper atmosphere, which is largely a function of changing amounts of neutrons bombarding the atmosphere at a given time (blame the sun for that), and changes in the relative amount of CO_2 stored in the atmosphere versus the ocean. Speed up or slow down how much CO_2 is squirreled away in the deep ocean, and one's radiocarbon-dated sample might have higher (or lower) amounts of ^{14}C—not because the sample is younger (or older), but because when it formed, the atmosphere had more (or less) ^{14}C to absorb.

To control for this variation, radiocarbon measurements are *calibrated* against objects whose ages are precisely known, such as the growth rings of a tree. Simplifying a bit: a tree adds one ring every year, and since most years differ from one to the next in rainfall and temperature, the rings are often different widths (wide and light colored if it's a good growth year, dark and narrow if not). The ring pattern becomes a fingerprint for a particular period in time. And like fingerprints, no two periods are exactly alike. By pushing the tree ring pattern back in time—thanks to some well-preserved and long-lived trees from the American Southwest, Ireland, and Germany (along with well-preserved wood specimens from archaeological sites)—a tree ring sequence has been compiled for the last 12,410 years.

By radiocarbon dating a specific tree ring of known age, one can measure how far the radiocarbon age diverges from the true age, making it possible to calibrate the radiocarbon result to bring it into line with a calendar age. When one sees an age listed as "cal BP," one is in the presence of a calibrated age.[3]

Unfortunately, the period of greatest interest to the study of the first Americans—the late Pleistocene—was also a window of geological time during which there were unusually rapid changes in ocean circulation (for reasons explained in Chapter 2), causing atmospheric ^{14}C to yo-yo. As a result, the radiocarbon clock at times ran too fast or too slow, and so a single radiocarbon age from this time period often corresponds to more than one calibrated age.[4]

TABLE 1   Approximate equivalence of radiocarbon and calibrated ages, from the Last Glacial Maximum to the Early Holocene.

Radiocarbon age	Median calibrated age
(^{14}C years before present or BP)	(calibrated years before present)
18,000	21,285
17,500	20,635
17,000	20,120
16,500	19,665
16,000	19,170
15,500	18,815
15,000	18,320
14,500	17,475
14,000	16,690
13,500	16,040
13,000	15,350
12,500	14,625
12,000	13,865
11,500	13,340
11,000	12,945
10,500	12,465
10,000	11,485
9,500	10,840
9,000	10,085
8,500	9,440
8,000	8,860
7,400	8,200

NOTE: As derived by OxCal 3.10 (http://c14.arch.ox.ac.uk/oxcal.php).

Because calibrating radiocarbon ages for this time period is neither straightforward nor certain,[5] calibrated ages are not used here; instead, all ages are given in radiocarbon years BP. Although this can mean a slight loss of chronological precision, that won't particularly matter since I am, for the most part, speaking of ages in general. At some point in the future, calibration of radiocarbon ages in this window of time will be more precise, and then we can make the switch. Until then, using radiocarbon years BP has the ancillary benefit of making them comparable to the vast bulk of the literature on the Pleistocene and on the first Americans, and so will cause less confusion for those who wish to look into that literature.

One can, of course, convert the radiocarbon years given here to calibrated years. Readers can try this at home, either using web-based programs such as CALIB (http://calib.qub.ac.uk/calib/), or by downloading calibration share-ware such as OxCal (http://c14.arch.ox.ac.uk/oxcal.php). I provide in the accompanying table a set of radiocarbon-to-calendar age calibrations at 500-year intervals (with one exception) covering the period from 18,000–7400 BP. These were calculated using OxCal 3.10.[6] These are just rough cuts and imply a more straightforward relationship between radiocarbon and calendar years than actually exists. Real calibration is a complicated and messy business, especially for the late Pleistocene.

EARLIER THAN WE THOUGHT?

The idea the first Americans were highly mobile, wide-ranging, big-game hunters, whose arrival was tied to the final rhythms of Pleistocene glaciation, made perfect sense. For a time. But there were always nagging doubts, not least the persistent claims of a *pre*-Clovis presence in the Americas. As more archaeologists took to the field in the 1960s and 1970s, perhaps driven (more than they might care to admit) by the chance of finding America's oldest site, every field season promised a pre-Clovis contender. Some were heralded with great fanfare: the legendary Louis Leakey, fresh from his triumph at Olduvai Gorge, flew to California to proclaim the Calico site to be middle Pleistocene in age (several hundred thousand years old). Unfortunately, its supposed artifacts—pulled from massive gravel mudflow deposits—proved indistinguishable from the millions of naturally broken stones the site's excavators burrowed through and tossed aside in great piles, still visible on final approach to Los Angeles International airport.

Other pre-Clovis claims were made by lesser mortals, but in all cases the result was the same: a purportedly ancient site burst on the scene with great promise, only to quickly tumble down what I came to call the *pre-Clovis credibility decay curve*, wherein the more that was learned about a site—for example, that its supposed artifacts were likely naturally flaked stone, or that the dating technique was experimental and unreliable, or that its deposits were so hopelessly mixed that the allegedly ancient artifacts were found alongside discarded beer cans—the fewer the archaeologists there were willing to believe it.

Dozens, even scores of sites failed to withstand critical scrutiny. There were so many false alarms archaeologists grew skeptical, even cynical, about the possibility of pre-Clovis. And we have long memories—it's part of our business, after all. The response may not have been commendable, but it was certainly understandable, particularly in light of the fact that once artifacts are out of the ground, they can never again be seen in their original context. In effect, we "destroy" aspects of our data in the process of recovering it, and because our sites cannot be grown in a petri dish in a lab, replication and confirmation of a controversial claim is no easy task and independent experiments to check results are nigh on impossible (archaeology may not be a 'hard' science, though it can be a difficult one all the same).

Pre-Clovis proponents cried foul, claiming the demands made of their sites and evidence were unfair, their work chronically underfunded, and their task overdemanding. Critics replied with a sneer that those same demands were met easily enough at Africa's and Australia's earliest sites, and perhaps the proponents' eagerness to find pre-Clovis sites marked a basic flaw in the motivational structure of American academia. Bystanders wisely kept their heads down and declared neutrality. Opinion quickly outran and outweighed the meager facts, and in science disagreement moves in quickly to fill the void between fact and opinion. So controversy grew.

All of this was testimony, cynics smirked, that academic battles are so ferocious because the stakes are so low.

The cynics are partly right. Knowing that the first Americans may have arrived 14,250 years ago, as suggested by artifacts deep within Meadowcroft Rockshelter, Pennsylvania, only tells us American prehistory is a couple of thousand years older than we used to think. In the grand scheme of the last 6 million years of human evolutionary history, that hardly matters. People could have arrived in the Americas tens of thousands of years earlier still, and it would not radically alter our understanding of human evolution (though if they came here hundreds of thousands of years ago, the ante is upped considerably—but the odds that happened are vanishingly small).

Nonetheless, there is more here than an academic turf war. Hanging in the balance is an understanding of when, how, how fast, and under what conditions hunter-gatherers can colonize a rich and empty continent; insight into the population and biological history of New World peoples; a gauge of the speed with which the descendants of the first Americans domesticated a cornucopia of plants (some as early as 10,000 years ago) and became the builders of the complex civilizations here when Europeans arrived; a better and more precise calibration of the rates of genetic, linguistic, and skeletal change in populations over that time; and most unexpectedly, a deeper understanding of the often-tragic historical events that unfolded in the wake of the Europeans' arrival on the shores of what they mistakenly, if self-righteously, proclaimed a New World.

As the peopling controversy deepened, support for pre-Clovis got a boost from an unexpected quarter. Starting in the late 1980s, molecular biologists and human geneticists began to piece together histories of modern American Indians from their mitochondrial DNA (which is inherited mother to child) and from DNA in the non-recombining portion of the Y chromosome (inherited father to son). By determining the genetic distance between modern Asians and Native Americans, and assuming that distance marks the time elapsed since they were once part of the same gene pool, geneticists have a molecular clock by which they can reckon the moment the ancestors of these groups split from one another. By some estimates, it was upwards of 40,000 years ago.

The linguists spoke up as well. There were an estimated 1,000 American Indian languages spoken in historic times. If all those evolved from a single ancestral tongue, they argued, then the time elapsed since those first speakers arrived in the New World might be as much as 50,000 years. The Clovis chronology, one linguist proclaimed, was simply in "the wrong ballpark." Although geneticists and linguists were happy to go where right-thinking archaeologists feared to tread, they could not prove the existence of pre-Clovis. Neither genes nor languages can be dated: only archaeological materials can.

Then the site of Monte Verde, Chile, excavated and analyzed by Tom Dillehay, came along. Monte Verde is an extraordinary locality, and what makes it so is that soon after this creek-side spot was abandoned, the remains left behind were submerged and ultimately buried in waterlogged peat, thereby stalling the usual decay processes and preserving a stunning array of organic items rarely seen archaeologically. These included wooden artifacts; planks used in hut construction; burned, broken, and split mastodon bones and ivory, along with pieces of its meat and hide, some still stuck to the wood timbers, the apparent remnants of coverings that once draped over the huts; and *Juncus* reed string wrapped around wooden stakes (Figure 3). There were also human footprints; a wide range of plants, some exotic, others charred, still others apparently well chewed, as well as a complement of stone artifacts. All of which dated to 12,500 years ago.

At Dillehay's invitation, a group of Paleoindian experts visited Monte Verde in January 1997, having studied in advance the 1,000 pages of his massive, soon-to-be-published second and final volume on the site. We came away convinced of its pre-Clovis antiquity. This was news even the *New York Times* deemed fit to print.

Although just 1,000 years older than Clovis, Monte Verde's distance (approximately 16,000 km) from the Beringian entryway and its decidedly non-Clovis look, raises a flurry of questions about who the first Americans were, where they came from, what triggered their migration, when they crossed Beringia, how they came south from Alaska (given the ice-free corridor would not be open until after they had arrived in South America), whether Monte Verde and Clovis represent parts of the same colonizing pulse, how many migratory pulses there were to America in Pleistocene times, how

FIGURE 3.

Two of the more than eighty "tent stakes" found at the Monte Verde site. These stakes have flattened heads from being pounded into the ground, were set behind timbers hewn from a different kind of wood, and had wrappings of string made from a third type of plant (*Juncus* reed). (Photograph courtesy of Tom Dillehay.)

and how fast the first Americans traversed the continent, and why (at the moment at least) the oldest site in the New World is about as far from Beringia as one can reach, with no sites in between as old or older.

The good news is we have plenty of answers to all these questions. The bad news is we cannot tell which answers are right. But I'll try to sift through what we know and don't, and what we can say or not.

TRACING FIRST PEOPLES

The chapters that follow explore the origins, antiquity, and adaptations of the first Americans. When they arrived, which at the very least was by 12,500 years ago, the world was still in the grip of an Ice Age, and North America was a vastly different place than it is today. Chapter 2 sets that stage. It explores the causes of Ice Ages in the intricate links between changes in the earth's orbit, solar radiation, ocean circulation and salinity, and greenhouse gasses such as carbon dioxide (CO_2), and their consequences, not least of which were the immense ice sheets of higher-latitude North America (as well as at higher elevations in lower latitudes). These were glaciers large enough to have bulldozed landscapes, changed the course of rivers (including the Missouri and Mississippi), altered atmospheric circulation (creating the paradox of Ice Age winters that in places were no colder and possibly even warmer than those of the present), and frozen so much water on land that sea levels fell worldwide, creating land bridges across which people could walk from one hemisphere to another.

South of the vast continental ice sheets and beyond their immediate refrigerating effects, North America experienced climates and environments unlike any at present, comprised of complex plant and animal communities that were changing dramatically, or in some cases heading toward extinction. The first Americans were there to witness and experience some of those changes, as well as the end of the Ice Age, which refused to die quietly but instead went out in a rush of floodwaters of Noachian proportions and one brief, if failed attempt to reassert its glacial dominance.

But just when did the first Americans arrive? During the most recent glacial cycle, or earlier still? The next few chapters range widely over the efforts, historical and contemporary, archaeological and non-archaeological, to establish the origins and antiquity of the first Americans. This is a problem that's been around, as detailed in Chapter 3, for well over a century, and has been disputed almost from the very moment it was first posed. The initial round of controversy was prolonged in part because archaeology itself was in its adolescence; it hadn't well-established methods and techniques for finding, evaluating, or reliably determining the age of ancient artifacts or sites; and it was being tugged in different directions by practitioners who wanted to craft the discipline in their own images.

Demonstrating people had arrived in the Americas by Ice Age times came only after better chronological markers were established, and when a particular kind of site was discovered, namely a kill site—as at Folsom—in which the prey was an extinct Pleistocene animal. If the animal lived during the Ice Age, then so did the people who killed it. This enabled a site's antiquity to be assessed in the ground, a necessity in those pre-radiocarbon dating years. That demonstration at Folsom also taught archaeologists what to look for and how to look for Pleistocene-age sites. Soon dozens more such sites were found, including Clovis, which not only helped paint a picture of North American Paleoindians, but also had the more subtle consequence of creating expectations that guided much of the archaeological research into the Paleoindian period over the ensuing decades.

One of those expectations—that Clovis sites were oldest and therefore represented the first Americans—quickly became fact, and as Chapter 4 shows, sparked a decades-long effort to prove otherwise. The criteria for demonstrating a pre-Clovis presence were straightforward in principle—one needed unmistakable artifacts in a secure geological context with reliable ages from radiocarbon or some other dating technique—but they proved extraordinarily difficult to meet in practice. Nature was partly to blame: it has the mischievous ability to break stone and bone in ways that neatly mimic primitive human artifacts. But we archaeologists shoulder part of the blame for not recognizing nature's deviousness, or for using unproven dating techniques, or for misreading geological circumstances. Even so, much was learned in the decades of contentious debate over pre-Clovis and how best to meet the standards of proof—which were finally met at Monte Verde in 1997.

How resolution came about was in some ways reminiscent of events that took place seventy years earlier at Folsom—including the venerable tradition of a site visit by outside experts—but in important ways, the events were very different, not least in the way that Monte Verde gave fewer clues of how to find sites like it. But it has certainly redirected where we look. In Monte Verde's wake, archaeological attention has shifted to the coast as a possible entry route, which was available for passage well before the ice-free corridor opened. It has also redoubled efforts to find sites of comparable age here in North America, but so far these have proven elusive. It leaves us wondering: why are pre-Clovis sites so hard to find, and how do they relate to Clovis? Are they different parts of the same colonizing pulse into the New World?

Archaeology speaks directly to questions of *when* and *where*, and sometimes *how*, the first people came to the Americas, but struggles mightily with the question of *who* these people were, in tracing their population histories (forward or backward) or in ascertaining their relationship to contemporary American Indians. It is no easy task to measure the historical affinity between groups widely separated in space and time from the manner in which they crafted their stone tools. Accordingly, Chapter 5 turns to DNA, language, teeth, and skeletal remains to attempt to fill the gap between the most ancient and modern Native Americans. By grouping together similarities in the words and grammar of many hundreds of native languages, and by examining the diversity and patterning in mitochondrial and Y chromosome DNA, it should in principle be possible to unravel the complex relationships among American Indians, and then go the next step to infer the number and timing (using molecular clocks or inferences about rates of language change) of their ancestors' migration(s) to the New World.

Assuming, that is, there is an unbroken chain from the present back into the past, and that modern Native Americans are descendants of the first Americans, a matter that's now hotly disputed by some physical anthropologists. They see among rare ancient human skeletal remains skulls that do not resemble the crania of American Indians—the most famous (infamous) being Kennewick, which after its discovery was described at a press conference by the arcane term "Caucasoid," which on the notepads of the assembled reporters quickly morphed into "Caucasian." Could the Americas have originally been peopled by Europeans? Were ancestors of American Indians *not* the discoverers of America, but later arrivals? These are not innocent academic questions, but ones that inevitably take on a political character with real-life implications for modern-day American Indians. Even so, a couple of archaeologists blithely leaped on that bandwagon, and proclaimed that Solutreans from Stone Age Europe had paddled the iceberg-choked Pleistocene North Atlantic and landed on the east coast of North America several thousand years before Clovis. But are there traces of non-Asian ancestry in genes or language? How reliable are skulls for tracing the origins of populations? Just what do crania tell us about "race"—whatever that loaded term implies? That's why Chapter 5 aims to detail how all these methods work, what they can and cannot reveal, and the reliability of the conclusions drawn from them.

That chapter also shows that compounding the evidence and methods being brought to bear on the peopling of the Americas has in no small measure compounded the controversy. Now, instead of archaeologists arguing with one another—as we still do, even in these post–Monte Verde days—linguists, physical anthropologists, and geneticists are haggling among themselves, and all of us with one another. There's a good reason for that, as explored in Chapter 6: linguists, physical anthropologists, and geneticists speak with no more unanimity on this question than archaeologists, nor is it easy to reconcile such radically different kinds of evidence. Each of these disciplines approaches the central questions from very different angles. Linguists and geneticists view the peopling of the Americas backward from the present, through the languages or DNA of living American Indians. Archaeologists and physical anthropologists, working with ancient sites and skeletal remains, come from the opposite direction.

Naturally, there are advantages and disadvantages to each, and significant differences in data and method, such that linking modern languages or genes with Pleistocene archaeological or skeletal remains proves no easy task—not that we haven't tried. We have many scenarios for the number, relative timing, and antiquity of migrations to America. Although there is no consensus among them, we have begun to answer questions about who the first Americans were and where they came from, and can perhaps narrow down the window of time within which the migration (migrations?) occurred, and what our best chance is of more precisely resolving such questions. Even so, controversy remains.

Of course, the search for the first Americans is not just about origins and antiquity—it's also about adaptations. Once here, they apparently colonized the length and breadth of the hemisphere in less than a millennium. That's a stunning achievement for any human group, but especially for hunter-gatherers in a novel and changing setting. Chapters 7 through 9 look into how it is they moved so far so fast, what life was like in Ice Age America for the new arrivals, and what adaptive strategies keyed their successful colonization of a continent as diverse and dynamic as late Pleistocene North America.

Central to these issues is the matter of adapting to a new land, considered in detail in Chapter 7. As these bygone Siberians moved south into an ever-more-exotic New World, they surely possessed a general knowledge of animals and plants, but were increasingly encountering ones they had never seen before. Which would feed them, clothe them, cure them, or kill them? There was no one to greet them or provide helpful advice about, say, rattlesnakes or poisonous plants. Nor were there signposts at the gateway to America as there are today (tongue-in-cheek) in downtown Barrow, Alaska, pointing the way to New York City or Ayachucho, Peru.

Colonists in new landscapes face great risks, especially early on when their numbers are low and they know little of the availability, abundance, and distribution of plant and animal foods, or of how severe local climates might be, or of where (and what) potential dangers might lurk. To reduce that risk, it would have been to their advantage to learn about their new world as quickly as possible, a strong incentive to range widely and rapidly. Yet, doing so would have meant moving away from other people.

The first Americans are often stereotyped as manly hunters, Pleistocene versions of the mountain men and fur traders who boldly ventured across the American West in the eighteenth and nineteenth centuries. But if the goal was not merely to exploit but also to explore, adapt, and settle, "early man" would not get very far without early woman, and without producing early children. And when those children came of age, they needed spouses. Where were those to be found? Within their immediate band, or among distant kin who'd split off to find their own way? And how could or did groups maintain long-distance contacts with others with whom they could exchange information, resources, and mates, and do so over a vast and uncharted landscape with few known landmarks, across which they and others were possibly moving rapidly?

We have only recently begun to model the processes of colonization. Central to seeing if those models work is an understanding of the archaeological record and what it reveals of Paleoindian adaptations, the subject of Chapter 8. The first Americans surely hunted more than gathered: their long Arctic traverse from Asia to America had few other options. Those habits continued as they moved south of the ice sheets, where Clovis Paleoindians took down mammoth, mastodon, and giant bison.

But just how often were they out hunting big game, or better, how often were they successful at it? So successful they drove the Pleistocene megafauna to extinction? By 10,800 years BP, soon after Clovis groups appeared, that extraordinary assortment of large mammals (some thirty-five genera all together) had disappeared, vanishing in a geological instant from a world where they had thrived for tens and hundreds of thousands of years. Paleoindians are charged with killing—or more properly, overkilling—the Pleistocene megafauna, a wholesale slaughter routinely invoked today by conservationists as a grim homily of human destruction.

Yet, if Paleoindians are guilty as charged, then they behaved unlike any other hunter-gatherer groups known before or since, and then artfully covered up virtually all evidence of their wrongdoing. It is possible, of course, that we've not found their kill sites, or that we do not know what members of our own species are wont to do on a rich, virgin landscape teeming with game never before hunted by wily human predators. Perhaps the rules that govern hunter-gatherers in other times and places do not apply here. The first Americans were unique in many ways; this may be another.

Of course, those extinctions also coincided with the end of the Pleistocene. The sweeping climatic and ecological changes that marked that transition are just as likely (maybe even more likely) to be responsible for this massive extinction event. But if that's so, more questions remain: why did horses disappear from North America at the end of the Pleistocene, and yet flourish when reintroduced by the Spanish in the early 1500s? And isn't it odd that the plants that comprised the diet of the giant ground sloths are common today outside the very southwestern caves these now-extinct animals once frequented? These are good questions for which we have, as yet, no good answers.

What is certain is that during Paleoindian times, climates were warming, glaciers worldwide were in full retreat, sea levels were rising, plants and animals were shifting their ranges (or going extinct), and the end of the Ice Age was just over the horizon. But around 11,000 years ago, the world's climates took a sharp turn. According to one prevailing theory, when the retreating Laurentide ice sheet uncovered the St. Lawrence seaway that apparently diverted glacial meltwater—which to that point mostly drained down the Mississippi River into the Gulf of Mexico—into the North Atlantic. Flushing very cold, very fresh water directly into the northern ocean upset circulation patterns in the Atlantic and triggered a nearly instantaneous climatic response: the Northern Hemisphere was plunged back into near-glacial conditions that lasted a thousand years. The Younger Dryas, as it's called, was no Ice Age rerun, since by then many of the conditions that had put the earth under Pleistocene ice had changed. Even so, the sudden polar freeze of the Younger Dryas is blamed for the "fragmentation of Clovis culture" and even the extinction of the Pleistocene megafauna: it wasn't Pleistocene Overkill. It was Pleistocene Over*chill*. Or not.

Regardless, the Younger Dryas set the stage on which the final millennium of colonization was played out. It was during this time, as discussed in Chapter 9, that the Paleoindian descendants of wide-ranging and highly mobile Clovis groups began to settle in different regions. As they did, they developed distinctive adaptations: lifeways in the mountainous and semi-arid Great Basin soon became very different from those on the grasslands of the Great Plains, or in the rich forests of eastern North America.

This settling in inevitably severed ties among populations, and over the next ten millennia, their descendants developed new dialects and languages, along with distinctive genetic lineages, cultures, and material culture. Evolutionary pathways diverged and converged as populations sporadically reconnected (peaceably or not) and exchanged genes, words, or artifacts. By the time Europeans arrived, some 400 generations of intermittent isolation, migration, and gene flow had passed, and the descendants of what may have been a single band of colonists was now many hundreds of separate peoples, cultures, and languages, whose histories were hopelessly entangled in complex skeins.

All shared, however, a Pleistocene ancestry, and it was nearly their undoing—as explored in Chapter 10. For when more than 12,000 years of isolation ended in 1492 and the peoples of the Old and New worlds came into contact, the consequences were profound, not least in the devastating impact of repeated waves of Old World epidemic disease on American Indians. The worst was smallpox, and against it—as well as against measles, influenza, plague, and other contagions—Native Americans had little, if any, immunity. Mortality rates may have spiked at over 90% in native populations and, in so doing, arguably altered the course of American history. But to understand why American Indians were so extraordinarily vulnerable to introduced infectious diseases, and harbored none of their own (which could have slowed the colonization of the Americas by Europeans), the answer must be sought deep in their prehistory.

And something more: American archaeology has changed dramatically in the last decade, not least because of events well outside the shelter of academia where we've long cloistered ourselves. Federal legislation—the Native American Graves Protection and Repatriation Act (NAGPRA)—aimed at righting the often egregious wrongs of history, mandated that skeletal remains held in museums and universities receiving federal funding (that's just about all of them) must be returned if requested to the American Indian tribes that are biologically or culturally affiliated. It's easy enough to identify affiliation if the remains come from sites of no great antiquity, where there is clear continuity from past to present. The task is immeasurably harder when attempting to identify specific tribal descendants of the first Americans. That has sparked plenty of fights about how or even whether ancient skeletal remains can be linked with modern peoples. Legally they can be, ethically they should be, but scientifically they can't be (at the moment anyway). And so at times, as with Kennewick, it's gotten ugly.

But there have been positive steps, too, often made far from the harsh partisan limelight: archaeologists and Native Americans have become increasingly more aware and appreciative of the other's perspectives. And all sides now recognize that questions about the peopling of the Americas matter, and can matter deeply—even if for very different reasons among different constituencies.

Getting the answers to those questions is a long story, and to start the telling requires returning, ever so briefly, to where it all began.

GETTING TO BERINGIA ON TIME

The deep roots of human prehistory reach back to Africa, and a long evolutionary line of early hominids. When our very earliest hominid ancestors become recognizable about 6 million years ago (we cannot call them humans just yet), they were barely refined apes, and certainly were not in possession of the adaptive abilities necessary to venture into the far north, let alone make their way to Siberia and then on to America.

The first groups to do that were members of the genus *Homo,* of which there are various species that first appeared nearly 2 million years ago. Within a few hundred thousand years of their emergence, they had mastered fire and learned to build shelters, which enabled them to establish beachheads in colder climes outside Africa, even with the astonishingly primitive stone tools that mark the Lower and Middle Paleolithic cultural periods. *Homo erectus* and its evolutionary kin ranged widely over temperate Eurasia and lived during glacial times, yet do not appear to have expanded in any significant numbers into northern latitudes, at least not until a few hundred thousand years ago when they and their descendants occupied Pleistocene Europe. By then they were clothed, revealed by the fact that body lice (which feed on the body but live in clothing) have made their evolutionary appearance.[7] Still, there were limits to humanity's range. They spanned the distance from western Europe to China, yet few (if any) descendants of this first wave to leave Africa made it to the far northern or eastern regions of Asia or Siberia, nor were they ever within striking distance of the Americas.

It was, instead, descendants of the second major wave out of Africa who, bearing a more sophisticated stone tool technology (not to mention increasingly elaborate artifacts of bone and ivory), pushed into Europe and all the way across Asia. These were our earliest direct ancestors, the first modern humans—*Homo sapiens*—that, based on genetic and archaeological evidence, arose in Africa nearly 200,000 years ago, and from which they subsequently dispersed.

The degree to which these early moderns are related to the descendants of the first wave of humans who left Africa, such as Europe's Neanderthals, has for many years been hotly debated. Because they briefly co-existed on the same landscapes—the one using vintage Middle Paleolithic stone tools, the other the more elaborate Upper Paleolithic technology—some paleoanthropologists insist we are descended from a genetic mix of Neanderthals and early moderns. That claim has steadily lost adherents over the years, precipitously so after ancient DNA extracted from Neanderthal skeletal remains showed a genomic sequence very different from that of living humans. Based on the molecular clock, it is estimated *Homo neanderthalensis* and *Homo sapiens* went their separate ways over 500,000 years ago (which is to say, they are both descendants of a deep common ancestor, but we trace our evolution via the *Homo sapiens* line).[8]

Once *Homo sapiens* struck out on their own, they scarcely stopped. It is this species that first traveled beyond temperate Eurasia to colonize the distant corners of the globe, including Australia and the Americas, the last of the habitable continents of prehistory. Although brainy, innovative, and highly adept hunter-gatherers, getting to America was no easy journey even for *Homo sapiens*, not with Siberia in between and especially not during harsh, full-glacial times. As archaeologist Ted Goebel observes, it appears no one was in Siberia (even southern Siberia) during the LGM, and understandably so. Climates were cold and harsh though, ironically, glacial ice was no barrier: virtually all of central and western Siberia was ice free, even during the LGM.[9]

Not that there is evidence humans had reached Siberia much before then. There are only a few archaeological sites north of 55°N latitude and east of 80°E longitude (near present-day Novosibirsk) that possibly predate the LGM. And the oldest of these, Nepa I in central Siberia, which dates to 35,000 years ago, and Yana RHS in northern Siberia near the Laptev Sea, dated to 27,000 BP, are both still many thousand kilometers shy of the western edge of Beringia, the New World's entry point.

Humans more or less permanently colonized far northeastern Asia only after 18,000 BP. By then, they had reoccupied the Lena and Aldan river basins and left behind a number of sites, including Dyuktai Cave, occupied as early as 14,000 years ago, where bifacial knives, blades, scrapers, and points were found with a range of animal remains, including mammoth, bison, musk ox, horse, reindeer, and moose. Even then, they were still several thousand kilometers away and well shy of the latitude of Beringia, which is mostly north of 60°N, about the latitude of Seward, Alaska (Figure 4).

Over late Pleistocene time, humans moved further north and east, and finally approached the gateway to America. Archaeologist John Hoffecker and paleoecologist Scott Elias suggest that improved stone tool technology, more efficiently insulated clothing, and a post-glacial expansion of trees (to provide wood for hearth fires) likely aided that expansion. But humans were still sparse on the ground. Their presence is well documented and securely dated only at the sites of Berelekh on the Indigirka River close to the Arctic Coast (at 70°N latitude), and at Ushki in central Kamchatka. These two localities were relatively late in the grand scheme of prehistory: they are no more than about 14,000 and 11,300 years old, respectively. Importantly, they contain artifact types—including the distinctive Chindadn point—we will soon see on the Alaskan side of Beringia.

Otherwise, it has so far proven difficult to pinpoint archaeologically when and from where in Siberia the earliest Americans originated. Could it be that they did not come this way at all? More on this later (Chapter 6), when we confront a bold daylight attempt to rob Siberia of its role as the jumping off point for the colonization of the New World.

Taken at face value, it appears that far northeastern Siberia and Beringia were not occupied by humans until as late as 14,000 years ago. But let's not leap to that conclusion just yet. The Siberian archaeological record on the whole is sparse, and gets even more so as one approaches the Bering Land Bridge. The timing of the peopling of Siberia will

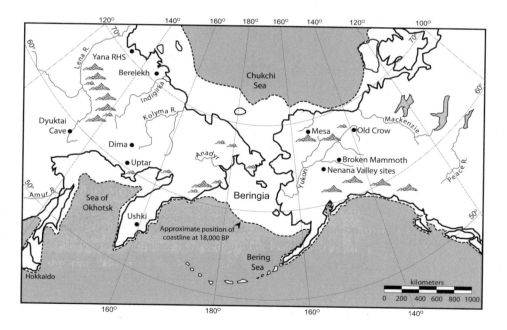

FIGURE 4.

Map of the extent of Beringia at the Last Glacial Maximum (LGM), showing the location of early Siberian and Alaskan archaeological sites discussed in the text.

not be known until Siberia is peopled by more archaeologists. Given the archaeological near invisibility of what must have been small and highly mobile human populations, the vast area to be searched for their sites, and the relatively limited archaeological work that's been done to date, this negative evidence is sure not to endure.

But prehistory in this region (or in the Americas) will likely not go too far back. No Neanderthals or any other earlier (non-*sapiens*) form of human has ever been found in far northeast Asia and Siberia, let alone in the New World. One should never say never in archaeology, but at this point, it seems exceedingly unlikely that premodern humans made it to the Americas, though that hasn't stopped speculation on this score—or claims from sites like Calico (California) or Old Crow (Canada) that the first Americans arrived some 200,000 to 350,000 or more years ago.

Of course, if such claims of deep antiquity (or European ancestry) are right, then our hard-won understanding of human evolution is badly wrong. But there's no need to rush a textbook rewrite just yet. More likely, such claims are simply flawed. We have long assumed, and have no reason to doubt, the Americas were colonized by anatomically modern humans, coming by way of Asia and bearing an Upper Paleolithic artifact technology, and arriving at some time during the latter stages of the Ice Age.

But what kind of place was this New World?

2

THE LANDSCAPE OF
COLONIZATION
Glaciers, Climates, and Environments
of Ice Age North America

In late June 1838, Charles Darwin set off on a geological mission to Scotland to examine the famous Parallel Roads of Glen Roy, several long, paired terraces that extended down both sides of this narrow mountain valley at matching elevations. Local legend pegged these as hunting paths of ancient Celtic warriors.[1] Geologists doubted that explanation, but were mystified all the same.

It was "admitted by everyone," Darwin later wrote, "that no other cause, except water acting for some period on the steep side of the mountains, could have traced those lines."[2] Yet, how could water have been raised to this elevation on land? Some surmised a lake had once filled the valley (Loch Roy, it was called), and that its water level had dropped over time. But this was an open-ended valley: where was the barrier that dammed the water to form the lake? Nature had to have built and destroyed several dams at successive periods, each at least a mile wide and minimally a thousand feet high, in order to form the several terraces. However, no evidence of any dams could be found.

Perhaps, Darwin suspected, instead of a dam pushing the water *up* into the land, the land itself had once been *down* in the water. He'd recently returned from a five-year voyage around the world on the H.M.S. *Beagle*, where on the earthquake-prone Chilean coast, he'd seen unmistakable evidence of a succession of wave-cut beaches that had formed as the land rose (episodically and sometimes violently) from the sea. He was certain the Parallel Roads of Glen Roy had been cut by a similar process, and were testimony Scotland had once been submerged, but had since risen from the sea.

Yet, as Darwin walked the Parallel Roads, he did not find a single seashell from those ostensibly wave-cut marine beaches. The quarrymen who worked in the area assured him they hadn't seen any shells either. Perhaps it was just a matter of poor preservation: not all the terraces on the South American coast were littered with seashells either. Besides, there was other evidence that helped convince him Scotland had emerged from the sea. Erratic boulders, so named because they were found far from outcrops of that rock type, were sprinkled throughout the upper reaches of Glen Roy. One giant granite boulder was perched 2,200 feet above sea level, yet it could only have come from a granite outcrop 6 miles away and nearly a thousand feet lower. Darwin knew no "rush of water so impetuous" that could wash a boulder that far and high up a steep mountainside. Surely its present perch was once closer to the sea, and the granite block had been deposited at a time when climates were much cooler, and icebergs could have rafted it here from its bedrock source.

Darwin spent a week tramping around Glen Roy, experiencing nature "looking as happy" as he felt. He then returned home and straightaway prepared a long paper, "Observations on the Parallel Roads of Glen Roy . . . with an attempt to prove they are of marine origin," which was published the next year in the *Philosophical Transactions* of London's Royal Society.[3]

Yet, scarcely two years later, Darwin was scrambling to defend his theory of the origin of the Parallel Roads. And he continued to do so for another two decades, though he was fighting a rearguard action and knew it. Finally he gave up, admitting to his friend and geological mentor Charles Lyell that he'd been "smashed to atoms about Glen Roy," his paper "one long gigantic blunder."[4] It was—one of the few gaffes Darwin made in an otherwise brilliant scientific career. But this one was understandable.

For Darwin in 1838 was unaware another geological agency was involved. Vast ice sheets—far larger than the mountain glaciers he knew—had once covered much of the Northern Hemisphere, northern Scotland included. His ignorance is no surprise. In those days, no one knew the extent, complex history, and role of glaciation in shaping the landscape, let alone that meltwater lakes could form behind dams of glacial ice and carve parallel terraces into valley walls, or that erratic boulders could easily be plucked and moved great distances by glaciers (ironically, Darwin supposed it was mountain glaciers reaching the sea during cooler periods that had released the floating ice that had rafted all those erratics).

Fully understanding the role of glacial ice and the history of glaciation on the planet—and for that matter, its role in the early peopling of the Americas and the adaptive challenges it posed for the first Americans—first required opening one's eyes to the geological signs that ice was once far more extensive than the glaciers that still sat atop Europe's alpine peaks.

STARTING AT THE TOP

Chamois hunters in the Swiss Alps were long familiar with the glaciers that perched high atop their mountains, and well understood that as these giant tongues of ice slid forward and back, they left telltale evidence of their movement. After all, they were living at the tail end of the Little Ice Age, a several-centuries-long (ca. AD 1550–1850) return to cooler conditions that activated many of the Alps glaciers (in a grim and indirect way, humans may be responsible for this episode of global cooling[5] as Chapter 10 explains). In the 1830s, Swiss geologist Jean de Charpentier began to record those signs: the smoothed and striated bedrock that ice had overridden, the piles of gravelly debris that had been pushed along and piled up (tills and moraines, we now call them), and the boulders that the ice had plucked from high peaks and transported dozens of kilometers down valley (erratics).

De Charpentier read from these signs that alpine glaciers had once extended over a much larger area, and in the summer of 1836, he showed those features to a fellow scientist there on holiday. Louis Agassiz was quite taken by what he saw, and though he'd won his scientific reputation as a fossil fish expert, he possessed a creative mind and realized what de Charpentier had not: many of these same features—albeit on a vastly larger scale—were present at lower altitudes in Europe's high latitudes.[6]

That included immense gravel deposits previously identified as water-laid diluvium by Oxford geologist William Buckland, which he attributed to "a transient *deluge*, affecting universally, simultaneously, and at no very distant period, the entire surface of our planet." When? Five or six thousand years ago by his estimate. The diluvium was not just from any old flood: it was from Noah's.[7] Granted, this deposit lacked the remains of humans who had perished in the flood, and in retrospect Buckland's Cambridge counterpart Adam Sedgwick admitted that should have given pause. Even so, Buckland found their absence not so hard to explain: in Noah's time, humans lived only in the "Asiatic" region. Their remains did not belong in floodwater deposits of northern latitudes.

Yet, in a classic argument by analogy, Agassiz reasoned otherwise. Since the diluvium on the low-lying plains of northern Europe was the same, albeit far more extensive, than the tills and moraines produced by glaciers at high altitude, then the tills and moraines across Great Britain and northern Europe surely had a glacial origin as well. Those glaciers, however, had long since vanished. Yet, they must have been, from the size and extent of their traces, of almost unfathomable size, and existed during a time of global cold. "Since I saw the glaciers," Agassiz crowed, "I am quite of a snowy humor, and will have the whole surface of the earth covered with ice."[8]

A rhetorical flourish, to be sure: the earth's surface was not entirely buried in ice back then. Still, in the main, Agassiz got it right. It was a bold, inferential leap that surprised de Charpentier and launched Agassiz, who was ambitious and charismatic (as de Charpentier was not), into the scientific limelight.

Agassiz's Ice Age was soon linked with Lyell's Pleistocene, a geological period defined independently by percentages of fossil shells of extinct species, but which seemed to coincide perfectly with glaciation. Buckland proved a willing convert, happily abandoning a Noachian flood for ice. Darwin too "gave up the ghost" of his marine subsidence theory once he was shown the traces of the glacier that had blocked Glen Roy and dammed the lake that cut those Parallel Roads.

The Ice Age soon jumped the Atlantic, when traces of continental ice sheets were spotted by paleontologist Timothy Abbott Conrad in 1839 in western New York. By the 1870s, American geologists realized that fossil plant layers sandwiched between sheets of glacial till, and the differential weathering of moraines, were clues there had been multiple episodes of ice advance and retreat. The Glacial Division of the United States Geological Survey (USGS), under the aegis of Thomas Chamberlin, spent the last decades of the nineteenth and first decades of the twentieth centuries mapping those separate episodes. By 1915, four distinct glacial stages, separated by long interglacials (warm, relatively ice-free periods), were recognized: from oldest to youngest, Nebraskan (followed by the Aftonian interglacial), Kansan (Yarmouth interglacial), Illinoian, (Sangamon interglacial), and Wisconsin (Holocene interglacial), the stages named for the states where their traces were especially well expressed.

Although Chamberlin believed he knew how many glacial episodes there were, he was certain he didn't know what caused them. He suspected the amount of carbon dioxide in the atmosphere played a role. Decades earlier James Croll had supposed it had something to do with changes in the shape of the earth's orbit and the wobble of its axis affecting the amount of heat received on earth. Chamberlin saw fatal flaws in both hypotheses, though perhaps their main flaw was being too far ahead of their time. But even as he wrote, Milutin Milankovitch, a Serbian mathematician and one-time concrete engineer, was laying the groundwork for resolving the cause of the Ice Ages and, in so doing, rendering the four-stage model of Pleistocene glaciation obsolete.

A NEWTONIAN COTILLION

Milankovitch had initially set himself the ambitious goal of creating a mathematical model of the world's climates. Well-meaning colleagues tried to dissuade him: who needs a model, when weather stations around the globe provide actual data? Milankovitch paid them no mind. With a model, he could predict—even retrodict—variation in climate across space, and perhaps even through time.

Driving the earth's climates, Milankovitch reasoned, was incoming solar radiation or *insolation*. Insolation varies by latitude and season, as the relative position of the earth and the sun shift over the annual cycle. It also varies over much longer time spans since, as Croll surmised, the earth's orbit is jostled by the gravitational tug of the sun and the other planets in a complex Newtonian cotillion. That alters the *eccentricity*, or departure, of our orbit from a near-perfect circle to more of an ellipse; it changes the *tilt*, or obliquity, of the earth relative to its plane of travel; and, it triggers the *precession* of the equinox, shifting the time of the year when the earth is closest to (perihelion) or most distant from (aphelion) the sun. All these changes occur in predictable cycles. Eccentricity varies over a 100,000-year period; the earth's tilt swings in 41,000-year cycles between 21.8° and 24.4° (during the most recent glacial episode it was 22.95°; it is 23.45° at present); the full precession circuit takes approximately 23,000 years (at the moment, perihelion occurs during Northern Hemisphere winters; 10,000 years ago, it was opposite).

These predictable cycles have, at least in principle, predictable results: they determine the total amount of insolation received in different seasons at different latitudes. So, for example, tilt the Northern Hemisphere just 1° away from the sun, and it reduces insolation by 3% to 4%, thereby cooling winters and summers. Similar effects come with changes in precession and eccentricity, though in the latter the climatic consequences are subtle indeed: reshaping the earth's orbit accounts for, at most, a 0.3% variation in insolation. Still, combined and compounded, these three orbital-forcing functions (as climatologists call them), can cause dramatic swings in insolation—upwards of 20%.

Long before computers made such calculations easier, Milankovitch calculated orbitally driven insolation changes by season and by latitude over long periods of earth history. The effort took years, though he worked steadily at it—even while a prisoner of war in Hungary during World War I (nowadays, one can instantly download the last 5 million years of orbital insolation changes from the Internet). When Milankovitch's work appeared after the war, it caught the eye of Wladimir Köppen, who saw that changing insolation might be the key to understanding glaciation. The link, he realized, lay in insolation received at high latitudes in *summer* since that determined how much of the previous winter's snow melted away. If it all vanished, it would be difficult to build an ice sheet, for glaciers are created and fed by snow that, by slightly melting, condensing, and recrystallizing, becomes ice. When summer insolation is reduced and temperatures cool, snow from the previous winter stays frozen on the ground (*firn*, it's called). The next winter, more snow falls, and if it does not melt away the following summer, the firn/snowbank grows larger. As it grows, the underlying layers of firn become steadily denser until ultimately they turn into ice as the mass above thickens.

Once the ice reaches a critical thickness, it begins to move a few centimeters to a few meters a day, either by internally deforming, or by sliding along on a slurry of water, gravel, and mud. Assuming temperatures remain cool, the ice can advance considerable distances from its zone of accumulation close to the poles or high in the mountains (see Plate 2). Massive, continental-scale ice sheets are primarily a Northern Hemisphere phenomenon since glaciers need land on which to grow—80% of the Southern Hemisphere is under water. Accordingly, the only ice sheet of consequence in the Southern Hemisphere is and was atop Antarctica, though there were relatively small glaciers high in the Andes and in Patagonia.

When the ice front reaches a latitude or elevation where temperatures are warmer, melting (ablation) overcomes the accumulation of ice, and the glacier stops in its tracks. Warm up the climate even more, and the glacier is forced to retreat. Where the glacier pauses for a time, it deposits a gravelly end moraine, marking a moment of geological time when climate was relatively stable and accumulation and ablation temporarily balanced one another.

For the most part, glaciers advance and retreat slowly: gravity and the sheer mass of the ice see to that. And it is often an uneven, stutter-step process of moving and pausing. However, in times of rapid climate change—especially global warming—glaciers can move quickly, even visibly, over the course of a few months or years. Not surprisingly, given the recent warming of the planet, most of the world's glaciers are currently in rapid and headlong retreat (which is not good news for many reasons, though with one very thin silver lining: melting glaciers have exposed a wondrous array of animals, plants, and people and their artifacts, preserved long frozen beneath the ice[9]).

ANSWERS FROM THE OOZE

At the behest of Köppen, Milankovitch ultimately pushed his insolation calculations back hundreds of thousands of years, and identified dozens of periods in the past when Northern Hemisphere summer insolation should have favored glacial advance. Yet, geologists had found evidence for only four such episodes. Either his model or the geologists' data were flawed. But which? An answer to this terrestrial puzzle first emerged in an unlikely place: the deep ocean floor.

Two different forms *(isotopes)* of oxygen bond with hydrogen to form water—one is the garden variety ^{16}O, and the other is the heavier isotope ^{18}O. (An isotope is a variant of a specific element possessing the same number of protons but differing in the number of neutrons, and hence has a different mass and slightly different physical properties). When water vapor from the ocean reaches the clouds, it contains a greater proportion of ^{16}O because these lighter water molecules evaporate more readily; the heavier ^{18}O is left behind. During nonglacial periods, rain or snow over land is returned to the ocean, and thus there is a relatively constant ^{18}O to ^{16}O ratio in ocean water. However, reduce insolation, cool summer temperatures, and freeze that precipitation in glaciers and prevent its return to the sea, and the ocean becomes progressively depleted in the lighter ^{16}O molecules. Thus, the ratio of ^{18}O to ^{16}O in the ocean reveals how much ^{16}O is locked up on land, and so provides a proxy for times of glaciation.

So how do we know $^{18}O{:}^{16}O$ ratios at different times in the past? Tiny planktonic animals, foraminifera such as *Globigerina bulloides,* precipitate calcareous shells that record the proportion of $^{18}O{:}^{16}O$ in the seawater at the time they formed.[10] After the foraminifera die, their shells drift to the sea floor and become part of the millions of years of mud that has accumulated on the ocean bottom. Over the past half century, geologists have extracted these shells from the mud and analyzed their isotopic composition ($\delta^{18}O$, measured in parts per million [‰] relative to a standard). The result is a lengthy record of changing ocean oxygen isotope ratios, and hence a history of ice on land. The details of that record have since been greatly enhanced by cores drilled through the Greenland and Antarctic ice sheets (and in thousands of smaller mountain glaciers), which yield a remarkably precise record that reaches back hundreds of thousands and even millions of years of oxygen isotopic composition ($\delta^{18}O$); greenhouse gases such as methane

(CH_4) and carbon dioxide (CO_2); and even traces of terrestrial dust, volcanic ash, pollen, and sea salt wafting about the atmosphere.[11]

The cores from sea and ice reveal there were at least ten major glacial-interglacial episodes (of varying intensity) over the last million years, and perhaps twenty over the entire Ice Age, the start of which is now fairly firmly dated to about 2.6 million years ago. Agassiz's Ice Age and Lyell's Pleistocene, tightly bound since the 1840s, have had to go their separate ways since the Pleistocene—as formally defined by the International Commission on Stratigraphy using criteria other than glaciation—began only 1.81 million years ago and thus, strictly speaking, covers only the later part of the Ice Age. And therein are the distant drums of battle brewing, for those of us who work in this time period are determined to reunite the two for reasons historical and otherwise. In the fall of 2006, the International Union for Quaternary Research demanded the International Commission on Stratigraphy reconnect the Pleistocene and the Ice Age, warning that their failure to do so would be considered a "unilateral and hostile" action. We demanded "nothing less than control over our period of geological time."[12] We Pleistocene types are a feisty lot.

Yet, if there were approximately twenty glacial episodes over the Pleistocene, why had geologists found evidence of only four? Lay the blame squarely on the geological record. The bulldozing action of later glacial advances removed nearly all vestiges of earlier ones. Even today, knowing there were more advances has not made finding them any easier: only traces of two more have been spotted (geologists have now also realized that glaciation isn't solely a function of Milankovitch forcing[13]).

Our present landscape was shaped primarily by the most recent of those, the Wisconsin—we can still use that name—which is subdivided as follows:

1. Early Wisconsin, a cooler period marking the onset of this last major glacial episode (80,000–65,000 BP; designated as Marine Isotope Stage 4 [MIS 4]);

2. Middle Wisconsin, a relatively warmer period, with glaciers present on the land but much reduced (65,000–35,000 BP [MIS 3]);

3. Late Wisconsin, a period of expanded glaciers (35,000–10,000 BP [MIS 2]), with ice volume reaching its maximum, and sea levels—though not necessarily temperatures—falling to their minimum at 18,000 BP, referred to as the Last Glacial Maximum (LGM). By virtue of this being the last episode of Pleistocene glaciation, it is by far the best known.

4. Holocene, the present interglacial (10,000 BP–present [MIS 1]), which we are in today.

The landscape, climate, and environment of the Pleistocene was very different than that of the Holocene. The details follow, but here are some broad themes to bear in mind. For one, the Pleistocene was not merely the modern world plus a few giant glaciers here and there. Its massive ice sheets changed the geography of land and seas (opening and closing routes to America in the process), as well as altered climates and environments in profound ways.

But it wasn't simply a matter of being colder and wetter (or drier) for long stretches of time. The Pleistocene was marked by cycles of climate change taking place on varying time scales, some at a stately hundred thousand–year rhythm, others running their course in millennia, centuries, or even decades (Figure 5). Transitions between those shorter cycles could be jumpy, with large and often-abrupt climate changes occurring in centuries or less. As a result, even though many Pleistocene plants and animals are familiar to our modern eyes, because they had to cope with a more-or-less constantly changing world, they were often joined in biotic communities that looked very different—not only from those of the present, but also over the millennia of the Pleistocene. Change, not stability, was their norm, save for those times when climates

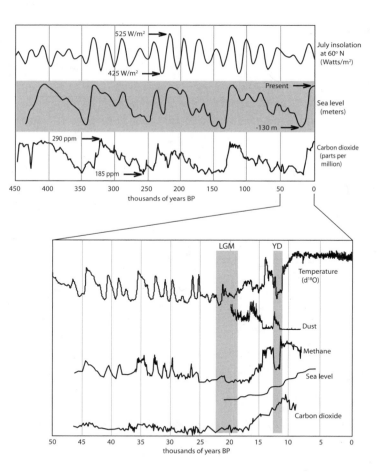

were at their coldest (as during the LGM) or warmest—which is why the climate and environment of the last few thousand years is quite tame by Pleistocene standards.[14]

How much of a challenge Pleistocene climate change would have posed for people depends on when they arrived, and how such changes affected resources vital to them. That the first Americans were adaptively light on their feet (they must have been: they made it here) was a very good thing as they moved out across this diverse and dynamic continent. This was especially true as the Pleistocene came to an end, for it went out with a bang (as we'll see, some take this literally), at a time when the first Americans were stretching their legs across the country. It was a challenging stage on which this human play was performed.

AN ICY STAGE

The first Americans arrived during the latter part of the Pleistocene, almost assuredly within the Wisconsin period. *When* in the Wisconsin is another matter. Most archaeologists put it in Late Wisconsin time; others push for a Middle or even Early Wisconsin

---

FIGURE 5.

Patterns and trends in Pleistocene climate and environment, 450,000 years ago to present and 50,000 years ago to present (all dates are in *calibrated* years BP).

The upper graph shows the period from 450,000 years ago to the present, and the trends over time—from top to bottom—of insolation, sea level, and atmospheric carbon dioxide. *Insolation,* as calculated for summers at 60°N latitude (in Watts/meter^2), is a significant—but not the sole—driver behind glaciation, with peaks roughly corresponding to glacial episodes. Global *sea level* presents something of a mirror image of the amount of glacial ice on land, varying from high stands that approximate the present to periods when global oceans were 120 ± 10 m below the present level, the latter depths reached during the Last Glacial Maximum (LGM). *Atmospheric carbon dioxide,* as measured in air bubbles trapped in the Antarctic ice sheet, ranged over that span between a low of 185 parts per million (ppm) to a pre-Industrial Revolution high of 290 ppm. The current figure of 385 ppm is well off the scale of this graph. (Adapted from Labeyrie et al. 2003:37; Overpeck, Whitlock, and Huntley 2003:83.)

The lower graph provides a close-up look at the last 50,000 years. Visible over that time in *average annual air temperature* (as inferred from δ^{18}O measured in the Greenland ice sheet) are the temperature lows during the Last Glacial Maximum (LGM) and Younger Dryas (YD), and the substantial warming that followed both periods. Also apparent are the wide swings of global temperatures throughout the Pleistocene, which narrowed significantly during the Holocene. The curves for *dust* and *methane,* also measured from the Greenland ice sheet, show the unmistakable signal of the Younger Dryas, when newly emerging peatlands—a significant source of methane—froze over, and the cold, dry, and windy conditions blew dust across Greenland. Likewise, the *sea level* curve (derived from the western Pacific) shows the two pulses of rapid sea level rise on either side of the Younger Dryas, as well as the flattening of the curve—that is, no significant change in global seas—during that period. Finally, *atmospheric carbon dioxide* began to increase after the LGM, and continued to do so through the YD, after which it stabilized for much of the Holocene, until more recent times. (Adapted from Liu et. al 2004; Oldfield and Alverson 2003:2; and Raynaud et al. 2003:22.)

appearance; a very few lobby for a pre-Wisconsin arrival. Unless radically new evidence is forthcoming, this last possibility seems remote. Playing the odds, our attention here focuses on the glacial geology, climate, and environment of the Late Wisconsin.

Late Wisconsin North America would be virtually unrecognizable today. Much of Canada and the northernmost United States were buried beneath an ice sheet comprised of three separate ice masses, the Laurentide, Cordilleran, and Innuitian. Little need be said of the small Innuitian ice sheet, save that it covered the islands of northernmost Canada (above 75°N) and spanned the gap between North America and Greenland. The other two were much larger, combining to cover about 15.8 million km^2 (the Antarctic and Greenland ice sheets today extend over approximately 15.4 million km^2), and accounting for more than a third of the approximately 42.5 million km^2 of glacial ice covering the LGM earth.[15] The Laurentide and Cordilleran played prominent roles in North America's Pleistocene climate, environment, and peopling.

The Laurentide, the larger by far, formed from the coalescence of glaciers flowing out from eastern Baffin Island, the Keewatin Uplands west of Hudson Bay, and Labrador. As the Laurentide expanded, it buried much of eastern and central North America, from above the Arctic Circle down to mid-Ohio (75°N to 40°N), and from the Atlantic coast to the eastern flank of the Rocky Mountains (64°W to 120°W). At the LGM, the Laurentide covered approximately 13.4 million km^2 and in places reached 3–4 kilometers in height (for comparison, the combined Antarctic ice sheets cover about 13.6 million km^2 and at their highest point in East Antarctica top out at just over 4.2 kilometers).

The Cordilleran ice sheet in the west was similarly a glacial composite, but of glaciers that originated in high elevation ranges of northwestern North America: ice expanded from the Alaska, Chugach, and St. Elias mountains in southeast Alaska, down the Coast Range of Canada, but also from the Rocky Mountains. When these fully coalesced, the Cordilleran covered an area of approximately 2.4 million km^2 and stretched from north of Anchorage, Alaska, to south of Seattle, Washington, blanketing much of the Yukon Territory and nearly all of British Columbia. The ice reached a thickness of 2.5 kilometers, overwhelming all but the highest summits, and flowed out into the Pacific Ocean.

South of the Laurentide and Cordilleran, isolated alpine glaciers formed on the Cascade, Sierra Nevada, and Rocky Mountain ranges, and on smaller ranges in places such as the Great Basin. The extent of these glaciers depended on topography, temperature, and moisture: northerly areas closer to the coast had more and more extensive alpine glaciation, while areas to the south and farther inland had glaciers only at the highest elevations. Thus, in the northern Cascades, glaciers descended to about 1,600 meters above sea level, whereas in the southern Sierra Nevada and Rocky Mountain ranges, glaciers formed only above approximately 3,000 meters in elevation. The impact of these relatively small and topographically restricted glaciers was insignificant—trivial, even—when measured against their continental-scale brethren.

The Late Wisconsin Laurentide ice sheet began expanding some 30,000–27,000 years ago, according to Arthur Dyke,[16] reaching the northern United States by about 26,000 BP. It reached its maximum southern extent at different times in different places: east of the Great Lakes region, the ice made it as far south as it would get very early—by roughly 21,000 BP—weaving a line from Georges Bank off Cape Cod, along southern Long Island, across northern New Jersey and Pennsylvania, and traversing Ohio, Indiana, and Illinois, whence it then turned sharply to the north up into Wisconsin. It toed that line for several thousand years. West of there, the Laurentide did not reach its maximum until about 14,000 BP.

Cordilleran ice probably began advancing no earlier than about 27,000 BP, with the expansion of ice in the high valleys of the Coast Range, and some 2,000–3,000 years later in the mountain ranges of southeast Alaska and interior northwest Canada.[17] As these glaciers spread out of their high valleys and coalesced, they initially moved west to the Pacific, approaching Vancouver Island by about 24,000 years ago. By 20,000 BP, the Cordilleran ice sheet had filled the lower-lying Fraser Plateau of south-central British Columbia, and by about 17,000 years ago had flowed south of Canada. It continued to expand in all directions, and by about 16,000 BP had reached its maximum extent along much of the Canadian coast, though it did not reach its southern extent in Montana, Idaho, and Washington until approximately 14,000 years ago.

Not surprisingly, these ice sheets had a profound influence on the topography of North America. Their advance disrupted and rerouted drainages, even reversing the flow of the Mississippi River, which in pre-glacial times exited North America via what's now Hudson's Bay. In retreat, they left behind moraines and other glacial landforms and lake-dotted landscapes (10,000 just by Minnesota's count), and in meltwater floods carried sediment that, when dry, was airlifted and deposited across large regions of interior North America (loess). Naturally, they had a significant impact on climate, which of course influenced the distribution of plant and animal species. Finally, this massive ice buildup helped set the stage for crossing over to the New World.

OCEANS AND SHELVES

Let's look first at what it meant for crossings: ice sheets the size of the Laurentide, Cordilleran, and their LGM counterparts require water. Nearly all of it must come from the oceans, where 97% of the planet's approximately 1.5×10^9 km^3 of water is stored. It's a zero-sum game. Building glaciers on land lowers sea level; melt the ice, and the seas rise again. That process is known as glacial *eustasy*, and it's ongoing: sea levels are currently rising, though at the subtle rate of 1–2 millimeters per year. (Of course, if the current melting of the Greenland and Antarctic ice sheets accelerates, so, too, will the rate of sea level rise, with potentially catastrophic consequences:

between them, these two ice sheets hold the equivalent of a whopping 60 meters of global seas. Low-lying island nations in the Pacific have good reason to be deeply concerned about global warming).[18]

During the LGM, roughly 5.2% of the world's water was frozen on land (about half in the Laurentide alone). That was enough to draw down global seas 120 ± 10 meters below their present level.[19] Shallow continental shelf became dry land, resulting in about 8% more land during the LGM than today. Small wonder the bones and teeth of mammals such as mammoth and mastodon are found in dredging the Atlantic floor: it's not because they swam out there. Since the Atlantic coastline was shifted hundreds of kilometers east of its present position, rivers had to extend further and cut deeper to reach distant Pleistocene oceans (today, their now-drowned channels are visible in bathymetric charts of coastal waters). The continental shelf on the Pacific side is much narrower, but in LGM times was still almost 50 kilometers beyond its present position. But it is to the far northern Pacific where our attention is drawn.

Today, Siberia and Alaska are separated by the Bering Strait, which even at its narrowest is still about 100 kilometers across. Seas only had to fall just over 50 meters for the continental shelf beneath the Bering and Chukchi Seas to become dry land. When sea levels were at their LGM minimum, a land bridge roughly 1,000 kilometers wide connected Asia and America. The Bering Land Bridge—or Beringia, which includes the portions of Siberia and Alaska it links—emerged in late Wisconsin times approximately 27,000 years ago, and was passable until severed by rising postglacial seas about 11,000 years ago. But from then until 10,000 BP, the breach was still narrow enough to readily freeze (and allow walking across) in winter. Soon thereafter, Pacific mollusks reappeared in the Chukchi Sea, an unmistakable sign the west-east Bering Land Bridge had once again become the south-north Bering Strait.[20]

Beringia was a vast, flat, ice-free, almost featureless highway trafficked by plants, animals, and people for millennia. That the same mammals inhabited Siberia and Alaska testifies to the ease of travel back and forth: only Siberia's woolly rhinoceros and a few Alaskan species (including the giant short-faced bear), failed to make the cross-Beringia commute. Traversable though Beringia was, hospitable it may not have been. From the absence of evidence for trees or lakes, and the presence of extensive sand dunes, loess, tracts of permafrost, and ice-wedge casts, it appears that the LGM climate was significantly drier and colder than at present (with precipitation estimated to have been 40%–75% lower, and average air temperatures lower by 7°F–11°F).

Still, the rich record of animal bones recovered from Alaska—sluicing for gold produces fossil wealth, too—bespeaks an abundant animal community. It was dominated, Dale Guthrie shows, by large grazers such as mammoth, horse, bison, saiga (an antelope), and musk ox, and their predators, the Pleistocene lion, saber-toothed cat, and giant short-faced bear. So many animals were present that Guthrie envisions LGM Beringia looking much like the African savanna with its vast game herds, though obviously one that was a great deal colder and drier. The *Mammoth Steppe*, he calls it.[21]

Once raised, the Mammoth Steppe notion was instantly criticized. Not by Guthrie's fellow paleontologists, but by palynologists studying the vegetation of Pleistocene Beringia. Their evidence, recorded in pollen-bearing sediments, pointed to a landscape shrouded in sparse, tundra-like vegetation. It was, they supposed, more polar desert than rich grassland, more polar Sahara than savanna. This contradictory testimony—lots of animals, which apparently thrived on no visible means of support—came to be known as the "Productivity Paradox." Yet, the burden of proof, Guthrie insisted, must be on the palynologists: Beringia had too many grazers for too long to have been as barren as they said.[22]

They're still arguing, for Guthrie has never demonstrated how abundant animals were (it's not an easy task), while palynologists have never resolved how much they can generalize about the Beringian landscape vegetation from their pollen cores. But reconciliation may be on the horizon. Sophisticated analyses of pollen records by Patricia Anderson and colleagues have shown that while LGM Beringia was indeed more tundra than steppe, it was a tundra unlike that in the Arctic today. Its productivity was much higher, and it included plenty of grasses for grazers.[23] Humans on this landscape may not have found many plants to eat, but there would have been animals to hunt as they made their way across from Siberia.

PLEISTOCENE PROMENADES

If migrating from Siberia to Alaska was relatively easy, traveling south to the Lower Forty-eight may not have been. The Laurentide and Cordilleran ice sheets grew toward each other, and for thousands of years around the time of the LGM, they coalesced along a nearly 2,000 kilometer stretch of western Canada (roughly today's British Columbia-Alberta border). Just how thoroughgoing a barrier the ice sheets formed is apparent in studies of ancient DNA from fossil bison and brown bears, which reveal an absence of gene flow between populations in Alaska and those south of the ice for much of the time between 25,000 and 13,000 years ago. Out along the coast during the LGM, Cordilleran ice flowed onto the Pacific Ocean's continental shelf, and along much of its length reached the Pleistocene sea, potentially blocking any coastal end runs.

People coming to America thus could have had an uninterrupted stroll from Siberia to Texas, if they arrived after Beringia emerged but before glacial ice had completely buried the landscape, or if they came after ice retreat but before Beringia was drowned. At times in between, it would have been a two-step shuffle: first to Alaska and later—after ice retreat—from Alaska to midlatitude North America. In either case, there was a choice of routes: through the continental interior along the eastern edge of the Rocky Mountains, or down along the Pacific coast (Figure 6). Timing, however, was everything.

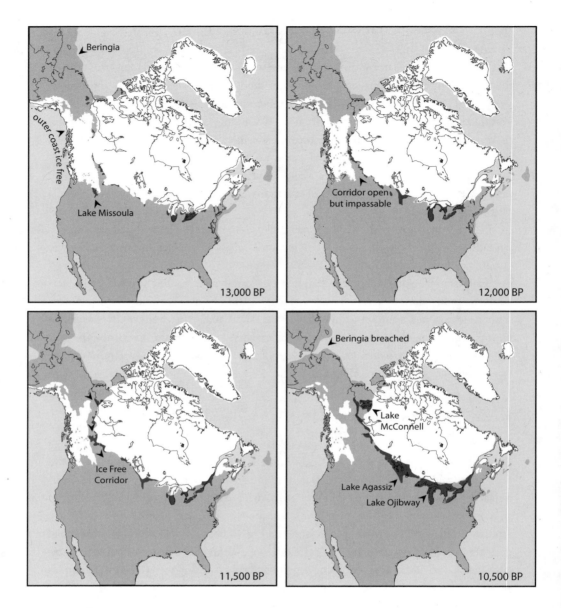

FIGURE 6.

Maps showing the position of retreating Cordilleran and Laurentide glacial ice at 13,000 BP, 12,000 BP, 11,500 BP, and 10,500 BP; the opening of the ice-free corridor; and the filling and draining of proglacial lakes. (Adapted from Dyke 2004 by David Willers.)

The interior route was effectively blocked by glacial expansion as early as about 28,000 years ago, and remained so until late glacial times. Yet, even after the Laurentide and Cordilleran glaciers retreated and an ice-free corridor began to open between them—which it did, zipperlike, from its northern and southern ends—the newly deglaciated terrain must have been an impenetrable mess. Low-lying areas would have been inundated by meltwater lakes and mud flats. The higher, drier terrain may have been traversable, but was likely inhospitable, at least while cold *katabatic* winds drained off nearby ice sheets. It was centuries or more—perhaps not until approximately 12,000 BP—that the land was fully restocked with plants and animals, and became a habitable passageway for human colonizers.[24]

A coastal route was open before about 23,000 years ago but was impassable during the LGM, as glaciers in the northern Gulf of Alaska blocked the route south. Were it possible to get around that barrier—either on foot or afloat—a trip down the LGM coast would still have challenged the hardiest Pleistocene pioneers. Travel would have been impeded by icebergs, sea ice, and glacier fronts hundreds of kilometers across, deeply crevassed, and with edges that calved into the ocean (see Plate 3). There were irregularly spaced segments of coast unburdened by Cordilleran ice even during the LGM, and though these may have provided safe refuge for plants and animals,[25] such areas would have been walled by ice and assuredly cold and harsh. It would have been risky for LGM travelers coming down the coast to count on finding such unpredictable landings. Better to travel once the ice had begun to retreat. But when was that?

Deglaciation of the coast was complex and took place at different times. Still, long reaches of the outer coast were ice free between 16,000 and 14,500 BP. Glacial ice blocked portions of the coast after that, but by 13,400 years ago those glaciers had retreated, and the coast was clear from Alaska to Washington State. The plants and animals necessary to human survival soon took root on land.[26]

But a caveat: our dates for the timing and availability of the different routes can change. Forty years ago, for example, the interior route was said to be *closed* from 27,000 to 12,000 BP; but then thirty and twenty years ago, the ice-free corridor was thought to have been *open* for much of Wisconsin time; ten years ago the pendulum of opinion swung back in the other direction, and once again the interior route was thought *closed* or otherwise impassable for much of the late Pleistocene (where matters now stand). This swinging to and fro was because of the continued radiocarbon and other chronological dating of glacial features and new interpretations that followed. It's a useful caution, suggesting as it does that although our current dates for the opening of interior and coastal routes might seem secure (and supported by DNA evidence), additional evidence on the timing of glacial advance and retreat and the environments along each route can change all that.

Regardless, once the first Americans got south of the North American ice sheets, glacial ice would not impact their movements until they reached the high Andes or the high latitudes of South America.

FORECASTING BACKWARD: A LOOK AT PLEISTOCENE CLIMATES

The climate that greeted the first Americans depends on when they got here, for climates were changing into, through, and out of the LGM.[27] Helping drive those changes were the ice sheets themselves, for these were mountain-sized weathermakers capable of diverting jet streams and air masses, influencing local climates as well as more distant land and sea surface temperatures (indirectly further altering air mass movement and behavior), and producing large anomalies in the earth's radiation balance and *albedo* (glaciers reflect more of the sun's energy back into space, compounding cooling).[28]

Overall, average annual air temperatures across the LGM globe were 9°F–12°F cooler than today. No surprise there: this was the Ice Age. However, the degree of cooling depended on location: it was only 5°F–9°F cooler in the tropics, 18°F–22°F cooler in areas closer to the ice sheets, and 37°F–41°F colder directly atop an ice sheet, were any human foragers hardy enough to venture atop one.[29]

In North America, an anticyclone (winds that rotate clockwise about a center of high pressure) developed over the Laurentide and Cordilleran ice sheets, and cold, katabatic winds flowed off the ice. Along their front, these produced a dry, periglacial band up to 200 kilometers wide, which left behind traces of permafrost such as ice-wedge casts, relict pingos (ice-mounded earth, since collapsed), and the scars of patterned ground. This would not have been a particularly hospitable place for people: annual air temperatures in permafrost regions average a chilly 21°F.

Yet, further south, temperatures might not have been so cold, nor were seasonal swings as pronounced as today. Winters, in fact, may have been relatively warmer. Here's why: today in North America, especially across its broad interior, we experience a *continental* climate marked by bitterly cold winters. These are a consequence of frigid Arctic air masses that sweep down from northern Canada and reach deep into the southern United States, plunging temperatures all the way. But cold air is low-lying air, and during glacial times would have been trapped in the Arctic behind the looming 3–4 kilometer-high Laurentide ice sheet, largely unable to penetrate midlatitude North America. Excluding that air mass dampened seasonal temperature swings, with the result that LGM climates were more *equable* than today, with cooler summers but relatively warmer winters.

Precipitation patterns during LGM times were likewise very different: it was wetter in some places, drier in others, and varied in different ways seasonally. Credit the ice sheet once more: it split the westerly jet stream into branches that flowed along its northern and southern edges. The northern, weaker branch brought relatively dry air across Alaska, where LGM climates were only a few degrees cooler than present. By the time the jet stream bent around to reach New England and the North Atlantic, it was feeding in cold arctic air, producing climatic conditions that were wetter than present, as storms formed along the border of glacial ice and ocean.

Further south it was not as cold or as wet. The presence of the ice sheet, coupled with the influx of Caribbean air (extending northward to a latitude of 40°N), strengthened westerly flow and pumped Pacific air into the Upper Midwest in summer, the air drying as it traveled eastward over the Rocky Mountains. Drier conditions also marked much of the southeast. Polar waters expanded south, cooling sea surface temperatures in the Atlantic 3°F–6°F (temperatures off coastal New England were as much as 25°F cooler than today). Cooler oceans give up less water to evaporation, reducing moisture available for precipitation on land. Florida was especially hard hit, suffering a steep decline in LGM rainfall.[30]

On the opposite side of the continent, the southern-shifted branch of the jet stream deflected moist Pacific air on a more southeasterly track, which brought onshore a strong flow of cool, moist air and rain to what is now the arid/semiarid Great Basin and American Southwest. Rain, of course, meant cloud cover, and that coupled with cooler temperatures lowered evaporation. By one estimate LGM precipitation was twice what it is at present, and evaporation about 25% less. Regardless of the exact figures, the result was lakes on the land: 25,000 dotted the Southern High Plains of Texas alone. *Pluvial* lakes, we call these to signify their formation under climate conditions wetter than today.

The largest of the LGM's pluvial lakes were in the Great Basin, and the greatest of them all was Lake Bonneville, which covered nearly 52,000 km² of northeastern Utah; it was about the size of present-day Lake Michigan. Today we see only its remnants: the Bonneville "salt flats" (the floor of the ancient lake); prominent wave-cut beaches (more parallel roads) etched into places such as the Wasatch Mountains east of Salt Lake City (Figure 7); and the Great Salt, Utah, and Sevier lakes, the puddles Bonneville left behind after it evaporated (which together represent no more than 6% of their ancestral lake). This vast inland sea hit its high water mark—a depth of about 372 meters—around 15,000 BP,[31] but the lake would have stayed higher longer had it not catastrophically drained following a breach in its sidewall. Ultimately it stabilized, and its later Pleistocene history is similar to that of other Great Basin pluvial lakes, such as Lake Lahontan and Mono Lake. All together, the Pleistocene Great Basin had about eleven times more water on its floor, and would have provided habitats for a wide variety of plants, animals, and people—were they in the neighborhood.

While the desert bloomed during the LGM, the rainy Pacific Northwest was much drier, owing to prevailing easterlies. Nevertheless, this was a time of catastrophic flooding in the Northwest, though for reasons that had nothing to do with precipitation. Instead, blame glacial ice. When Cordilleran ice reached its maximum extent, it dammed the Clark Fork of the Columbia River, forming proglacial Lake Missoula. With a volume of about 2,500 km³ (the size of lakes Erie and Ontario combined) and as much as 650 meters deep, Lake Missoula drowned the valleys of northwest Montana and etched a series of Glen Roy–like terraces into its mountains (travelers tip: when visiting Missoula, Montana, be sure to look up at the hillsides).

FIGURE 7.

The power of water: a wave-cut notch on the side of Fort Rock Cave in the northern Great Basin, Oregon. The pluvial lake that once filled this now desert area and eroded that notch was relatively small by Great Basin standards. (Photograph by George T. Jones, courtesy of Charlotte Beck.)

Lake Missoula existed from 15,700 to 13,500 BP. But the water was held in check by glacial ice, a massive but notoriously unstable dam that was no match for the tremendous pressure exerted by the lake water rising up behind it. Repeatedly, the water bored a tunnel through the base of the dam, burst through, and reached full flood within a few hours (helped by the relatively warmer water rushing through and enlarging the tunnel), then stopped flowing when pressure from the overlying ice squeezed the tunnel shut. Overall, the Lake Missoula 'spigot' was opened some 90 times over the lake's existence, though at progressively shallower lake levels as the ice dam thinned.[32] At its flooding peak, Lake Missoula waters burst out across Idaho and into Washington (Figure 8), discharging at ~30 million meters^3/second—the Amazon River, in comparison, can barely muster 200,000 meters^3/second—reaching the Pacific Ocean within a few days.

The floods carved islands in the loess, rafted 100-ton boulders like corks, and tore deep gashes in the bedrock, one of which— the tandem of Lower and Upper Grand Coulee, is nearly 80 kilometers long, as much as 7 kilometers wide, and nearly 300 meters deep. When Lake Missoula's waters plunged back into the Columbia River in central Washington, they created waterfalls that in places were twice the height and width of Niagara Falls. The sight and sound of one of the Missoula floods must have been awesome, were anyone on the ground to see it, hear it—and survive it.[33]

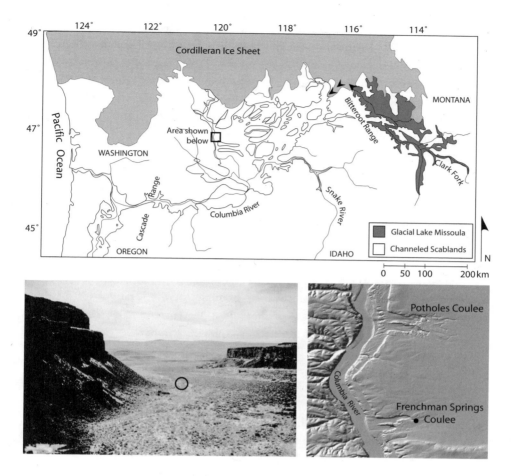

FIGURE 8.

Map of glacial Lake Missoula and the Channeled Scablands of the Pacific Northwest (arrows between these areas denote the exit path of the lake water). The digital elevation model (DEM) on the lower right shows the topographic scars left on the landscape of central Washington (just west of George, WA) as the floodwaters of Lake Missoula reached the Columbia and tore gashes into weak areas of bedrock, creating brief but spectacular waterfalls, which are now Potholes and Frenchman Springs coulees. The photograph on the lower left was taken from what would have been the lip of the waterfall at Frenchman Springs Coulee (near the dot on the DEM), looking toward the Columbia River. For scale, there is a full-sized white pickup truck (circled) on the floor of the coulee more than 100 m below where the photograph was taken. (Photograph by David J. Meltzer; map and DEM by Judith R. Cooper.)

ENVIRONMENTS WITHOUT ANALOGUE

Late Wisconsin plant and animal communities are well known from some areas, less so in others. But in all cases, they were arranged very differently than they are today: it's because of the effects of the ice sheets once again. Nowadays, plant and animal distributions in eastern North America are largely determined by factors such as seasonal temperature and growing season. The northern and southern boundaries of the boreal forest across Canada, for instance, roughly mark the average positions of summer and winter Arctic air (respectively). Beyond the one-hundredth meridian, in semiarid and arid western North America, moisture plays the more dominant role in shaping ecological communities (though at high elevations, temperature reasserts its importance). But bring on continental glaciers and change the climate ground rules, and plant and animal communities change dramatically.

Importantly, they did not merely shift south en masse in response to advancing ice, but instead these communities came apart: individual species set off on their own and at their own pace, moving (or not) depending on their specific climatic and ecological tolerances. Some went far: during the LGM, spruce trees grew in Louisiana and Florida, while arctic caribou grazed in northern Alabama.

The northern species that migrated south did not run roughshod over temperate species already present, but instead were integrated with them in diverse communities often far richer than those of today. Living in Peccary Cave (Arkansas) and Cheek Bend Cave (Tennessee) during the Pleistocene, for example, were several species that now live far apart: among them, *Microtus xanthognathus* (the taiga or Yellow-cheeked vole), which today never strays south of the Canada-United States border; *Geomys bursarius* (the Plains pocket gopher), which today inhabits western prairie and plains; and *Dasypus bellus* (the beautiful armadillo), an extinct animal that, like its modern counterpart, inhabited subtropical regions. These species could co-exist because LGM climates were equable: summers south of the ice were cool enough to support northern species, while winters were not so very cold that they drove southern species away.

In the mountain west, much of the shifting of species' ranges was according to altitude rather than latitude. Plants and animals that are today confined by heat and aridity to higher elevations, during glacial times were able to migrate downslope in response to cooler temperatures and greater moisture at lower elevations. The consequences of that shift are dramatically illustrated in the Great Basin, with its long, high, spiny mountain ranges separated by low valleys. The elevation difference between valley floor and mountaintop can be as much as several thousand meters. Today, it's a difficult trick for many animals adapted to the cool climate and alpine vegetation of the high peaks to scramble down and across the harsh desert lowland—think Death Valley—and scamper up into an adjacent mountain range, even in the cool of winter. And, yet, many of the same species (small mammals mostly) occupy adjacent mountains, suggesting to ecologist James Brown these isolated "islands" were connected during the cooler and wetter Pleistocene when intervening lowlands were more habitable and enabled easier passage. Subsequent work by Brown, zooarchaeologist Donald Grayson, and others showed—by documenting species common to separate mountaintops—that was mostly the case, though the details vary by species.[34]

All this shifting and integration of plants and animals made for Pleistocene biotic communities that lacked modern analogues. Although that's partly because those communities also contained many now-extinct animals, even those species still present today occurred during the Pleistocene in numbers and distributions that differed from the present. Complicated as it was, painting a picture of the LGM landscape is best done with broad brush strokes.

Along the southern edge of the ice sheet, from New England to Montana, and south along the summit of the Appalachians, there was a discontinuous treeless zone of tundra plants and animals—arctic and alpine beetles included—intermingled with southern and prairie forms. South of the treeless zone over much of eastern and central North America was a diverse forest, dominated by spruce in the northern Midwest and pine in the northeast.

Still further south in North America, forests increasingly included oak, ash, and hickory, and other temperate and deciduous elements. In these, cold-loving shrews, lemmings, and voles lived alongside warm-loving armadillos and tapirs, chipmunks and ground squirrels. Naturally, the ratio of cold- to warm-loving species decreased as annual temperatures increased from north to south. These forests were also the domain of a variety of now-extinct large mammals, such as mastodon, sloth, and mammoth, whose browsing and grazing were in part responsible for these Pleistocene forests being much more open than the dense and closed forests of the historic period.[35]

In the Pacific Northwest, areas along the ice margin were dry and largely treeless, while further south, an open parkland of spruce and lodgepole pine grew in the unglaciated portions of the Puget lowland. This setting supported a mix of animals, in this instance denizens of alpine tundra, subalpine coniferous forest, and grassland species.

In the far west and southwest, snow lines and tree lines dipped—nearly 1,000 meters in places—below their present altitude. In the Great Basin, bristlecone and limber pine grew in places where in today's hotter and drier climates only desert scrub (e.g., creosote) can survive. On the Colorado Plateau, complex forests developed as species of pine, spruce, hemlock, and fir shifted downslope. In the deserts of the Southwest, piñon-juniper forests shrouded much of the landscape, intermingled with sagebrush, with jackrabbits and pocket gophers scrambling about in the open woodland.

On the plains and prairie of the midcontinent, a mixed spruce/pine forest pushed west from the Great Lakes into the eastern plains. How far it extended, and at what times (its front moved) is unknown, partly for want of sites of the proper age. Likely, however, the grassland/forest boundary was further west than it is today: greater LGM precipitation on the plains would have enabled trees to grow in areas now too dry for their survival. Still, the plains then—as the plains now—was largely open grassland.

It's no easy task to get excited about what *kinds* of grasses were present on that landscape, but it is worth brief notice, because not all grasses are alike. There are the warm-season types that grow in the summer, and cool-season forms that undergo their growth in the fall, winter, and spring. Which type is in an area is driven primarily by summer temperature and precipitation, but also by atmospheric CO_2: warm-season forms are more common in southerly areas, cool-season in higher elevation and northerly areas.

During glacial times, cool-season forms should have been more widespread, yet may not have been. Knowing their distribution over the last few millennia of the Pleistocene is potentially quite important because warm-season grasses contain potent anti-herbivory toxins, possibly bad news for Pleistocene mammals that may have grazed on those, especially since some of those mammals soon went extinct.

MEGAFAUNAL MENAGERIE

Thirty-five genera of mammals ultimately vanished from the North American Pleistocene landscape (see Table 2). It's a stunning loss, made all the more so because the roster of the disappeared is a spectacular zoo of mostly (though not entirely) very large animals—megafauna—of highly diverse adaptations and habitats, and for which humans are a prime suspect in their demise.[36]

The tally includes four species of mammoth, animals that stood upwards of 4 meters at the shoulder, weighed nearly 4,500 kilograms (5 tons), and carried massive ivory tusks (overgrown incisors, really) that sometimes reached lengths of 4.5 meters. Mammoth herds grazed the vast grasslands of Pleistocene North America, Europe, and Asia, and some of their freeze-dried carcasses have been unearthed in Siberia and Alaska (Figure 9), where in the last century, their flesh was fed to dog teams (none of the dogs were reportedly worse off for the experience). More food, this time for thought: bits of ancient DNA have been successfully recovered from tissue and bones of the woolly mammoth *(Mammuthus primigenius)*.[37] Although fragmentary, these reveal that the mammoth genome is virtually identical to that of African elephants. It's too early to bet the ranch on the chances of cloning a mammoth inside an elephant.

In the same taxonomic order (Proboscidea), but a different family than the mammoth and elephant, was the American mastodon *(Mammut americanum)*. These were shorter (up to 3 meters at the shoulder) and stockier than the mammoth, and were likely relatively solitary animals that dined on twigs and cones of conifers and other browse in forests, where their bones have been unearthed for centuries. One very early find was presented to Cotton Mather who, evidently worse at human anatomy than he was at identifying witches in seventeenth-century Salem (Massachusetts), mistook mastodon bones as proof of the Biblical passage, "There were giants in the earth in those days." Right size, wrong animal.

Also on the list: four genera of giant ground sloths (see Plate 4). The smallest of these was about the size of a modern black bear; the largest, *Eremotherium*, reached a length of 5.5 meters (tall giraffes top out at under 5 meters), and probably weighed 2,700 kilograms (3 tons). Then there was *Megalonyx jeffersonii*, named for Thomas Jefferson, who first mistook this "animal of the clawed kind" as a giant carnivore three times larger than the modern lion (the Latinized term *Megalonyx* comes from the Greek as "great claw"). It was, in fact, the smallest of the ground sloths but still about

FIGURE 9.

Dima, the freeze-dried baby woolly mammoth—note the hair around the ankles—shortly after emerging from the permafrost in 1977 on a terrace of the Kirgiliakh River, a tributary of the Kolyma River in Siberia. Dima was perhaps just 4 months old when he walked into and was entrapped in a muddy pit. Ultimately, he sank deep into the mud, slowly starved (preserved in his stomach were bits of plant detritus and sediment), then died from inhaling sediment into his lungs. (Photograph courtesy of David M. Hopkins.)

the size of a modern black bear (the sloths' modern, distant relatives, the tree sloths of South America, might weigh 9 kilograms soaking wet).

Then there was the glyptodont, a mammal encased in a turtle-like shell with massive limbs and an armored tail and skull. Stretching nearly 3 meters in length, 1.5 meters in height, and weighing upwards of a ton, this cumbersome animal appears to have lived along lakes, streams, and marshes, and may have been semi-aquatic. Rounding out the mammalian herbivore list were horses and three genera of camel (one of which looked like a large version of a dromedary; the other two were more like present-day llamas); pronghorn; musk ox (the latter including a genus longer limbed and able to live in warmer climes than the modern musk ox); multiple genera of peccaries, deer, and tapir; and assorted bit players such as Harrington's mountain goat, the Aztlan rabbit, and the giant beaver *(Castoroides ohioensis)*, which weighed over 90 kilograms and packed incisors 20

Order and family	Genus and species	Common name
North American genera that went extinct in the Late Pleistocene		
Cingulata		
Pampatheriidae	*Pampatherium* sp.	Southern pampathere
	Holmesina septentrionalis	Northern pampathere
Glyptodontidae	*Glyptotherium floridanus*	Simpson's glyptodont
Pilosa		
Megalonychidae	*Megalonyx jeffersonii*	Jefferson's ground sloth
Megatheriidae	*Eremotherium rusconii*	Ruscon's ground sloth
	Nothrotheriops shastensis	Shasta ground sloth
Mylontidae	*Paramylodon harlani*	Harlan's ground sloth
Carnivora		
Mustelidae	*Brachyprotoma obtusata*	Short-faced skunk
Canidae	*Cuon alpinus*[a]	Dhole
Ursidae	*Tremarctos floridanus*[a]	Spectacled bear
	Arctodus simus	Giant short-faced bear
Felidae	*Smilodon fatalis*	Saber-toothed cat
	Homotherium serum	Scimitar cat
	Miracinonyx trumani	American cheetah
Rodentia		
Castoridae	*Castoroides ohioensis*	Giant beaver
Hydrochaeridae	*Hydrochoerus holmesi*[a]	Holmes's capybara
	Neochoerus pinckneyi	Pinckney's capybara
Lagomorpha		
Leporidae	*Aztlanolagus*	Aztlan rabbit
Perissodactyla		
Equidae	*Equus* spp.[a]	Horses
Tapiridae	*Tapirus* spp.[a]	Tapirs
Artiodactyla		
Tayussuidae	*Mylohyus nasutus*	Long-nosed peccary
	Platygonus compressus	Flat-headed peccary
Camelidae	*Camelops hesternus*	Yesterday's camel
	Hemiauchenia macrocephala	Large-headed llama
	Paleolama mirifica	Stout-legged llama
Cervidae	*Navahoceros fricki*	Mountain deer
	Cervalces scotti	Stag-moose

TABLE 2   *(continued)*

Antilocapridae	*Capromeryx minor*	Diminutive pronghorn
	Tetrameryx shuleri	Shuler's pronghorn
	Stockoceros spp.	Pronghorns
Bovidae	*Saiga tatarica*[a]	Saiga
	Euceratherium collinum	Shrub ox
	Bootherium bombifrons	Harlan's musk ox
Proboscidea		
Mammutidae	*Mammut americanum*	**American mastodon**
Elephantidae	*Mammuthus* spp.	**Mammoth**

North American species *that went extinct; members of the* genus *survived here or elsewhere*

Xenartha		
Dasypodidae	*Dasypus bellus*	Beautiful armadillo
Carnivora		
Canidae	*Canis dirus*	Dire wolf
Ursidae	*Temarctos floridanus*	Spectacled bear
Felidae	*Panthera leo atrox*	American lion
Artiodactyla		
Bovidae	*Oreamnos harringtonii*	Harrington's mountain goat
	Bison antiquus	**Bison**

Select North American large mammal genera *and* species *that survived late Pleistocene extinction*

Artiodactyla		
Cervidae	*Alces alces*	**Moose**
	Cervus elaphus	**Elk**
	Odocoileus spp.	**Deer**
	Rangifer tarandus	**Caribou**
Antilocapridae	*Antilocapra americana*	**Pronghorn**
Bovidae	*Ovibos moschatus*	**Musk ox**
	Ovis spp.	**Mountain sheep**

NOTE: Animals listed in bold type are ones for which there is secure archaeological evidence they were preyed upon by human hunters. Taxonomic data from E. Anderson 1984; Grayson 2007; Kurten and Anderson 1980. Information on archaeological occurrences from Frison 1991; Grayson and Meltzer 2002.

[a] Other members of the genus survived outside of North America

centimeters long (judging by its teeth, this was no dam builder, but nonetheless frequented lakes and ponds).

Preying on these herbivores were some formidable carnivores, including the giant short-faced bear *(Arctodus)*, long limbed and 33% larger than the largest living carnivore, the Alaskan brown bear. *Arctodus* was likely omnivorous, but may have fed on bison and other large grazers. Also on the carnivore list was a large dog (the dhole), the short-faced skunk (larger and more carnivorous than its modern relative), and several species of big cats, including the most spectacular carnivore of all, the saber-toothed cat.

The saber-toothed cat, aptly named *Smilodon fatalis* (who says taxonomists have no sense of humor?), had 15-centimenter (6 in)-long upper canines, which were saw-toothed front to back and dagger thick side to side. This lion-sized predator probably fed on young mammoth or other slow-footed herbivores, its sword-like teeth used to slash the throat or abdomen of its prey to produce suffocation, heavy bleeding, and death. The saber-toothed cat had short, powerful legs: it was not built to run fast or far, but was probably an ambush hunter. Thousands of its remains have been found in Los Angeles's La Brea tar pits, suggesting *Smilodon* haunted the area, waiting for prey to become mired in the tar pits. Studies of *Smilodon* teeth indicate they rarely gnawed on or cracked open the bones of their prey, which explains why their abandoned kills attracted scavengers such as the dire wolf *(Canis dirus)*, which seemingly made a career out of cleaning up saber-toothed cat leftovers.

A common feature of most of these extinct animals was their large size: virtually all, with the exception of the rabbit *Aztlanolagus* and the diminutive pronghorn, had an adult body weight of at least 50 kilograms (100 lbs), and in the elephant-sized animals, of many tons. There were by some estimates over 100 million of these aircraft carriers of the animal kingdom inhabiting the diverse landscape of Pleistocene North America. Some of those genera were evolutionarily homegrown (such as the horse and camel), many had been in the Americas hundreds of thousands if not millions of years, and all had weathered multiple glacial-interglacial cycles. Yet, by 10,800 BP, all were extinct—which is not to say that all went extinct at that moment, for current evidence indicates only half of the thirty-five genera had survived until then.

Even so, because their disappearance appears to coincide roughly with two very distinctive events—people coming in and the Pleistocene going out—disentangling what caused these extinctions has proven no easy task, and is a matter hotly debated. Whether humans were to blame is taken up in Chapter 8 when the archaeological record of Clovis hunters is examined; here, let's stay with the climatic and ecological changes wrought by the end of the Pleistocene.

A WARMING WORLD

We officially mark that end at 10,000 BP. That particular date is arbitrary and was mostly chosen, as geologist David Hopkins cheerfully admitted, because it's a nice, round number. In fact, the changes that signaled the beginning of the end of the Ice Age began earlier, and the end—at least in northerly areas—came later.

Around 16,000 years ago, the first hints global climates were changing appeared. Higher summer insolation (those orbital cycles never stop) began steadily warming the Northern Hemisphere. Winter snow and ice were no longer staying frozen through the summer, and glaciers slowly started to melt. It was the Antarctic glaciers and those in the midlatitudes that began retreating earliest (around 14,500 years ago). Deglaciation did not begin in earnest in the Northern Hemisphere for another millennium or so, and it would take some 10,000 years: it's not easy to melt 54,000,000 km³ of ice.[38]

Sometime after 14,500 BP, the Laurentide and Cordilleran ice sheets began to pull back, though with occasional, brief re-advances of ice lobes and even-briefer surges (the latter are triggered by the slippery internal mechanics of glaciers[39]). By 13,000 years ago the Laurentide had decreased in area by about 25%, and in volume by about 50%: it thinned more rapidly than it shrank. By 11,000 BP, glacial ice had abandoned New England for Quebec, opening the St. Lawrence Seaway, an event that may have had profound climatic consequences (of which, much more below). By 10,000 BP, the Laurentide was half its former self, and had vanished from western Canada and the upper Midwest. By 7700 BP, it made its last stand over Hudson Bay, then split in two (an event also with climatic consequences). By 5,000 years ago, its surviving remnants huddled over northern Quebec and on Baffin Island.[40]

On the other side of the continent, Cordilleran ice lingered for a brief time at its maximum position after 14,000 BP, then departed Seattle and the Puget lowland. It melted more rapidly than the Laurentide, and by 10,000 BP had nearly vanished, save in the mountains of southeast Alaska and at high spots on the Coast Range. Robert Fulton and colleagues suspect its fast-disappearing act was the result of geography: the bulk of the ice sheet rested in the low-lying region between the Coast Range and Rocky Mountains. Once post-LGM warming slowed and then stopped the replenishing flow of ice down from those mountains, the low-lying ice sheet was doomed. The interior portion, isolated and starved, was unable to retreat gracefully and wasted in place. On the Pacific side, coastal ice rapidly calved into the sea.

As ice sheets withered, meltwater returned to the oceans. Much of the Laurentide drained down the Mississippi River into the Gulf of Mexico, swelling the river far wider and deeper than any historically known flood and presenting a formidable challenge for crossings. And it was choked with sediment and flanked by extensive mud flats, which when dried and scoured by winds fanned aloft chokingly thick clouds of fine-grained sediment (silt) that fell onto the surrounding uplands. In places, these loess deposits are 40–50 meters thick.[41]

Of course, once those waters returned to the ocean, sea levels began to rise, flooding river valleys and trenches carved by glaciers that once had extended to Pleistocene shores, and creating large estuaries (such as the Chesapeake Bay) and deep fjords (such as the port of Seattle). By one estimate, some 110 meters' worth of sea level was returned to the oceans from 16,000 to 6,000 years ago (modern sea levels were reached around 4000 BP, the timing varying by area). That's an average of 1.1 meters per century, fast by modern standards—the seas are rising today at the rate of 10–20 centimeters per century. But rates and averages mean little here, at least in terms of glacial and climatic history, for they mask the fact that this overall, upward trend was mostly slow and steady, but occasionally punctuated by pulses of very rapid sea level rise (two major pulses occurred at 16,000 and 12,500 BP).[42]

Had human colonizers come down the coast sometime after 16,000 BP, they likely would not have noticed the rising seas—water levels weren't coming up *that* fast. Surely, however, they would have encountered the milky deltas formed at the mouths of sediment-choked rivers draining interior ice fields. It likely took several centuries before these rivers ceased muddying tide pools and choking off near-shore environments.

Rising sea levels, of course, must have drowned many early sites that were once located close to the coast—but it did not necessarily drown *all* of them. For as sea levels were rising, so, too, were portions of the newly deglaciated coast of Alaska and British Columbia: when freed from beneath their weighty ice burden, they rose upward (a phenomenon geologists call *isostatic rebound*[43]). What goes down, must come up. And because this is also a tectonically active area, occasionally the land comes up very quickly indeed. All that's good news for archaeology because along the Pacific northwest coast, where we might hope to find evidence of early colonizers making their way south, segments of the Pleistocene shoreline on which they may have walked are now well above sea level. Finding the sites that might be on them in an area now covered in lush temperate rain forest is a challenge, but not an impossible one (Chapter 4).[44]

COLD WATER AND HOT AIR

Rising seas did more than just swallow coastline; they also drove climate on land by toying with one of the principal controls on the earth's thermostat: carbon dioxide. As Wallace Broecker explains,[45] the world's oceans—particularly the Atlantic—contain a grand system of watery "conveyor belts" that distribute and equilibrate global heat between the equator and poles. Lurking less than 1 kilometer below the waves of the Atlantic, a current of warm, salty water (the Gulf Stream) flows north from the tropics along the North American east coast. As it approaches the vicinity of Iceland and Greenland, strong winds sweep the surface water aside, and the newly arrived and now-exposed warm equatorial waters rise and cool (by about 8°C), and in so doing release water vapor and heat that presently bring wet, mild winters to western Europe. Those erstwhile tropical waters are now colder, saltier, and denser than when they arrived, and so they slowly sink back into the abyss. There, the deep water rides a southbound belt of the conveyor: its next destination, the southern ocean. Meridional overturning circulation is its official name, but the process is known more colloquially as *thermohaline circulation,* and it moves with astonishing power: below the relative calm of the Atlantic surface are currents with a flow equal to 100 Amazon Rivers.[46]

Thermohaline circulation operates very efficiently nowadays because of the large temperature difference between the equator and poles that provides its energy, and because of the relative salinity of the North Atlantic, which is critical to the density-driven sinking that drives deep water formation.[47] But if there were a large injection of freshwater into the northern gears of the conveyor from, say, a melting ice sheet (an experiment nature has performed multiple times), the seas can freeze in winter (saltwater has a lower freezing point). And if frozen, the surface water will fail to sink, and when that happens thermohaline circulation slows to a halt or simply collapses.[48]

When that happened in the late Pleistocene—the Younger Dryas, it's called, for a tundra flower (*Dryas octopetala*) that flourished in Europe during this time—there were direct and geologically instantaneous climatic consequences. Across the northern hemisphere, especially downwind of the now-frozen North Atlantic, temperatures plummeted: the Greenland ice cores record an *average* annual air temperature during the Younger Dryas of about −52°F. But the effects of the Younger Dryas were felt elsewhere as they rippled through Earth's climate system, and at a moment when the first Americans were dispersing across the continent (more on this below, and in Chapter 9).[49]

Switching thermohaline circulation on and off also influences global climates in a less obvious but no less consequential way, for the oceans harbor vast amounts of carbon dioxide, CO_2, a greenhouse gas that looms large in our story of past climate (and, not inconsequentially, future climate). Although CO_2 resides in both atmosphere and ocean, it's not shared in equal measure: the oceans contain sixty times more CO_2. Much of that CO_2 is stored deep in the abyss, where it can be sequestered for thousands of years (and so effectively the U.S. Department of Energy is pursuing the possibility of injecting excess anthropogenic CO_2 into the deep ocean—a sort of CO_2 government witness protection program). Thus, the amount of CO_2 in the atmosphere at a given time is strongly influenced by how much is absorbed into the ocean and/or photosynthesized by phytoplankton, and then sent to the deep. That, in turn, depends on a host of factors, the most relevant to our purposes being changes in deep water formation, which of course are tied to changing thermohaline circulation.[50]

Why does that matter to climate, or even to archaeology? Because atmospheric CO_2 falls when thermohaline circulation stalls, and rises when the conveyor re-starts. And with each increase of approximately 70 parts per million (ppm) of atmospheric CO_2, global temperatures rise about 1°C(1.8°F). Over the last million years or so, CO_2 levels overall have moved only within a narrow range: between 180 and 290 ppm. The lower end marks cooler, glacial periods, the higher end the warmer, interglacial periods. The efficiency with which the earth's climate system has locked down that range is in no small measure why the very recent, rapid rise in atmospheric CO_2 to its current level of about 385 ppm is cause for great concern.[51]

Not surprisingly, turning the thermohaline circulation switch off and on also wreaks havoc on radiocarbon dating, since doing so also scrambles the ratio of atmospheric:ocean radiocarbon, making the radiocarbon clock sometimes run too slow (because of a decline in atmosphere's $^{14}C{:}^{12}C$ ratio), or too fast (because of a rise in the atmosphere's $^{14}C{:}^{12}C$ ratio).[52] It's little wonder that radiocarbon dating sites of the first Americans that were occupied during the Younger Dryas is complicated (Chapter 1).

Thermohaline circulation remained weak (and atmospheric CO_2 levels flat) following the initial pulse of meltwater (16,000 BP), perhaps because waters of the North Atlantic stayed cool, and the overall temperature difference between equator and pole remained relatively small. But as the planet warmed, and especially following the second major pulse of meltwater pulse (at 12,500 BP), thermohaline circulation strengthened (oceanographer Peter Clark and colleagues suggest that was the result of the melting of the Antarctic ice cap, which sent freshwater flooding north from the Southern Hemisphere, thereby jump-starting North Atlantic deep water formation).[53]

Stronger thermohaline circulation sped widespread warming—the Bølling-Allerød warm period, it's called—and an increase in atmospheric CO_2.[54] This further hastened the demise of North American ice sheets, and as their climate-modifying effects steadily diminished, the southern branch of the jet stream and the winter storm track shifted north, and the glacial anticyclone began to weaken. Swamps and peatlands formed on newly deglaciated landscapes, which helped cause a spike in another greenhouse gas—methane (prior to the modern era, swamp gas was the major source of methane).[55]

Correspondingly, there was on land a "major reorganization" of biota in response to such changes.[56] Northern areas once in tundra were replaced by spruce forest, while spruce-dominated woodlands in turn became hardwood-dominated forest (in the Midwest), or pine and birch forest (the Northeast and New England). Further south, deciduous trees (maple, oak, and beech) expanded their numbers. Throughout, the amount of tree cover generally increased, and yet was still marked by species associations that lack modern analogues.

This process played out in complicated ways across space and through time because plants and animals left their Pleistocene communities the same way they had entered: it was every species for itself, finding their own ways to new ranges and habitats (or, in some cases, to extinction), their speed and direction tied to their individual tolerance limits, dispersal abilities, interspecies competition, location during the

LGM (oak trees were already lurking in low numbers in protected settings in North Dakota, for example, ready to spring forth—metaphorically speaking—when climates improved), the availability of suitable habitats, and a variety of other factors. Had the process been recorded in time-lapse photography and then viewed at high speed, it would have resembled a "Keystone Kops" routine, a free-for-all of species shuffling in different directions at different rates at different times to colonize new habitats.

Estimates vary, but some shuffled faster than others. Most tree species moved 100–500 meters per year, but some pines virtually sprinted across the landscape at upwards of 3,000 meters per year.[57] In some cases vegetation was *almost* able to keep pace with climate change, lagging behind by only a century or so. But then, Sitka spruce in the Pacific Northwest is still expanding west in response to deglaciation. Animal species, also moving at their own individual paces, presumably kept up with climate and vegetation change, but again depending on the species: some small mammals are still adjusting to postglacial climatic conditions. The central point stands, however: conditions were, as Daniel Mann describes them, a "biogeographic free-for-all."[58]

What of humans? Were these changes happening fast enough they would have been detectable to them? The Pleistocene came to a screeching halt on a geological time scale, but would grandparents have noticed a difference they could tell their grandchildren? Knowing just how fast climate and environment changed could help us better understand the challenges faced by newly arrived colonists on this unfamiliar landscape (Chapter 7).

THE PLEISTOCENE'S LAST STAND

Not all glacial meltwater returned to the sea. Some from the Laurentide filled bedrock depressions left exposed by the retreating ice, creating in the Upper Midwest a land of lakes. All together, there are hundreds of thousands, including the five Great Lakes, the earliest versions of which appear about 14,000 years ago. The waters of these lakes would initially have been lifeless, turbid, and very cold, perhaps hovering just above freezing. If humans were in the area, it would have been to hunt game; fishing had to wait on the natural post-glacial stocking.[59]

Great as the Great Lakes are—and combined, they contain 22,700 km³ of water—they are little more than kiddy pools compared to the now-vanished glacial Lake Agassiz. This vast freshwater lake formed in the wake of the Laurentide ice retreat, first appearing around 11,700 years ago (its basin was ice-filled before then). At its maximum the lake covered 840,000 km² (which included much of western and central Canada, and a prong that reached down to the far northeastern corner of South Dakota), contained some 163,000 km³ of water, and drowned low-lying regions beneath more than 770 meters of icy water. Over its 4,000-year life span, Lake Agassiz rose and fell, but overall it was a downward trend as water made its way back to the sea. The water exited via different routes, as ice retreated and opened different outlet channels. Early on, the water went south down the Mississippi River into the Gulf of Mexico. Later it flowed east into the Lake Superior basin and out through the St. Lawrence Valley into the North Atlantic. After that, water moved northwest into the Arctic Ocean via the Athabasca and Mackenzie valleys, then east again down the St. Lawrence; finally, a dying pulse went north into Hudson's Bay and out into the North Atlantic.[60]

Ultimately, most of Lake Agassiz drained into the sea, though it left behind an arc of remnant lakes in western Canada, including Great Bear Lake, Great Slave Lake, Lake Athabasca, and Lake Winnipeg. Few humans would have paddled Lake Agassiz or fished its shores, but Lake Agassiz looms large in our story since for a couple of geologically brief moments, it played a singularly important role in Late Glacial climate, and possibly in Paleoindian adaptations.

When Lake Agassiz's water drained south, down the Mississippi River into the Gulf of Mexico, the earth's climate paid it no mind. But by 11,000 BP, the Laurentide ice sheet had retreated just far enough that a drainage outlet opened across what is now the northern Great Lakes and Upper Midwest. Lake Agassiz's waters suddenly flowed east rather than south, down to the newly opened St. Lawrence lowlands and out into the North Atlantic (the Mississippi continued to flow, though much diminished).

At that moment, Lake Agassiz's drainage was rerouted and some 9,500 km³ of very cold, very fresh water was flushed into the North Atlantic, shutting down thermohaline circulation and plunging the Northern Hemisphere back into *near* glacial conditions for the next thousand years.

Although there is considerable evidence that a shutdown in thermohaline circulation caused the Younger Dryas, there is also considerable controversy over what caused that shutdown. The problem? There ought to be geological evidence of Lake Agassiz's eastward drainage—an outlet channel, perhaps, or flood debris dating to 11,000 BP. But it hasn't been found. Without it, several alternative ideas have been proposed to account for why thermohaline circulation suddenly stopped—most controversially, that it was caused by an extra-terrestrial impact.[61]

Leaving aside matters of cause, let's consider the Younger Dryas consequences. Most important, this was no mere LGM rerun, for by then the climatic influence of the North American ice sheet was much reduced, atmospheric CO_2 had climbed to nearly interglacial levels (approaching 265 ppm compared to ~190 ppm at the LGM), and summer and winter insolation was higher and lower (respectively) than at any time in the preceding 70,000 years. To be sure, it grew chilly in the higher latitudes of the Northern Hemisphere, with temperatures in Greenland falling 27°F in only decades. That's cold, but hardly the 41°F plunge that marked the LGM. Younger Dryas climates were different in another important respect: because of the seasonal difference in insolation, this was

THE YOUNGER DRYAS: IT CAME FROM OUTER SPACE?

In 2001 the *Mammoth Trumpet*, a newsletter for a lay audience on happenings in Paleoindian studies, carried an unusually long, highly technical article declaring there'd been a Pleistocene doomsday. A supernova-caused neutron bombardment centered over the Great Lakes had fried the earth 12,500 years ago, Richard Firestone and William Topping announced.[62] That nuclear catastrophe heated the atmosphere to over 1,800°F, and radiated plants and animals at the equivalent dose of "a 5-megawatt reactor for more than 100 seconds." Megafauna died en masse because they were—as the authors reported on the good authority of the *Saturday Evening Post*—especially susceptible to radiation. The explosion purportedly rearranged maize genes, readying the plant for human domestication; gouged out the Carolina Bays (oval depressions in the coastal southeastern states); and so spiked atmospheric radiocarbon concentrations that ages on Paleoindian sites were thrown off by up to 40,000 years.

The claim was so far out literally and figuratively—we'd been using radiocarbon dating since 1950 and no one had noticed a 40,000 year error?—it was met with bemusement, or simply ignored. Only two scientists—experts in radiocarbon dating—bothered to reply. They concluded Firestone and Topping's claims were "at best, highly problematical and, at worst, difficult to take seriously."[63] Few did.

Later, when Firestone and others published *The Cycle of Cosmic Catastrophes: How a Stone-Age Comet Changed the Course of World Culture,*[64] the supernova had become multiple comets that had struck the earth repeatedly, with apocalyptic results. Maybe so, but the book was long on anecdote, short on evidence, weak on fact, and jumped to more than a few bizarre conclusions (extraterrestrial events caused variation in human skin color and blood types?). The scientific community's response: cold. But the book got a warm reception among (as Amazon.com reported) those who also purchased *The End of Days: Armageddon and Prophecies of the Return* and *Forbidden History: Prehistoric Technologies, Extraterrestrial Intervention, and the Suppressed Origins of Civilization.*

Usually when an idea slips into the pseudoscientific netherworld, it rarely returns. But this one did. E-mails started flying in the spring of 2007, with word of a symposium at the annual meeting of the properly scientific American Geophysical Union (AGU). There, oceanographer James Kennett—who has sterling scientific credentials—joined with Firestone and others to present evidence that a comet had struck the earth 11,000 years ago, and to outline its consequences—not least, that it was now blamed for triggering the Younger Dryas. In announcing the symposium, Kennett admitted there had been missteps in the past. This time, he said, it would be a different ball game.

Perhaps, but it began familiarly enough. The newly constituted team held a press conference at the AGU meeting, and over the following weeks, the supposed Pleistocene extraterrestrial catastrophe was hyped as fact from FOX News to the *Economist*. You can even watch their press conference on YouTube. The staid National Science Foundation, which that spring awarded Kennett and Luann Becker $53,000 to test the ET impact idea, issued a press release on the team's conclusions before the team had even completed their research, let alone presented their hypothesis and evidence in enough detail for evaluation and testing by others (turning the usual procedure of releasing scientific results upside down).[65]

It's unfair to blame scientists for how their words, however carefully crafted, are hijacked and sensationalized: caution doesn't make headlines or sell magazines. Still, when phrases such as "the entire continent was on fire" are spoken,[66] the fault is not entirely with the journalists. It's important to admit, as well, that scientists are not immune from the lure of public acclaim, or fail to recognize it's the public that ultimately supports our endeavors, and as such is entitled to know what we're up to. Yet, extraordinary claims require extraordinary proof, especially when the extraordinary is extraterrestrial.

To be sure, invoking ET impacts to explain earthly events is not unprecedented. In the 1980s, it was announced dinosaurs were destroyed by an asteroid that hit earth 65 million years ago. But skeptics became convinced only *after* there was independent confirmation of an impact's telltale signs: the finding of high levels of iridium (an element rare on earth) and quartz grains scarred by impact shock in rocks of that age, and the discovery of the "smoking gun"—the impact crater (Chicxulub, off the tip of Mexico's Yucatán Peninsula).[67] That an asteroid had collided with earth was beyond dispute. Even so, how the asteroid killed off the dinosaurs—or if it did—is still contested. The lesson: we must keep a firewall between two distinct questions: Is there indisputable geological evidence of an impact? If there is, then what were its consequences?

The first scientific publication on a late Pleistocene ET impact appeared in the fall of 2007. Much of the argument and evidence were familiar to readers of *The Cycle of Cosmic Catastrophes*, though some of that book's more colorful claims had been quietly locked away. Firestone and his twenty-five co-authors drew attention to a Younger Dryas–age "carbon-rich black layer" (the Black Mat) found at Clovis sites in North America,[68] which reportedly contains high levels of iridium (though in far lower amounts than in 65-million-year-old rock), metallic grains high in titanium, glass-like carbon studded with nanodiamonds, fullerenes high in extraterrestrial helium, and soot and charcoal—the last from the intense wildfires ostensibly ignited by the blast. Evidently, Europe caught fire, too: the authors report a charcoal-rich layer at Lommel, Belgium, and imply such are widespread on the Continent (for the record, Europe did not experience the kind of Pleistocene-ending extinctions seen in North America).[69]

Firestone and colleagues could not identify an impact crater, though by their reckoning, the comet must have been more than 4 kilometers in diameter, large enough to make a substantial dent in the earth's surface (about 50 km wide, by one estimate). Perhaps, they say, it hit the Laurentide ice sheet, which cushioned the blow and prevented cratering, or maybe it disintegrated in the atmosphere before hitting the surface. Perhaps its exploding debris caused the "enigmatic" depressions reported beneath Hudson Bay or even the Carolina Bays. Of course, Hudson Bay was still buried beneath ice at that time, which by their logic should have prevented cratering, and no meteoritic material has ever been recovered from the Carolina Bays, nor are all the bays the same age or date to the geological moment of the supposed impact.[70]

There's not been enough time for geologists and impact scientists to assess the merits of the case, but early reviews have been harsh. Impact specialist Christian Koeberl finds the whole scenario "contrived," since "their data don't agree with anything we know about impacts." But what of the iridium and other ET indicators? Geochemist Paolo Gabrielli, whose analysis of the Greenland ice core was cited as evidence of a large increase in iridium during the Younger Dryas, reported his data showed no such thing, while impact geologist David Kring is likewise skeptical, since iridium can be concentrated by algae (which happens to be a principal constituent of the Black Mat), and the reported levels of titanium and nanodiamonds embedded in melted carbon "make no sense to him."[71] They don't make sense to Nicholas Pintar and Scott Ishman either, since the identified ET elements cannot all come from the same type of extraterrestrial body: some (magnetic grains) require iron-rich meteorites, for example, others a stony meteorite, while still others (carbon spherules) suggest a carbon-rich source. If something did hit the earth, they say, it must have been a "Frankenstein" of an ET.

And could an explosion over North America set Europe afire? Thermal radiation, Pintar and Ishman point out, is zero below the horizon. But even leaving Europe out of it, there are tens of thousands of lakes in the Upper Midwest near the proposed ground zero. Since the impact ostensibly caused horrific wildfires, these lakes—where charcoal, soot, pollen, and other airborne particles settle and preserve—ought to have recorded the debris of a soot-filled atmosphere. Odd, then, that palynologists examining late Pleistocene-age sediments from thousands of lakes have not noticed traces of the inferno. Could it be they didn't look for it, or didn't know what to look for? That's unlikely: they often examine charcoal in sediments cores to help understand fire history. No matter. Several teams are returning to their field sites and stored sediment cores to look for a soot/charcoal spike at the time of the purported impact.

Until these and other studies are completed, it's best not to get ahead of ourselves exploring the consequences of an impact that may never have happened. But Firestone and colleagues cannot resist the urge. This impact, they proclaim, "explains three of the highest-debated controversies of recent decades": the cause of the Younger Dryas, the extinction of the Pleistocene fauna, and the precipitous decline of post-Clovis human populations.[72]

It's hard to see the last of these as controversial, since there's no evidence there was a population collapse immediately following Clovis times.[73] On the Great Plains, where the archaeological record is well documented and dated, Folsom groups follow Clovis, and their populations seem to have boomed (Chapter 9). So, too, invoking an ET impact to explain extinctions is another case of a solution in search of a problem. It's not been shown that Pleistocene extinctions were simultaneous, or even coincided with the supposed impact (Chapter 8). If they were and did, then why didn't this apparently global conflagration burn out far more of life on earth, compounded as the fires were by "increased deadly UV radiation," greater "chemical toxicity," and "diminished photosynthesis"?[74] This is not a rhetorical question.

As for the Younger Dryas, Firestone and colleagues believe that when the comet hit (or exploded above) the Laurentide ice sheet, it instantly melted or broke off large portions of glacial ice, releasing freshwater and icebergs into the North Atlantic, thereby—as in the currently accepted explanation—weakening thermohaline circulation and causing abrupt cooling. Once again, they're asking a great deal of what they must admit is geologically invisible.

There's no doubt that the past sometimes requires creative leaps to comprehend, and this certainly is a creative idea. But there needs to be much more evidence in order to make the Younger Dryas extraterrestrial impact a leap of science, and not a leap of faith.

a time of strong annual arcs between cold, harsh winters, and summers not nearly as cold as in LGM times.[75]

During the Younger Dryas, northern hemisphere glaciers stopped retreating and in places even began to re-advance. There was a sharp decline (from 680 to 460 parts per billion) in atmospheric methane as peatlands froze over, and increases in the amount of dust and salt from higher winds and greater storminess over Northern Hemisphere lands and seas occurred. It was a time of fierce nor'easters in the North Atlantic.[76]

But such changes were not true of the rest of the hemisphere (and possibly not of the southern hemisphere, including Antarctica which apparently warmed at this time).

Across North America, Younger Dryas climates varied. In the northernmost United States Younger Dryas temperatures were cooler than present, with a drop of about 9°F being about the maximum. It was drier as well—in places. Yet, in the southeastern United States, the Younger Dryas was relatively warmer and wetter also, a result of thermohaline circulation shutdown, which left warm, tropical waters stalled off the Atlantic coast. Further west, much of the interior of midlatitude North America stayed relatively warm and temperate, since cold, arctic air was still mostly trapped north of the remnant ice sheet, and warm, Caribbean air was able to penetrate northward and amplify the insolation-driven highs of summer. Younger Dryas climatic conditions had "no modern equivalent in North America."[77]

Plants and animals that had been busy responding to Bølling-Allerød warming scrambled to adjust to Younger Dryas reversal. Cold- and dry-adapted sedges abruptly returned to prominence in northern areas; spruce trees that had migrated north now turned south, while pine trees expanded west from New England into the southern Great Lakes region, and with astonishing speed for something rooted to the ground: nearly 300 kilometers in little more than a century. In high elevations of the Rockies, tundra and timberline shifted downslope, along with cold-loving trees such as Engelmann spruce and bristlecone pine.[78]

The most dramatic vegetation changes occurred in the more northern and higher-elevation areas. Yet, across North America, distinctive, albeit short-lived vegetation communities emerged during the Younger Dryas. That was presumably true of Younger Dryas mammalian communities as well, though their fossil record is not as precisely known.

However distinctive, these communities would not last more than a millennium, and would dissolve when the Younger Dryas came to an abrupt end.[79] Ironically, the Younger Dryas may have been the instrument of its own demise. The colder temperatures it brought to the Upper Midwest, coupled with lake-effect snows from nearby Lake Agassiz (think Buffalo, New York, in winter), triggered the re-advance of a glacial lobe that by 10,000 BP had pushed far enough south that Lake Agassiz's eastern outlet once again was blocked. Lake Agassiz's meltwater once more drained down the Mississippi to the Gulf of Mexico, or north into the Arctic Ocean. With that, thermohaline circulation was re-established, this time for good, and the Younger Dryas came to an end: above the Greenland ice sheet, temperatures shot up 15°F in a matter of a decade, while global methane amounts rose by 50% in a century as the world's wetlands expanded.[80]

Live by the ice, die by the ice. It's hard not to agree with Broecker that the Younger Dryas—if indeed caused by a freshwater flood that sucker punched thermohaline circulation—was a "freak event."[81] That, coupled with its occurrence amid an unusual combination of solar insolation, ice sheet extent, and greenhouse gas concentrations, triggered complex climatic and biotic responses, and lead inescapably to this next conclusion: the Younger Dryas was unique in the annals of glacial history. It has traditionally been thought the events ending this last glacial cycle were no different than those ending previous glacial cycles. That's now doubtful, but that's good news: it might help explain why many of the Pleistocene mammals, which so successfully had survived previous glacial-interglacial cycles, succumbed to this one.

The climatic and ecological changes of the Younger Dryas were the most dramatic millennial-scale changes recorded since the LGM.[82] Yet, viewed over a longer time frame, the Younger Dryas was merely a detour en route to the much more profound changes that occurred as the earth switched from glacial (Pleistocene) to interglacial (Holocene) mode. That switch was marked most obviously by the virtual disappearance of the great ice sheets, which unleashed the polar air masses and repositioned the jet stream; by substantially higher sea levels; and of course, by a much warmer world as air temperature and its fellow travelers, atmospheric CO_2 and methane, rose quickly after the Younger Dryas.

Such changes triggered the wholesale rearrangement of biotic communities, which included mass extinctions (which were complete by about 10,800 BP). At the close of the Younger Dryas, eastern North America supported a complex mosaic of boreal and deciduous forest: think plaid. Yet, the modern environment is one of latitudinal stripes: a Canada-to-Florida road trip passes through tundra, boreal, mixed conifer/hardwood, temperate deciduous, and southern pine forests, a pattern that reflects increasing temperature as well as the seasonal positions of the arctic air mass that had been effectively excluded from North America during the Pleistocene. Changing from Pleistocene plaids to Holocene stripes took several thousand years, but was largely finished by 5,000 years ago.

On the plains, as precipitation went the way of the northward-shifting jet stream, lake levels fell, forests contracted, and grasslands expanded in all directions, even into the Upper Midwest. By early Holocene times, it came to resemble the plains of Hollywood stereotype: a sea of grass, on which vast herds of bison grazed. Bison had been on the plains all along, but had to share the grasslands with mammoths, horses, camels, and other large grazers. Once those were extinct, bison had the plains to themselves, and as a result, their numbers skyrocketed (*competitive release*, it's called), an explosion fueled in part by the increasing expanse of warm-season C_4 grasses, which bison love (there's a reason one of the most widespread of these grasses is named *buffalo* grass).

Further west in Early Holocene times, alpine and boreal species began their march upslope, replaced down below by species better able to tolerate increasingly warmer and drier conditions (in the Southwest), and warmer and wetter conditions (in the Northwest). The process was relatively drawn out, and of course involved a reshuffling of the vegetation deck: the vertical zones characteristic of modern times would not form up for several thousand years, depending on the area. But soon enough, harsh desert filled in the low-lying, intermountain valleys of the southwest and Great Basin, forever stranding on mountaintops those mammals that in cooler and wetter times had been able to scurry from peak to peak.

The Pleistocene would not end, however, without one final death rattle. As the Laurentide ice sheet steadily shrank, Lake Agassiz followed it north. By 7700 BP, it had ballooned to enormous size, having joined Lake Ojibway, and was held in check by a horseshoe-shaped dam of remnant Laurentide ice hovering just south of Hudson Bay. As all ice dams do, this one eventually failed, and what was left of the world's largest freshwater lake flooded into Hudson Bay, out Hudson Strait, and into the North Atlantic via the Labrador Current. An estimated 163,000 km³ of freshwater was flushed into the North Atlantic. That's seven times the volume of all five Great Lakes put together, and seventeen times more water than the trifling 9,500 km³ that triggered the Younger Dryas. This was a flood worthy of Noah's; the Reverend Buckland would have been thrilled.

Yet, the climatic cooling that followed this last flood—termed the 8200 BP event (for its *calibrated* calendar age)—was a mere blip compared to the Younger Dryas, dropping temperatures in Greenland only about 10°F for about a century or so.[83] Why was it so climatically inconsequential? Perhaps, as Dyke suggests, such is the difference between releasing freshwater directly into the North Atlantic, where it could T-bone the thermohaline conveyor (the Younger Dryas scenario), compared to sending it through more northerly, already-cold and less-saline waters, thereby diminishing its impact on arrival. The very largest of the large post-glacial superfloods, Richard Alley observes, is one that didn't stick.[84]

People living in North America at the time may scarcely have noticed the 8200 BP event. Besides, their ancestors had experienced climatic and environmental diversity and change of much greater magnitude, such as the Younger Dryas and perhaps even the Last Glacial Maximum. But that depends on just when those ancestors first arrived, and how fast they moved through this new land.

3

FROM PALEOLITHS TO PALEOINDIANS

In the fall of 1781, the governor of Virginia received a letter from the secretary of the French Legation, inquiring about the political institutions, natural history, and native peoples of his state. Times being what they were (hostile British forces were advancing on the governor's home) his answer was postponed until the following summer, after Cornwallis had surrendered at Yorktown and the governor had retired from office. "Great question has arisen," Thomas Jefferson then replied in his now-classic *Notes on the State of Virginia*, "from whence came those aboriginal inhabitants of America." For that matter, who were they, and when had they arrived?[1]

These questions were deeply unsettling to a nation that viewed the past through a biblical lens, and which well realized American Indians were not mentioned in the chronicles of Moses. Still, on the presumption of monogenesis, the idea that "all the varieties of the human race were descended from a single pair, and that after the flood the earth was indebted solely to the ark of Noah for the replenishment of man and beast,"[2] American Indian origins were sought among historically known or imagined groups, among them wandering Egyptians, Phoenicians, Mongols, Welsh, Hindus (those *other* Indians), survivors of the Lost Continents of Mu or Atlantis, and the Ten Lost Tribes of Israel. The possibility that American Indians were Israelites had two great virtues: it explained who the Indians were, and where the Israelites had been lost all those years.

Yet few capable thinkers, and Jefferson was an extraordinarily capable thinker, paid these any mind. Diverse as the Native Americans were, Jefferson realized they shared a common ancestry. Judging by their appearance, he supposed their ancestral homeland must have been eastern Asia. That being the case, how had they reached the New World? The answer, anticipated in the late sixteenth century by the Jesuit José de Acosta, was that they came by land, possibly via northern Asia and northwestern North America. But it was not until 200 years later that Jefferson was able to report that "the late discoveries of Captain Cook, coasting from Kamchatka to California, have proved that, if the two continents of Asia and America be separated at all, it is only by a narrow streight [sic]."[3]

Having found an easy migration route from the Old World to the New still left unsettled the question of when that migration had taken place. Jefferson had his suspicions. Native Americans spoke a Babel of languages (few of which were mutually intelligible or bore any resemblance to an ancestral Asian language), they did not all look alike—the usual stereotypes notwithstanding—and they had diverse cultural practices, all of which implied a long period of divergence from what was assumed to be a common ancestor. Could this variability have arisen in the biblically allotted 6,000 years? Jefferson was doubtful: such linguistic, physical, and cultural divergence from a common source seemingly required "an immense course of time; perhaps not less than many people give to the age of the earth."[4]

As old as the earth? Jefferson's political enemies howled, accused him of heresy, and branded him an atheist. The earth itself could turn out to be very old, perhaps tens of thousands of years old as astronomers and geologists were already insisting, but Jefferson's critics demanded human history go no earlier than the 6,000 years allotted in the scriptures. And it didn't.

At least not for another seventy-two years.

THE DISCOVERY OF DEEP TIME

Before 1859, the Bible was history, chronology, and ethnography all rolled into one: it was a detailed and sacred account of human genealogy and lifeways from Adam and Eve on down. Compiled by people who had either been present or had access to a supernatural informant, it linked modern life back to the very creation of heaven and earth. Well, almost to the Creation.[5]

By the late seventeenth century, scientists such as Isaac Newton began lobbying for more time, at least for earth history. They knew from physical evidence the earth had to be more than 6,000 years old, but not daring to reject Genesis entirely, they placated themselves by treating the first days of Creation as allegorical only, supposing that Moses was simply "accommodating his words to the gross conceptions of the vulgar," as Newton put it, and that the Biblical chronology "relates only to the human race." This was an argument easy to make. After all, as Edmund Halley (of comet fame) observed, if the sun was created only on the fourth day, how could the time before then be measured in "days" in the literal twenty-four-hour sense?[6]

So geologists got busy. From the strange fossils brought to light by exploring expeditions and the rapacious mining that fueled the Industrial Revolution, Georges Cuvier and others revivified the wonderfully exotic plant and animal life of earlier periods of earth history. The most recent of those earlier periods was inhabited by fossil elephants, including North America's mastodon and mammoth. Cuvier had it on the good authority of American Indians—"nomadic peoples who move ceaselessly around the continent in all directions" and thus were in a good position to know—that these "creatures no longer existed."[7]

The "unconsolidated . . . layers of the earth" in which the fossils of these animals were found lay close to the surface, in what appeared to be sand and gravel laid down by water—Buckland's diluvium. It would be several decades before those deposits were recognized as the residue of once-vast continental ice sheets (Chapter 2) and, in turn, linked with Lyell's Pleistocene. Nonetheless, even by the early 1800s, Cuvier realized these were not truly ancient deposits, for there were still more primitive fossil elephants and other animals in deeper layers of the earth. He concluded that earth's history was one of multiple periods of creation of different kinds of animals, each of which came to a cataclysmic end. The animals of the diluvium represented merely "the last or one of the last catastrophes of the globe." But if these animals were not old in geological time, they were assuredly from the pre-modern world. The Bible made no mention of mammoths.[8]

Accordingly, no one was looking for human remains alongside mammoth fossils in Pleistocene deposits, nor expected to find any. This was a period unknown to human history, beyond range of the Mosaic chronicles, and thus a baseline against which human antiquity could be measured. The uncoupling of earth history from human history was a splendid compromise in a world fast becoming inconceivably old: humanity's best hope for divine origins and its own uniqueness in the animal kingdom lay in the affirmation of the Bible. Here, fortunately, Genesis and geology seemed to agree: humans were the most recent creation, or so it seemed on good authority.[9]

Still, through the first half of the nineteenth century, human remains were found alongside Cuvier's extinct animals at an increasing number of sites. But why accept those finds at face value?[10] Most came from continental Europe and especially France, and for this reason immediately lost credibility with the more theologically conservative British who, since the French Revolution, had suspected the French of atheism and harbored a lingering distrust of their latter-day Enlightenment notions. Much of the evidence came from excavations by provincial amateurs who were looked upon, in John Lubbock's charitable Victorian parlance, as mere "enthusiasts."[11] They were hardly a trustworthy source of evidence on such a momentous question. No scientist was going to reject long-held beliefs on their say-so.

Compounding resistance, the bulk of finds were made in caves, settings regularly churned by burrowing animals that mix deposits of different ages. Caves are ill suited to demonstrating whether artifacts found alongside animal bones were the same age, or just accidentally associated. When the "Red Lady" of Paviland, an ocher-covered human skeleton, was found in a cave with the remains of a mammoth, it was all too easy for the Reverend Buckland to explain (with a knowing wink) that a nearby Roman camp site threw "much light on the character and [antiquity] of the woman under consideration." Ocher apparently wasn't the only reason Buckland supposed the Lady of Paviland was red (as it happens, she was no lady either, but an Upper Paleolithic male, now radiocarbon dated to slightly over 26,000 BP).[12]

Finally, none of the evidence fit prevailing expectations of human history based on the Bible or geology, and thus there was no compelling reason to accept it. This isn't dogmatic; it's good science: one doesn't reject a long-established worldview without compelling reason. As Lyell later confessed, "I can only plead that a discovery which seems to contradict the general tenor of previous investigations is naturally received with much hesitation."[13] Only when the counterevidence becomes too weighty to ignore does the model of the world get reassessed and, if needed, rejected.

That reassessment began in the summer of 1858 at Brixham Cave in southwestern England where, beneath a nearly impenetrable layer, a crack team of excavators under the supervision of Britain's finest geologists (Charles Lyell among them) had uncovered stone tools in direct association with Pleistocene fossils. Either the team had failed to observe that stone tools had been mixed in from higher, younger deposits and had botched the excavation (a conclusion no one was keen to admit), or the theologically disconcerting was true: humans had occupied Brixham Cave during the Pleistocene, alongside now-extinct animals, and thus had an ancestry that predated biblically recorded history.

Brixham Cave's revelations prompted another look at the long-standing claims of Jacques Boucher de Perthes, who for decades had been collecting stone tools and Pleistocene fossils in the Somme Valley of northwest France. Until that moment, he'd been mostly ignored because many of the "artifacts" he illustrated were clearly not artifacts at all, and what genuine evidence he did have was embedded in arcane theories that had long since been rejected. Charles Darwin was hardly alone in reading Boucher de Perthes' work and concluding "the whole was rubbish."[14]

Yet, some of the stone tools Boucher de Perthes illustrated looked a great deal like those from Brixham Cave and, better, had come from floodplain deposits, which are less prone to the mixing of deposits that occurs in caves. A procession of geologists and antiquarians made the pilgrimage to see Boucher de Perthes and his sites and collections, including Joseph Prestwich and John Evans, who in April 1859 witnessed and photographed a handaxe in situ at a locality in Amiens.

At a meeting of London's Royal Society a few weeks after their return, Prestwich read a paper on the glacial age and stratigraphy of the Somme Valley gravels, and Evans spoke extemporaneously on the stone artifacts found in them. Listening to them, Evans recalled, "were a good many geological nobs . . . Sir C. Lyell, Murchison, Huxley . . . Faraday, Wheatstone, Babbage, etc. so [we] had a distinguished audience." That's an understatement. The audience included not only the geological elite (Lyell, Thomas

Henry Huxley, Roderick Murchison), but also one of the greatest experimental scientists of all time (Michael Faraday); the inventor of the stereoscope and a pioneer researcher in acoustics, electricity, and telegraphy (Charles Wheatstone); and one of the trio who revolutionized English mathematics in the nineteenth century and a dabbler in crypt-analysis, probability theory, geophysics, astronomy, and computing machines (Charles Babbage). It might not have been possible to gather in one place a more influential group of scientists in all of England. The question of a deep human antiquity was out of the closet and on center stage, and the favorable reception their audience accorded Prestwich and Evans ("Our assertions as to the findings of the weapons seemed to be believed"[15]) immeasurably helped the cause.

In the summer of 1859, Charles Lyell himself, foremost among those who had long challenged all claims of great human antiquity, made the pilgrimage to Abbeville. He too returned from France a convert, announcing, "I am fully prepared to corroborate the conclusions which have been recently laid before the Royal Society. . . . I believe the antiquity of the Abbeville and Amiens flint instruments to be great indeed if compared to the times of history and tradition."[16] Humans had seen Agassiz's glaciers and preyed on Cuvier's fauna.

Only months before, the idea of a deep human antiquity had been the dubious claim of provincial amateurs. Now, it was almost universally accepted fact. As one contempo-rary joked, people were no longer insisting "it was not true" or that "it was contrary to religion," but that "it was all known before."[17]

The Paleolithic, or Stone Age, occupants of Brixham Cave and the Somme Valley predated history, and lived at a distant time "when man shared the possession of Europe with the Mammoth, the Cave bear, the Woolly-haired rhinoceros, and other extinct animals," using "rude yet venerable weapons" of stone, as John Lubbock put it.[18] This was a human past about which the Bible said absolutely nothing, and which suddenly rendered it obsolete as the story of humanity's past. The Paleolithic would be knowable only through its silent artifacts and skeletal remains. To investigate that past required a new discipline of *pre*history, with its own body of theory and methods, kin to geology and not sacred history.

As dozens more Paleolithic sites were found across Europe, humanity's roots were pushed back ever deeper in time: hundreds of thousands of years, some suggested, per-haps millions of years, others supposed. What had transpired over that span was poorly known, testament to the vast "chasm which separates the flint folks from ourselves." Julia Wedgwood, Darwin's niece, saw "something dreary in the indefinite lengthening of a savage and blood-stained past,"[19] but however dreary or theologically unnerving, those rude artifacts were nonetheless vivid testimony of the savage depths from which humanity had climbed. They became for many Victorians a triumphant demonstration of social progress.

The demonstration of a deep human antiquity came at virtually the same moment as Darwin published his *Origin of Species,* which laid out the theory of evolution by natural selection. Although the two had independent origins, they soon and ever after were linked. The *Origin* had no more to say about the evolution of the human species from our animal forebears than the understated one-liner, "Light will be thrown on the origin of man and his history." But everyone knew what *that* meant: ancient human ancestors and ultimately a pedigree shared with other primates. One could scarcely accept the implications of Darwin's views of human evolution without the deep past that prehistory provided. Darwin well recognized that, and a decade later when he finally tackled the topic of human evolution, he knew precisely where to start: "The high antiquity of man has recently been demonstrated by the labors of a host of eminent men, beginning with M. Boucher de Perthes; and this is the indispensable basis for understanding his origin. I shall, therefore, take this conclusion for granted."[20] It was rubbish no more.

And that's when the search began on the other side of the Atlantic.

THE RISE AND FALL OF THE AMERICAN PALEOLITHIC

The discovery of the European Paleolithic fired American scientific imaginations: why couldn't American prehistory be just as old? After all, the geology of the two continents seemed so very similar. But archaeologists quickly realized that here in America there were no deeply stratified river valleys or caves with human artifacts indiscriminately mixed with Pleistocene fauna.[21]

No matter, they soon had another inspiration: if artifacts were ancient, they ought to look the part, so ones similar to the tools of Paleolithic Europe must be the same age. The catalyst for these studies was Charles Conrad Abbott, a New Jersey physician (and coincidentally, nephew of Timothy Abbott Conrad, who first found traces of the Ice Age in America). By all accounts—including his own—Abbott had a dreadful bedside manner and was unable to earn a living as a physician. But a living had to be made, for although he came from a prominent, land-holding family, his was not a wealthy branch of the family tree. Abbott made a half-hearted stab at farming and the occasional odd job to help support himself and his family, but his livelihood came to revolve around authoring popular books on the natural history and archaeology of the Trenton area. Abbott soon realized the artifacts he was finding near his Trenton home appeared to be as "rude" as those of the European Paleolithic (Figure 10). Better still, some came from geological deposits that hinted of considerable antiquity, and perhaps they were even Pleistocene in age.

We now know, based on archaeological work done at the Abbott farm since, the artifacts he was finding were not very old at all, and certainly not glacial in age. But Abbott didn't know that, nor did most of his peers. In February 1877, Harvard University geologist Nathaniel Shaler studied the Trenton ground with Abbott and pronounced the artifact-bearing gravels to have been deposited during the Ice Age. In his diary that night, Abbott recorded his triumph: "*I have discovered glacial man in America*"[22] (emphasis in the original).

FIGURE 10.

Charles Abbott searching for paleoliths in the
Trenton gravels, New Jersey, ca. 1880s. (Photograph
courtesy of the Peabody Museum, Harvard
University. © Harvard University, Peabody Museum
2004.24.31046.)

Heady stuff, this, and it inspired Abbott's compatriots. Harvard archaeologist
Frederic Ward Putnam hired a local man to walk the freshly dug railroad right-of-ways
and sewer trenches of Trenton watching for more "Paleolithic" artifacts. As Abbott's
work in the Delaware Valley gained prominence, so, too, did the American Paleolithic.
Throughout the 1880s, reports came in of other paleoliths from the eastern seaboard
into the Upper Midwest. Some were found under geological circumstances that suggested a Pleistocene antiquity, but even in the absence of secure geological evidence,
these artifacts so readily mimicked European paleoliths of undeniable antiquity, they
must be as old. As Reverend George Frederick Wright, an Oberlin College theologian
and geologist, put it, in Paleolithic times, as today, American fashions followed the
Paris line.[23]

By the end of the 1880s, Abbott and his colleagues possessed the proof they needed
that the first Americans had arrived thousands of years ago, when northern latitudes
lay shrouded in glacial ice. The American Paleolithic spawned a formidable literature of symposia, feature articles, and books. Abbott was lionized here and abroad as
"America's Boucher de Perthes." Everyone believed the first Americans had arrived in
the Pleistocene; the only question remaining was how much further back in time that
prehistory extended. It had to be deep, Abbott insisted, for were it only 10,000 years old,
he would be "compelled to crowd several momentous facts in American archaeology
into a comparatively brief space of time."[24]

But the consensus surrounding the American Paleolithic proved short-lived. Scarcely
a year later, it was under withering fire from USGS geologists and archaeologists at the
Smithsonian Institution's Bureau of Ethnology (later, the Bureau of American Ethnology,

FIGURE 11.

William Henry Holmes sitting in "an ocean of paleoliths" at the Piney Branch quarry, in Washington, D.C., ca. 1890. Holmes was only kidding about those being paleoliths. He used the cobbles and workshop debris from Piney Branch to show that the "primitive-looking" and supposedly ancient artifacts of the American Paleolithic were mere quarry refuse. (Photograph courtesy of the Smithsonian Institution Libraries, Washington, D.C.)

or BAE). By late 1889, the archaeological community was humming with rumors that BAE archaeologist William Henry Holmes's excavations at a prehistoric stone quarry along Piney Branch Creek in Washington, D.C. (Figure 11), were set to demolish the conceptual foundations of the American Paleolithic.

At the Piney Branch site (which, astonishingly, remains intact just a few miles north of the White House), prehistoric groups found quartzite cobbles suitable for making their artifacts. Scattered across the site surface were the debris of tool manufacture, ranging from cobbles with just a few flakes struck from them, to nearly finished items that had broken or otherwise been rejected in the final stages of preparation. Many of the rejects bore an uncanny resemblance to Paleolithic artifacts. Holmes had an *aha!* moment: the manufacturing sequence at Piney Branch mimicked the long evolution of stone toolmaking from primitive to refined. Thus, just as the most ancient stone tools were little more than barely sharpened cobbles, so, too, were stone tools in the early stages of manufacture. That being the case, he reasoned, an artifact jettisoned prematurely during manufacture would naturally resemble a primitive stone tool, even if it was scarcely a few hundred years old.

In late January 1890, Holmes hurled his opening salvo in the *American Anthropologist*: artifact form had no inherent chronological significance.[25] Proponents of an American Paleolithic were mistaking primitiveness for antiquity, failing to realize their artifacts were merely manufacturing failures. An artifact's age, he insisted, must be determined by its geological context—and not because it happened to look like one illustrated by Boucher de Perthes or John Evans. That included Abbott's Trenton paleoliths, which to Holmes's eye looked very much like manufacturing failures. Still, at that moment, he was feeling charitable, and he wasn't suggesting the Trenton specimens were recent in age. Not yet, anyway.

Word soon reached Abbott of Holmes's paper, and he dashed off a letter to Henry Henshaw, editor of the *American Anthropologist,* requesting a copy. He got one, along with an invitation. Would Abbott like to come to Piney Branch, preferably immediately? "It seems to me," Henshaw wrote, "that you would be particularly interested in the matter since you have done so much work at Trenton, and a visit here just now could not fail to prove instructive. Mr. Holmes would be very glad to see you, and we will all do what we can to make your visit pleasant and instructive."[26]

"Instructive," Henshaw said. Twice. But Abbott wasn't looking for instruction. He was looking for confirmation. But he went to Washington anyway, and toured Piney Branch with Holmes in early 1890. A smug Holmes later recalled that

> on parting Abbott said, "I have learned more arch[aeology] in three hours than ever before in three months." This I was content to think of as a pleasant compliment but from my subsequent studies and increased wisdom I concluded that he probably meant what he said.[27]

If Abbott indeed said that, he hardly meant it, for he left Washington unswayed. As he later wrote, Holmes may well have been correct in his archaeological interpretation of Piney Branch, but Abbott failed to see that the story at Piney Branch had any relevance to either Trenton or the larger question of human antiquity in America. After all, Holmes's "so-called failures" were "not identical with the true American Paleolithic implements of the Delaware River Valley," which were found in what Abbott deemed undisturbed Pleistocene glacial deposits. Even if geologists did quibble over their precise age, Abbott was sure "no verbal jugglery" could make the Trenton gravels that much younger.[28] If Holmes aimed to reject Trenton as a Paleolithic site, he would have to go to Trenton.

The lines were drawn in the sand (and gravel). The Great Paleolithic War began.

Holmes spent the next several years examining the alleged Paleolithic sites in eastern North America, including Trenton.[29] In each case, Holmes solemnly proclaimed that mistakes had been made. All alleged paleoliths were manufacturing failures having no appreciable antiquity, and were merely the debris of historically known Native Americans. None, in his view, were Pleistocene in age, and those allegedly found in glacial deposits must have fallen down rodent burrows or cracks in the earth, and

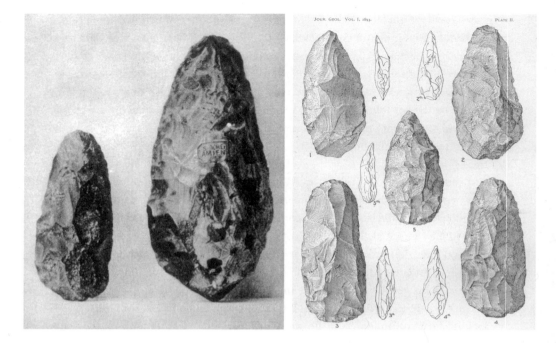

FIGURE 12.

Two perspectives on the Newcomerstown, Ohio, "paleolith." Left, G. F. Wright's (1890) composite of the Newcomerstown paleolith alongside a European paleolithic biface (reduced to one-half size) from Amiens, France. Right, Holmes's (1893b) depiction of the Newcomerstown specimen alongside "four ordinary rejects." Holmes left it to the reader to decide which of the five specimens was from Newcomerstown, and which were quarry rejects.

fortuitously settled in those older deposits (Figure 12). And if the specimens were actually in situ in "Pleistocene" gravels, Holmes could say, on the supreme authority of the USGS's Thomas Chamberlin and WJ McGee, that those gravels were not Pleistocene age at all.

Abbott was furious. If the critics were correct, why were only primitive-looking specimens found in gravel deposits, and not the more refined artifacts of the American Indians? If these were quarry rejects, where was the flake debris resulting from their fashioning? And who were these self-styled "expert geologists" to lecture *him* about the Trenton gravels? None had spent a fraction of the time he had walking the banks of the Delaware, and assuredly could not have greater knowledge of its geology than he possessed. "I lay a claim to a smattering of gravel-ology," Abbott declared in *Science,* and if "up pops some 'authority' and declaims the possibility that the ground was washed from beneath the big stone and the implement slipped in. Well, we can go on supposing till the crack o'doom, but as to proof, that is another matter."[30]

Perhaps. But as Holmes snorted in reply two weeks later, demonstrating that arti-facts were truly found in glacial gravels required the skills of "competent and reputable observers of geological phenomena." That Holmes believed no glacial-aged artifacts had yet been found made it quite clear he did not consider Abbott's claims to "gravel-ology" equal to competence. A testy Abbott immediately fired back with doggerel virtually guar-anteed to offend the humorless Holmes:[31]

The stone are inspected,
And Holmes cries "rejected,
　　　They're nothing but Indian chips"
He glanced at the ground,
Truth, fancied he found,
　　　And homeward to Washington skips.

They got there by chance
He saw at a glance
　　　And turned up his nose at the series;
"They've no other history,
I've solved the whole mystery,
　　　And to argue the point only wearies."

But the gravel is old,
At least so I'm told;
　　　"Halt, halt!" cries out WJ [McGee],
"It may be very recent,
And it isn't quite decent,
　　　For me not to have my own way."

So dear WJ
There is no more to say,
　　　Because you will never agree
That anything's truth
But what issues, forsooth,
　　　From Holmes or the brain of McGee.

Holmes was not amused (though the editor must have been, since it was highly unusual for *Science* to publish verse). The American Paleolithic, he snarled, was little more than the blunders and misconceptions of "amateurs" with little scientific under-standing of stone toolmaking, let alone of geological age and context. If Abbott still didn't get the message, Holmes spelled it out in a lengthy, sharply pointed critique, which asked the decidedly nonrhetorical question, "Are there traces of man in the Trenton gravels?" Holmes's answer was "No."[32] And then the controversy turned ugly.

In the fall of 1892, USGS and BAE scientists attacked George Frederick Wright's just-published *Man and the Glacial Period*. Wright's wrongs were obvious enough: he had spoken in favor of the American Paleolithic; he had challenged one of Chamberlin's (and by extension the entire USGS Glacial Division's) intellectual monuments—the demonstration there had been multiple glacial events; and, worst of all, he had put those heresies in a book written for a general audience. Wright's critics, orchestrated by Chamberlin, set out to destroy his credibility as a glacial geologist, as an archaeologist, and especially as a public spokesman for science. They launched a barrage of vicious reviews of *Man and the Glacial Period*, which were unprecedented in number and savagery. "No one," Chamberlin thundered, "is entitled to speak on behalf of science who does not really command it." WJ McGee, a one-time staunch supporter of the American Paleolithic but by now a zealous convert, was especially bloodthirsty, labeling Wright's work absurdly fallacious, unscientific, and an "offense to the nostrils," then dismissing him as "a betinseled charlatan whose potions are poison. Would that science might be well rid of such harpies."[33]

The maliciousness of the attacks appalled Wright's colleagues, as well as many who hardly knew him or his work. To Wright's allies, the near-simultaneous appearance of the reviews and their "sameness of tone" smelled of a conspiracy. They were right. It was. Yet, the assault on Wright was more than just personal. It was a thinly veiled proprietary dispute in which BAE and USGS scientists sought to impose their vision of archaeology and geology on those fields, and contrast their brand of science against those, such as Wright and Abbott, they deemed rank amateurs.

Because the BAE and USGS scientists were richly funded and backed by the power of the federal government, at a time when the government dominated American science (the balance of scientific research and power would shift to universities only in the twentieth century), the atmosphere was charged with accusations that arrogant, heavy-handed federal scientists were conspiring to crush state and local practitioners. Those accusations reached Capitol Hill in 1893, when Congress was deciding the annual budgetary fate of the USGS and BAE. Neither agency could afford such bad press, since their appropriations were already threatened by the economic fallout from the Panic of 1893. Paleolithic proponents got a measure of revenge: the BAE and USGS budgets were slashed.

After savaging each other in meetings and in print, the warring parties suspended hostilities in late 1893, though more from battle fatigue than by truce, for both sides had hardened beyond compromise or retreat. They met again in the summer of 1897 in Toronto at a joint meeting of the American and British Associations for the Advancement of Science. There, Paleolithic proponents suffered a devastating blow: Putnam handed Sir John Evans a set of Trenton artifacts. With barely a glance, the dean of Stone Age archaeology dismissed the lot as not Paleolithic at all. Putnam desperately tried to

A MAMMOTH FRAUD IN *SCIENCE*

The boldface headlines of the September 8, 1894, *Philadelphia Inquirer* nearly screamed aloud: "Dr. Hilborne Cresson Takes His Own Life." As the paper indelicately put it, Cresson "blew his brains out in a park in New York City." He had gone insane, the *Inquirer* reported, his mind disordered as a "result of scientific study" and the "too close application of esoteric principles."

Cresson was an archaeologist.

He began his career an expatriate student of art and archaeology at the École des Beaux Arts and École d'Anthropologie in Paris in the 1870s. Apparently, he had a talent for recreating prehistoric art. He returned to the United States in 1880 and began collecting artifacts and excavating around Philadelphia. By 1887 he'd attracted the attention of Frederic Ward Putnam, who hired Cresson as an archaeological field assistant.

In December 1889, Cresson appeared in Putnam's office bearing a whelk (*Busycon*) shell pendant, on which was engraved the figure of a mammoth (Figure 13).

FIGURE 13. The Holly Oak pendant (Delaware), and as drawn by W. C. Sturtevant. Note that there is ample room on the shell for feet; that they are not shown was one of the hints that the mammoth image was made from a book, and not from real life. (Adapted from Griffin et al. 1988.)

Cresson reported he'd dug up the specimen as a schoolboy years earlier near Holly Oak, Delaware. Putnam, a staunch champion of the American Paleolithic, would have instantly understood the pendant's significance: if the mammoth had been engraved from life, it would be a tidy solution to the question of human antiquity in the New World, and would show that Paleolithic Americans had created works of art to rival in age and beauty those of Paleolithic Europe. How timely that Cresson should appear with a specimen that could tip the balance of the Great Paleolithic War, which Putnam knew was looming on the horizon.

Even so, one searches Putnam's papers in vain for the happy announcement his cherished beliefs had been vindicated by this remarkable evidence. At Cresson's request, Putnam showed the Holly Oak pendant at the February 1890 meeting of the Boston Society of Natural History, yet was ever after silent about it. But Putnam left behind a telling clue of his opinion: a photograph of the Holly Oak pendant alongside a drawing of the La Madeleine mammoth, which had been engraved on a segment of mammoth tusk that was discovered by Eduoard Lartet in 1865 in the Dordogne region of France.

Any good archaeologist, and Putnam was a good archaeologist, would have instantly seen the similarities between the Holly Oak and La Madeleine engravings. To be sure, in Putnam's day there were few mammoth depictions known from Europe, and none from America. Possibly, he may have concluded the similarity between the two merely reflected the fact both depicted mammoths, as opposed to both depicting the *same* mammoth (nowadays we know of thousands of European Paleolithic depictions of mammoth—still none from America—and those depictions vary greatly, as one might expect).

Putnam's suspicions might have been further aroused when Cresson reported his discovery had been made in 1864, conveniently predating La Madeleine's by a year, and claimed that his French tutor, allegedly involved in the Holly Oak discovery, had been a student of Lartet's. Cresson never satisfactorily explained why it took him twenty-five years to reveal the pendant's presence.

Cresson suggested to Putnam that he look at Charles Rau's book, *Early Man in Europe,* to see how well the Holly Oak mammoth resembled the mammoth skeleton it illustrated. In doing so, Putnam would have seen that the book also provided an illustration of the La Madeleine engraving, one that differs from better reproductions of the original in ways strikingly similar to Holly Oak: the contours of the back and tusks and trunk can be overlaid; they have the same orientation and overall posture, especially the leg positions; they treat the feet in a similar way. Most mammoths are shown with bulbous feet, but the La Madeleine feet are cut off by a break in the original specimen, and the Holly Oak feet terminate in precisely the same way, despite the fact that there is ample room on the shell where feet could have been engraved. Putnam, almost certainly suspecting the Holly Oak pendant was a fraud, quietly ignored it.

For that matter, there was hardly any mention of the pendant in the thousands of pages published during the Great Paleolithic War. That silence was more damning than even the rare voice condemning the pendant, that of archaeologist Henry Mercer who, busy peddling his own mastodon-engraved forgery, sneered that Holly Oak was a fraud and Cresson a liar. Since the archaeological community in those days was so very small, its members must have known of the pendant. One can safely surmise they doubted its authenticity, and were just too polite to say anything about it. And why bother? Everyone knew the story.

Unfortunately, that story died with its participants, though the pendant itself survived in obscurity in a drawer at the Smithsonian Institution, only to emerge in 1976 on the cover of *Science*.[34] An accompanying article by John Kraft and Ronald Thomas reported that geological work in and around what is now Claymont, Delaware, revealed deposits that were at least 10,000–40,000 years old. They asserted, in the absence of any evidence that the pendant actually came from those deposits, it must be just as old, thereby making it "definite evidence of association of early American man with the woolly mammoth" and the only known example of North American Paleoindian art.

The *Science* cover caught the eye of Bill Sturtevant, a curator at the Smithsonian, who was instantly reminded of the La Madeleine mammoth. When he read the pendant came from Delaware, not the Dordogne, he straightaway suspected fraud. Knowing of my interests in Paleoindians and the history of American archaeology, he asked me to help investigate.

We quickly established there were concerns about unrelated but equally suspicious archaeological discoveries of Cresson's that, like Holly Oak, could never be field checked. We learned, too, Cresson was not trusted by his peers, and for good reason. In late 1891, Cresson was working in Ohio on Putnam's archaeological field crew when his supervisor, Warren Moorehead, caught him stealing artifacts from the excavations and shipping them to his home to Philadelphia. Cresson was fired on the spot.

Reading the details of Cresson's pilfering in Moorehead's diary, I was reminded of a comment made to me a few months earlier by James B. Griffin, one of the deans of American archaeology. Griffin thought it odd that the Holly Oak pendant, ignoring its engraved mammoth, bore such a striking resemblance to shell pendants found in Fort Ancient period sites, which generally postdate AD 1000. In virtually the same moment, I remembered that prior to revealing the existence of the pendant in 1889, Cresson had worked for Putnam on Fort Ancient–age sites and museum collections. What better way to pass off a forgery than to carve a mammoth on a genuine archaeological specimen stolen from a collection that at least *looked* old, but in the late nineteenth century could not be independently dated?

Of course, in the late twentieth century it could be, and when accelerator radiocarbon dating came on line, we submitted a tiny fragment of the pendant's shell. While we waited for the dating results, Smithsonian archaeologist Bruce Smith organized a pool to guess the pendant's age. The resulting radiocarbon age, when calibrated, came to approximately AD 875 (with a one standard deviation range of AD 760–990). Our prediction of its age was off by just a couple of centuries, probably attributable to the margin of error in dating marine shell; more precise calibration would require knowing where the shell came from in order to account for the local carbon reservoir, but on this point the trail is cold.

No matter. The Holly Oak pendant was engraved from looking at an image in a book, and not at a live mammoth. Bruce Smith won the pool.

Archaeology, not being a hard science but a difficult one all the same, rarely yields the kind of unequivocal results obtained in the Holly Oak case. When it does, it is gratifying—regardless of whether one happens to be right or wrong. But let's be honest: most of us want to be right. Science is like that. After we published the radiocarbon date on the shell, many colleagues claimed they had known all along the Holly Oak pendant was a fake. Science is like that, too.

Cresson had been desperate to place his name on the rolls of American science, and evidently believed that if he made a discovery so wonderful that it resolved the most bitter dispute then facing American archaeology, the archaeological world would beat a path to his door. The flaw in that strategy is that spectacular finds cannot be made on command, unless one is blindly ambitious and utterly dishonest. So it was that a stolen Fort Ancient shell pendant in the hands of a skilled artist schooled in French prehistory and with an illustration of the La Madeleine engraving in a book in front of him, became for a moment the first American Paleolithic art object. But only for a moment.

Whether it was the disgrace of being summarily fired, or the dismissal of his pendant (more by stony silence than public censure), or some other event that started Cresson on his downward spiral into insanity, we will never know. But perhaps Cresson knew. Among the items in his pockets at the time of his suicide was a note in his handwriting that he feared he was "suspected of counterfeiting and that Secret Service detectives were continually on his track."

change Evans's mind and undo the damage, but failed. Critics gloated. Proponents were badly shaken. Only the faithful remained undaunted. A bitter Abbott blamed Putnam for the "unfortunate Toronto business."[35]

Nevertheless, the active search for the deep past continued, though following that pivotal summer, no more claims were made on behalf of American paleoliths. However, the new century brought a new kind of evidence to center stage.

NEANDERTHALS IN AMERICA?

In December 1899, Putnam's hired hand in Trenton found a human femur (upper leg bone) deep in what appeared to be Pleistocene-age gravels at Trenton. Hope for a Pleistocene human antiquity was renewed, but Putnam was worried. "We must not make any blunder about it," he warned, and for good reason. Government critics quickly scorned the fast-traveling news of the find. Putnam turned the femur over to Aleš Hrdlička, a young physician turned physical anthropologist, whom he hoped would show that the bone had some antiquity. But Hrdlička was none too impressed by the specimen, which looked no different from femurs of recent American Indians. Perhaps, he supposed, the geology of the site might shed light on its age. Wright had examined the geology, and thought the specimen's Pleistocene age was so clear that there was scarcely anything to discuss. Wright, however, was hopelessly optimistic. By then, the Trenton gravels had such geological notoriety that any agreement on their age was impossible.[36]

The Trenton femur proved to be the first of many human skeletal parts found over the next twenty-five years in apparent Pleistocene-age deposits. Over that period, Hrdlička would emerge—after 1903 in the employ of Holmes at the Smithsonian—to challenge each and every claim, "like Horatio at the land bridge between Asia and North America, mowing down with deadly precision all would-be geologically ancient invaders of the New World," as one of his contemporaries put it.[37]

Hrdlička's position was this: if the earliest Americans had arrived in the Pleistocene, they should look like a Pleistocene-age fossil human—like Neanderthals, say, and not like the American Indians who inhabited the region. In structure this was no more than Abbott's argument (if it's old, it should look primitive) applied to skeletons. But there was one significant difference: the argument worked for Hrdlička. He was fast becoming the premier physical anthropologist of his day, and few could challenge his considerable knowledge of human variability and evolution. To claim a human skeleton was Pleistocene in age, one had to play by Hrdlička's rules. So were there Neanderthals in America?

On a farm along the Missouri River just outside Lansing, Kansas, in February 1902, brothers Michael and Joseph Concannon were digging a tunnel to store fruit and vegetables on their father's farm. Seventy feet into the hillside and twenty feet beneath the surface, they shoveled into two human skeletons. The bones were pushed aside—there was a tunnel to be dug—but after a few months, word of their discovery reached the City Public Museum in nearby Kansas City.

The museum's curator visited the Concannon farm, and seeing that the bones came from under a layer of apparent glacial loess, he alerted geologists Warren Upham and Newton Winchell of the Minnesota Historical Society, who rushed to Lansing. The two men were stunned by what they saw: Pleistocene loess atop the bones. Within days, Upham prepared a paper announcing that Lansing proved a human presence in the New World prior to the last episode of ice advance in North America: in round numbers, perhaps 30,000 years before the present.

To his friend Wright, Upham happily chirped that Lansing vindicated the claims made in *Man and the Glacial Period*, then sent his hastily written paper to Chamberlin and Holmes, inviting "verification or correction of our view."[38] Verification, naturally, was what Upham preferred and briefly thought he had. Chamberlin and Holmes (and later Hrdlička) visited Lansing, and soon afterward Upham heard rumors they had endorsed the site's great antiquity—rumors Upham vigorously fanned.

But the rumors were false. Unbeknownst to Upham, even before Chamberlin left for Lansing, he was already grousing about Upham's "fundamental untrustworthiness." Chamberlin's mood hardly improved at the site, where he concluded Upham—his former USGS employee—had badly misread the stratigraphy and geology. Back in Chicago, Chamberlin blistered Upham in a string of letters, lecturing him on the attributes of loess, scientific ethics (like Wright, Upham had "gone public" with his claims about Lansing), and even ominously accused him of "direct falsification."[39]

Battle lines were drawn once more: Upham, Winchell, and Wright fighting for a Pleistocene age for the loess overlying the skeleton, with Chamberlin and Bohumil Shimek, the Midwest's leading loess expert, dismissing the Lansing deposit as neither true loess nor Pleistocene in age. Holmes and Hrdlička joined Chamberlin, chiming in that the Lansing skulls were no different than crania of American Indians of the region, and surely no more than a few hundred or a few thousand years old. To Abbott watching from the sidelines, it all recalled "the merry old days of earnest work . . . [and] the controversial days that embittered me."[40] Just like in the old days, neither proponents nor critics backed down.

Scarcely four years later and 195 kilometers up the Missouri River, the exercise was repeated. At the Gilder Mound, just outside Omaha, Nebraska, human crania—Neanderthal-looking to some—were plucked from apparent Pleistocene loess. Eight years later, attention shifted to the tar pits of Rancho La Brea, where a human skull was extracted from the asphalt ooze along with bones of an extinct (Pleistocene) condor. Two years after that (1916), it was Vero, Florida, where human remains rolled out of sand deposits yielding the bones of extinct mammoth and sloth. Nearly a decade later, it was a similar story at Melbourne, Florida.

In each case, a swarm of archaeologists and geologists descended on the site to inspect the skeletal remains and their geological context, Hrdlička usually leading the charge (Figure 14). Testifying to the duration of the dispute, there was now a second

FIGURE 14.

Aleš Hrdlička examining the stratigraphy in the Gilder Mound, Nebraska, January 1907. (Photograph by E. Barbour, courtesy of the Nebraska State Museum.)

generation of participants. Literally, a second generation: geologist Rollin Chamberlin visited Vero in the place of his father, Thomas.

There were the usual disagreements over whether the human skeletal remains at each site were in primary context, or contemporary with the apparently ancient loess or extinct mammal–bearing deposits in which they were found. Hrdlička took a cue from Holmes: since human beings bury their dead and because bone is so easily broken and moved in the earth, the odds were that any bone in ancient deposits came from later times. "Perhaps," paleontologist Oliver Hay tartly replied, "we get a clue here to the reason why civilized people nail up their dead in good strong boxes."[41]

Besides, Hrdlička continued, even granting human remains had not moved in the ground, what assurance was that of their great age? Geologists' opinions were utterly divided. The Lansing and Gilder Mound specimens were either in true loess or not in true loess, and thus either Pleistocene or post-Pleistocene in age, while the Vero and Melbourne fauna was either Pleistocene or post-Pleistocene in age—it all depended on which geologist one heeded.

Hrdlička naturally followed the Chamberlins, but his true allegiance was to the bones. What did they have to say? Only if they spoke of an anatomically distinct pre-modern human could they be Pleistocene in age, never mind the geology. But they said no such thing.

Chamberlin (senior) was disappointed in Hrdlička's low opinion of geological testimony, but no more so than in many of his geologist colleagues. In fact, the irreconcilability of interpretations soured relations all around. Anthropologists and archaeologists bickered among themselves over what a Pleistocene-aged human should look like, then argued with paleontologists about the timing of mammal extinctions. Paleontologists wrangled with geologists about where to draw the line between Pleistocene and post-Pleistocene formations. Geologists fought each other over the number, timing, and evidences of glacial history. Even linguists got in the act, clucking with disapproval at everyone's failure to provide them sufficient time to evolve the great diversity of native North American languages (Chapter 5).

Once again, the situation reached an angry impasse. Holmes darkly pronounced the evidence from Vero "dangerous to the cause of science." Journalist Robert Gilder, who had found the Nebraska "Neanderthals," called Hrdlička a "liar."[42]

So it went. Over the decades, scores of purportedly Pleistocene-age sites were championed, some with stone tools, others with human skeletal remains, but all were suspect, and all faced withering criticism from Holmes, Hrdlička, and others. In this wide-open field, there were few rules of engagement, and the dispute exposed deep rifts over what constituted legitimate proof of human antiquity. At its worst, Frank Roberts darkly admitted, "the question of early man in America [became] virtually taboo, and no anthropologist, or for that matter geologist or paleontologist, desirous of a successful career would tempt the fate of ostracism by intimating that he had discovered indications of a respectable antiquity for the Indian." Archaeologist Nels Nelson advised his colleagues to "lie low for the present." Shrewd advice and many, such as Alfred Kidder, followed it: we "comforted ourselves," he later confessed, "by working in the satisfactorily clear atmosphere of the late periods."[43] Ironically, the key to resolving the dispute had, since 1908, quietly lain exposed in an arroyo in New Mexico.

IN THE BELLY OF THE BEAST

On August 27, 1908, torrential rains fell on Johnson Mesa in a remote corner of northeastern New Mexico. The thunderstorm broke in early evening, yielding a beautiful sunset, but then, unusually, fired up again. After dark, from a ranch just below the mesa, a frantic phone call went out to Sarah Rooke, the local telephone operator: the Dry Cimarron River was rising fast. Everyone downstream in the town of Folsom needed to head to higher ground. From her switchboard, Rooke began calling the townspeople. Many heeded the warning. Others could not be saved, including Sarah Rooke, who stayed at her post sounding the alarm until the floodwaters tore her small operator's shed from its foundations and washed it away. Her body was found in a field the next spring, buried in mud.[44] The flood forever changed the town of Folsom. So, too, American archaeology.

Sometime after the flood, George McJunkin, the foreman on the Crowfoot Ranch below Johnson Mesa, went out to check his cattle and fences, and came across a new and deeply incised portion of Wild Horse Arroyo, a tributary of the Dry Cimarron. Looking down, he noticed bones jutting out near the base of the arroyo. Most cowboys would have passed them by: bones are hardly an uncommon sight in ranch country. But by all accounts, McJunkin was no ordinary cowboy. Born a slave in pre–Civil War Texas, he was befriended at an early age by his plantation owner (Jack McJunkin), who taught him to read and supplied him with books. In his teens, George moved to Midland, taking a ranching job using the McJunkin name. By the time he was in his twenties, he was working on the Crowfoot Ranch.

Precisely what McJunkin, a self-taught naturalist, thought of the bones in the bottom of Wild Horse Arroyo is not known. But they obviously piqued his curiosity, for he told others about them. One was Carl Schwachheim, the blacksmith in nearby Raton, who had built a fountain in front of his home using the antler racks of two bull elk that had become entangled in a mortal contest. McJunkin had passed by the fountain, saw that its builder was a kindred spirit, and on trips to Raton would stop in to talk to Schwachheim—at some point, he must have described the bones in Wild Horse Arroyo.

Schwachheim did not visit the Folsom site until December 1922 (after McJunkin died). He and Raton banker Fred Howarth collected a few of the bones, which they took to the then Colorado Museum of Natural History in Denver. Jesse Figgins, the director, turned the bones over to the museum's paleontologist, Harold Cook, who identified them as being from an extinct species of bison. Figgins and Cook visited the site in March 1926 and decided to excavate, with the aim of acquiring a bison skeleton for museum display. They were not looking for, nor did they expect to find, any archaeological remains.

Still, they were well aware of the human antiquity controversy. Cook was the discoverer and namesake of *Hesperopithecus haroldcookii*, a fossil he had found in the early 1920s near his family ranch in western Nebraska, which was identified as a pre-Pleistocene form said to resemble what was then the oldest-known human ancestor, *Homo erectus* (Java Man, as it was then known).[45] Sadly for Cook's hopes of taxonomic immortality (Cook Man?), *Hesperopithecus* proved on closer inspection to be a fossil pig that had become extinct millions of years before our ancestors appeared on the plains of Africa. But Cook paid that little mind—the tooth might well be from a fossil pig, but he was convinced some of the bones found with it had been broken by humans. Ever the optimist, Cook was sure humans had been in America a very long time.

He saw further evidence of that at the site of Lone Wolf Creek, in Colorado City, Texas. Figgins had hired workers there in 1924 to extract the bones of an extinct bison for display (Figure 15). Unfortunately, they made a mess of the task, lopping off the ends of bones to make them fit packing crates, rather than just getting larger crates. Worse, Figgins learned only afterward that three artifacts had been found with the bison. The artifacts were neither left in place nor photographed; one was missing.

The following spring, Figgins dispatched Cook to the site to assess the geology, and even though the evidence was long out of the ground (and badly mangled), Cook still confidently concluded that Lone Wolf Creek provided "good, dependable definite evidence of human artifacts in the Pleistocene in America," perhaps as much as 350,000 years old.[46]

That was a daring claim at a time when most were unwilling to push human antiquity in America back to 10,000 years ago. But because of the sloppiness of the discovery,

FIGURE 15.
H. D. Boyes and Nelson Vaughn at their excavations at the Lone Wolf Creek site, Texas, 1924. Note the bison remains in place. (Photograph courtesy of the Heart of West Texas Museum.)

there was little reason for confidence in Lone Wolf Creek, no matter how vigorously Cook tried to promote it. And he tried very hard indeed. But that hardly convinced skeptics like Holmes, who immediately asked about Cook's scientific reliability. The answer from Cook's old mentor was none too flattering: "Harold has a somewhat optimistic temperament, and I find it necessary to discount his geological conclusions more or less." Holmes and Hrdlička did as well.[47]

Cook persisted. In a broadside attack on Hrdlička published the next year, Cook invoked Lone Wolf Creek and *Hesperopithecus* to bolster his claim of a Pleistocene or even earlier human presence in the New World. In Hrdlička's angry eyes, Cook's latest paper was just "another head of the hydra,"[48] and he moved swiftly to decapitate it. But like a hydra, new heads kept popping up.

In early 1927, Cook was called to a gravel quarry in Frederick, Oklahoma, following a report of mammoth and other extinct mammal bones found alongside grinding stones in apparent Pleistocene gravels. "Strangely enough," Cook remarked, "these implements show a degree of culture closely comparable with that of the nomadic modern Plains Indians."[49] He assessed the geology and concluded the site was about 365,000 years old. But as at Lone Wolf Creek, crucial details of what was found, and where, rested on the unreliable testimony of an inexperienced collector. Worse, an independent assessment of the site's geology concluded the deposits were "not necessarily more than 10,000 years old, and might be somewhat younger."[50] That was followed by a searing critique of the archaeology that ridiculed the absurdity of Pleistocene-age grinding stones (even today, such are rare).

Yet, Cook and Figgins paid the skeptics little mind, writing that *Hesperopithecus*, Lone Wolf Creek, and now Frederick pushed human antiquity back "by hundreds of thousands of years."[51] By then, however, few were taking them or their sites very seriously. It was in this harshly skeptical climate that their Folsom work emerged.

Schwachheim was hired by Figgins to excavate at Folsom, and began in May 1926. By early July, he was down to the level of the bison bones, and in mid-July, an artifact was found. This was no "rude" paleolith, but a delicately made spear point with a distinctive central groove or *flute*. Unfortunately, the point was out of the ground before he spotted it, so whether it was associated with the bison bones was uncertain. Notified in Denver, Figgins instructed Schwachheim to watch "for human remains and then in no circumstances, remove them, but let me know at once."[52] He wanted to inspect the remains in place. The remainder of the summer, Figgins waited anxiously for word. None came.

Nonetheless, he and Cook were convinced this was another Pleistocene archaeological site. That fall they wrote matching papers for *Natural History* magazine that, as Figgins boasted to Oliver Hay at the Smithsonian, were "a deliberate attempt to arouse Dr. Hrdlička and stir up all the venom there is in him." As Figgins explained, "Everyone seems to think Hrdlička will attack [and if he] tears a chunk of hide off my back . . . there is nothing to prevent my removing three upper and two lower incisors, black one eye and gouge the other, after I have laid his hide across a barbed wire fence. I am daring the whole miserable caboodle of them."[53]

Brave words, and they inspired Hay to march down the hall to Hrdlička's office to arrange "a showdown" with Figgins in Washington. When Hay reported what he'd done, Figgins backpedaled fast, declaring it would be much better if Cook went to Washington to "be the [sacrificial] goat." In the end, Figgins checked his faltering bravado and boarded a train for Washington. By the time he arrived at Hrdlička's office, he was in a fearful lather. Yet, much to his astonishment and relief, Hrdlička was courteous, "extremely pleased" to see the Folsom point Figgins carried with him, and even offered some advice: if additional points appeared during the coming excavation season, they should be left in place and telegrams should be sent around the country inviting "outside scientists" to come and examine them in the ground.[54]

Good advice, Figgins thought, and he left with newfound respect for Hrdlička. What he didn't appreciate, however, were Hrdlička's motives for offering that advice. Hrdlička didn't trust Figgins or Cook for a moment, and would not be convinced by anything they said about Folsom's age, or any possible association of artifacts with extinct animals. He wanted others called in to judge the evidence.

Schwachheim resumed excavating at Folsom in the spring of 1927. In late August, he uncovered a Folsom point, this time firmly in place between a pair of bison ribs. Figgins was alerted, and immediately broadcast telegrams around the country announcing, "Another arrowhead found in position with bison remains at Folsom, New Mexico. Can you personally examine find." Schwachheim was commanded to guard the point night and day until the visiting dignitaries arrived. He dutifully awaited the parade of "Scientists, Anthropologists, Archaeologists, Zoologists, or other bugs."[55]

It began a few days later with the arrival of Figgins, paleontologist Barnum Brown of the American Museum of Natural History, and archaeologists Alfred Kidder of the Carnegie Institution of Washington and Frank Roberts, a young colleague of Hrdlička's at the Smithsonian (Hrdlička was invited, but was in Alaska at the time; Holmes by then had retired) (Figure 16). All agreed this was no accidental association of artifact and bone: human hunters had killed this now-extinct Pleistocene bison. Here, finally, was proof in those pre-radiocarbon days of an Ice Age human presence.[56]

That certainty, however, gave way to uncertainty about the site's absolute age, since the taxonomy and timing of bison extinction were still not altogether clear. Even so, within weeks Kidder announced publicly what he'd always hoped for privately: the first Americans had arrived some 15,000–20,000 years ago (now he could rebut those who claimed the Americas hadn't been occupied long enough for its civilizations to have developed on their own). The announcement, subsequently elaborated by Brown and Roberts, electrified the scientific community.

Brown, in fact, was so taken by what he saw that he returned to Folsom in 1928 to expand the excavation. That July, when fluted points were again found alongside bison

FIGURE 16.
Carl Schwachheim (left) and Barnum Brown posing with the first in situ Folsom point, September 4, 1927. The point is the one shown in close-up in Plate 1 of this book. (Photograph courtesy of American Museum of Natural History.)

remains, telegrams were once more broadcast across the country, and in response, the find was seen by several of "the best men in the country,"[57] including some of a new breed of USGS geologists who mapped the region, and who independently affirmed Folsom's late Pleistocene antiquity (based on my own investigations at Folsom seventy years later, which involved extensive radiocarbon dating, I can report their age estimates were off by only a couple thousand years: that's quite an achievement, given they had no techniques for absolute age dating[58]).

The decades spent searching for paleoliths or pre-*sapiens* fossils were over. In retrospect, that effort seemed strangely misguided. Folsom was late Pleistocene in age, but it looked nothing like what anyone expected a Pleistocene human occupation to look like.

HISTORICAL HOMILIES

Proponents of a great human antiquity in the Americas, recalling with fondness Holmes and Hrdlička apparently getting their comeuppance at Folsom, see in this history a vindication of their belief that critics hinder the recognition of bona fide early sites and retard the progress of science. That's a serious charge, but it needn't be taken too seriously.

Historian of science David Hull observes that "the least productive scientists tend to behave the most admirably, while those who make the greatest contributions just as frequently behave the most deplorably."[59] Without question, Chamberlin, Holmes, and Hrdlička often behaved deplorably: they were merciless in attack and engaged in no small amount of behind-the-scenes skullduggery.

Nevertheless, far from hamstringing inquiry, these critics actually sped it along toward resolution. Their tough questions and criticisms crystallized debate on stone tool technology (the means of making tools), the relationship between tool form and age, human evolution, glacial geology, and dating techniques—critical issues all. Much was learned in the half century of dispute, not least what a solution to the human antiquity controversy had to look like. The swift acceptance of Folsom attests to that.

Besides, look at the sites the critics rejected. Today, Trenton, Newcomerstown, Lansing, Gilder Mound, Vero, and all the other allegedly Pleistocene sites are significant because of the battles fought there, not because of the great age of their archaeological remains. None of them were what they were claimed to be. Skepticism about a Pleistocene human presence was not arbitrary, but rather forged in the face of repeated cases that failed to withstand critical scrutiny (circumstances that will arise again).

The antiquity of the Folsom site, like that of Brixham Cave before it, was based on artifacts in close association with the remains of an extinct Pleistocene animal, the only secure means then available for telling time. That made Folsom unlike virtually every previous contender, where it had been impossible to demonstrate that the artifacts (or human skeletal remains) had been deposited at the same time as the Pleistocene-aged deposits or fauna with which they were found.

Equally important, Folsom was a kill site that would ultimately yield the remains of nearly three dozen bison and almost as many points. Consequently, its excavation provided several opportunities for scientific visitors to witness newly discovered points in place. Site visits to evaluate claims of great antiquity had been common since the 1890s. Yet, they were not always successful nor, for that matter, welcome. After a joint visit to Trenton in the late summer of 1897, Abbott griped, "I cannot say, looking back over the past four days, that I have enjoyed it. There is too much assumption of extra-carefulness, as they call it, which is simply a lot of childish twaddle. They cannot grasp the subject in its entirety and see the facts. . . . They may all be very eminent men, but it took me a good deal less time to learn that we had here evidences of man's antiquity."[60]

Those earlier site visits were often exercises in incompatibility, never achieving consensus and serving largely to highlight differences in interpretation. This was hardly surprising: archaeological methods and techniques in those years were uneven, many discoveries were made under dubious conditions, training was spottier, more amateurs were in the mix, there was considerable disagreement about how to recognize Pleistocene-age deposits, the criteria for evaluating evidence were less explicit, and a site's age had to be assessed in the field after careful examination of the stratigraphy and geology, artifact context, and the nature of the associated remains. At Trenton, Lansing, Gilder, and Vero, proponents and critics visited the sites, looked at the very same evidence, but came away with radically different views of what it meant. The 1927 visit to Folsom, where everyone agreed on what they were seeing, was the exception far more than the rule. But then by 1927, archaeology, geology, and vertebrate paleontology had become more professional sciences.

It was not inevitable that resolution of the human antiquity controversy would occur at Folsom, only that a site such as Folsom was needed, where the association of points and extinct animals was indisputable and could be repeatedly witnessed. That's critical, for the evidence at Folsom was seen—on Hrdlička's good advice—by members of the scientific elite.

Historian of science Martin Rudwick has shown that controversy in science, or at least nontrivial controversy, is not resolved by consensus across the community.[61] Rather, resolution is brought about by a core elite within the field who are recognized as experts, even if they are not particularly involved in the research within that area. These elite scientists regard themselves, and are regarded by others, as competent arbiters of the fundamental questions of a discipline: for example, the antiquity of people in America.

That Kidder examined Folsom, then publicly announced his acceptance of the evidence, carried enormous weight, for in the 1920s, he was at the height of his considerable power and influence (like Lyell in the 1860s). He was not being immodest when he explained to Figgins that "as an archaeologist, I am of course not competent to pass either upon the paleontological or the geological evidences of antiquity, but I have paid great attention for many years to questions of deposition and association. On these points I am able to judge, and I was entirely convinced of the contemporaneous association of the artifact which you so wisely had left 'in situ' and the bones of the bison."[62]

Wise, indeed. Figgins, of course, had done so in order to convince two more members of the elite: Holmes and Hrdlička. We know their opinions mattered, and not just because they thought so. Cook and Figgins thought so, too. In every paper they wrote, they wrote for—or rather, against—the Smithsonian duo. They recognized, however much they disliked the idea, that it was only "right and proper [that Holmes and Hrdlička] should not take without question such basic evidence as may seem necessary to establish a given fact beyond reasonable question."[63]

Holmes and Hrdlička did accept the evidence from Folsom, though on their own terms. When asked directly, the eight-one-year-old Holmes understandably replied he was finally content to leave judgment to others (the old lion passed away a half-dozen years later). Hrdlička quietly probed for weaknesses in the Folsom case, but did so more out of habit than hope, for he respected Kidder's authority in such matters. It was not true, as one paleontologist teased, that Hrdlička would not accept that the Folsom bison were speared by humans unless he had "fired the arrow himself."[64] But Hrdlička did want to know who was there when the "arrow" was unearthed. When later confronted on the larger question of human antiquity in America some years later, Hrdlička stuck close to his original script: there was still no evidence of a pre-*sapiens* skeleton in America. In that, he was correct.

Holmes and Hrdlička for decades had been the scourge of claims for a deep human antiquity. That they said nothing about Folsom spoke volumes. No mea culpa was offered, but then no one expected them to admit to being wrong for so long.

But if Holmes and Hrdlička lost the war over a Pleistocene human antiquity in America, it was not because Cook and Figgins won the battle at Folsom. Much like the situation at Brixham Cave, there was a sharp divide between those who made the discoveries (Boucher de Perthes; Cook and Figgins), and those who were called upon to judge their significance (Evans, Lyell, and Prestwich; Brown, Kidder, and Roberts). In both instances, the opinions of the discoverers were largely ignored because of their propensity to make absurd claims about what they'd found. Figgins understood their place, privately admitting "our opinions are valueless."[65] He was not being humble; he was being honest. At a time when virtually all archaeologists were skeptical of a Pleistocene presence in the Americas, he and Cook were campaigning for several spectacularly weak cases. Nelson lectured Figgins that if everything he and Cook said were true, "we shall have to revise our entire world view regarding the origin, the development, and the spread of human culture."[66] Nelson was not ready to do that. Few were.

Even worse, it wasn't obvious to Cook and Figgins, as it was to everyone else, that Folsom was the pick of the litter. In fact, Cook judged Folsom the "weakest and least conclusive of our localities," and Frederick the strongest.[67] That they couldn't properly evaluate their own evidence destroyed any remaining shreds of their credibility.

Cook and Figgins fell victim to what's been called the *Matthew Effect:* "for whosoever hath, to him shall be given . . . but whosoever hath not, from him shall be taken away even that he hath." Hrdlička's reputation may have been roughed up by Folsom, but only slightly, and he continued to be a sought-after authority regarding human antiquity in the Americas. Not all scientists are created equal; some are more equal than others. And inequality is most visible during episodes of controversy, when the stakes are highest. In the end, and despite their crucial role in the discovery at Folsom, neither Cook nor Figgins was asked to interpret the meaning of what he had found, nor given the opportunity to participate in any of the half-dozen major symposia devoted to human antiquity that followed in the next decade, let alone receive the acclaim for his contributions. Unfair, perhaps, but at least there were these kind, albeit private words to Figgins from the ever-gracious Alfred Kidder: "Anthropology owes you a very great deal for having handled this material so carefully and so intelligently, and I think the researches of yourself and Dr. Cook will go far towards opening a new era in the study of the question of Pleistocene man in the New World."[68] They did.

A MAMMOTH BARRIER

The Folsom find did not result in an antiquity comparable to the remote ages of Paleolithic Europe. But by 1927, it was obvious American prehistory did not extend that far back. Nonetheless, Folsom did show that the first Americans arrived at least by the end of the Pleistocene. They were not Paleolithic peoples, but assuming them to be ancestors of modern American Indians, they came to be called *Paleoindians*.

The Folsom find also taught archaeologists how to look for Paleoindian sites. The strategy was simple: look in arroyo channels or ancient lake beds, or track down reports of large and easily spotted bones of extinct Pleistocene mammals (Figure 17), then carefully comb those localities for any associated human artifacts. So it was that Clovis came to light. A road crew, mining gravel from an old pond near Clovis, New Mexico, had struck immense fossils bones. Word of their discovery reached Edgar B. Howard of Philadelphia's Academy of Natural Sciences, who began excavations there in the summer of 1933, and soon uncovered mammoth bones alongside fluted points that were longer, broader, and less finely made than Folsom points (Chapter 8). These "generalized Folsoms" soon became known as Clovis points.

Naturally, how archaeologists looked for Paleoindian sites predisposed what they found: almost all of the nearly two dozen found in the decade after the Folsom discovery were kill sites with bones of extinct animals and artifacts. "Boneless" Paleoindian sites were rare.[69] Not fully realizing how much this pattern was biased by their search strategies, archaeologists saw in the many sites littered with bones of bison

FIGURE 17.

A bison tibia (the larger of the two lower leg bones) exposed on the wind-swept floor of a now dry Pleistocene lakebed in far west Texas. Spotting bones eroding out in this manner led to the discovery of many Paleoindian kill sites, particularly during the 1930s Dust Bowl. (Photograph by David J. Meltzer.)

and mammoth the testimony that Paleoindians were top predators, who specialized in the killing of big game. An inspiring vision of gutsy hunters, holding a trumpeting and mortally wounded animal at bay, came to embody North American Paleoindians, and often established expectations of Paleoindian sites in areas environmentally and climatically different from the Great Plains, where virtually all the iconic kill sites had been found. Steven Simms, who works in the Great Basin, gives voice to many when he complains of "stereotypes of [Paleoindian] lifeways applied uncritically from the Plains."[70] Big-game hunting makes for good copy, but the appearance masks a different reality (Chapter 8).

Sites with Folsom points soon proved to be limited in their distribution to the western plains, but Clovis points were more widespread—across the continent, in fact. They were also found in deposits below Folsom points, evidence the Clovis archaeological complex was older than Folsom. How much older would only be learned a couple of decades later, following the advent of radiocarbon dating (it would also become clear that by Folsom times, there were many other archaeological complexes in other parts of the continent, as detailed in Chapter 9).

But were Clovis groups the first Americans, or had people arrived earlier still? In the aftermath of the Clovis discovery, archaeologists sought traces of more ancient Americans, and soon found themselves again at loggerheads over the question of antiquity. As early as 1953, Alex Krieger warned his colleagues that having overthrown the Holmes-Hrdlička "dogma," they were now in danger of replacing it with another. The first Americans would be permitted a late Pleistocene entry, but he feared that 10,000 years was fast becoming the new "allowed antiquity."[71]

Yet, in 1953 he tallied a half-dozen sites that "may and probably do" break that barrier. In 1964 he upped the total to fifty sites in North and South America that he thought pointed to a human presence predating Clovis. Not all sites were likely genuine, as Krieger well appreciated. But what impressed him most was how many sites *looked* old. As he saw it, where there's smoke there's fire. The sites on his list included some with radiocarbon ages ranging from 21,000 to more than 38,000 BP. Others had bones of extinct animals that appeared split, burned, or broken by human hands. And then there were the Malakoff heads—giant sandstone boulders from deep in a Texas gravel quarry that had crude "faces" carved into them, looking like enormous versions of Mr. Potato Head toys. Krieger even tossed in several American Paleolithic sites, Trenton included, for they "cannot all be set aside as insignificant."[72]

In fact, many of the sites on Krieger's list recalled the American Paleolithic, for they contained crude stone or bone artifacts. Krieger insisted he was not making Abbott's mistake of equating artifact form with age, merely raising the possibility of a "pre-projectile (pre-Clovis) point stage." Perhaps. But few were eager to follow Krieger out on his speculative limb. Others were busy sawing it off behind him.

The same year Krieger published his pre-Clovis compendium (1964), C. Vance Haynes reported the first secure radiocarbon ages for a half-dozen Clovis sites (including the type site). Their ages fell in a very narrow slice of time, between 11,500 and 11,000 BP.[73] None were more than 12,000 years old, and none yielded evidence of a precursor or pre-Clovis population, despite having underlying sediments of the right age, and with ecological conditions that should have been favorable to occupation—had people been present in the area.

But if there was no pre-Clovis, how might one explain the "apparently sudden appearance" of Clovis over much of North America 11,500 years ago? Geology provided a clue. By the late 1950s, there were a raft of radiocarbon ages available on the timing of deglaciation, and in them Haynes spotted a striking concordance between the geological and archaeological records. It appeared that 12,000 years ago the ice-free corridor had opened, linking Alaska with the rest of the continental United States for the first time in 15,000 years. Was it mere coincidence Clovis appeared south of the ice sheet at this moment?

It all made perfect sense: the land bridge connecting Siberia and Alaska only emerged during glacial cycles, but once migrants reached Alaska, ice sheets had blocked their path south. Either the first Americans came before the Late Wisconsin ice advance, in which case they had to contend with crossing the open Bering Sea, or they came later and walked across the land bridge, then cooled their heels in Alaska waiting for the Cordilleran and Laurentide ice to retreat. The splendid chronological correlation between the disappearance of the ice and the appearance of Clovis surely favored the latter hypothesis.

The way Haynes saw it, Clovis progenitors probably were in Alaska some 12,500 years ago, moved out across the Arctic slope soon thereafter, and then down the ice-free corridor fast on the heels of its opening. The migration from Siberia into northwestern North America might have taken 1,500 years altogether, but once south of the glaciers, Clovis groups could have spread rapidly east and west, colonizing the entire continent in fewer than 1,000 years.

For Haynes the pieces were falling neatly into place. For Krieger, his worst fears were coming to pass (Figure 18). The notion that Clovis was one of the *older* occupations of North America was steadily losing ground to the idea that it was the *oldest* occupation in North America.

Still, in 1964 Haynes saw "good indications" there were people in America before 12,000 years ago. He just didn't think they were related to Clovis, or perhaps there were just very few of them. But by 1969, he was losing enthusiasm for the idea of pre-Clovis.[74] His newly found skepticism was understandable. In the intervening years, he had learned the hard way that sites too good to be true often weren't.

Tule Springs, located near Las Vegas, Nevada, had everything going for it: genuine artifacts, bones of Pleistocene megafauna seemingly broken by human hands, and a radiocarbon date of 28,000 BP. Anxious to learn more of this occupation, a team of archaeologists, paleoecologists, and geologists embarked on an ambitious excavation program, overseen by a blue-ribbon panel of scientists. Almost immediately Haynes,

FIGURE 18.
C. Vance Haynes and Alex Krieger examining specimens at the 1970 Calico Conference. (Photograph courtesy of David J. Wilson.)

the site geologist, noticed that the "hearths" at Tule Springs were nothing more than organically blackened deposits. That was welcome news, solving as it did a longstanding puzzle: if these were hearths, why were they full of unburned snail shells? Then the curious fracturing of the megafaunal bones turned out to be restricted to remains found in the spring vents, and likely had been broken by the frantic trampling of animals trapped in those quicksand-like sediments, and not by human hands. Their "burning" was merely groundwater staining. The only indisputable artifacts at Tule Springs were from deposits much younger than 12,000 years old.[75] Tule Springs was not pre-Clovis. Still, neither Haynes nor anyone else categorically rejected the possibility of an older human presence in the Americas. Just so, the seeds of skepticism were sown.

At the outset of the Great Paleolithic War, Charles Abbott had expressed the fervent hope that "the 'doubting Thomases' [would] be fewer by the year 2000."[76] As it turns out, Abbott finally got something right. But it was no easy road getting there.

4

THE PRE-CLOVIS
CONTROVERSY AND ITS
RESOLUTION

In the late 1980s, about the time the pre-Clovis dispute in America was nearing a boil, archaeologist Jack Harris visited my university, bringing with him the just-excavated artifacts from the Senga site in Zaire (now Democratic Republic of Congo). Judging by the animal bones found with them, he inferred the specimens were more than 2.3 million years old—making them nearly 100,000 years older than any previously known artifacts from Africa, and thus the earliest human tools ever found. At the end of a brief talk about his excavations at Senga, Harris laid the specimens on the table.

It is surprisingly humbling to see a human culture starter set, even if it is several hundred shattered quartz flakes, most no more than 2 centimeters long and all utterly nondescript, even by forgiving Lower Paleolithic standards. But what I really noticed was the silence. No one in that audience of mostly Old World archaeologists asked Jack how sure he was of the artifacts' age, whether their geological context was secure, or why he was certain they were even artifacts—questions I was asking myself. After all, I'd been in rooms where artifacts on display from the pre-Clovis age sites of Monte Verde, Chile, and Meadowcroft Rockshelter, Pennsylvania, dated to 12,500 and 14,250 BP, respectively, triggered noisy debate.

So why does Senga, plunging the birth of cultural behavior 100,000 years deeper into the past—almost to the doorstep of the Pliocene—scarcely elicit a murmur of skepticism, while sites just 1,000 years older than Clovis come under withering fire? Senga helps redefine what makes us human. Monte Verde and Meadowcroft only tell us American prehistory began slightly earlier than we thought.

I asked my colleague, South African archaeologist Garth Sampson, why Senga was so readily accepted. He tried not to be condescending: "My dear fellow," he said, "because it's obvious Senga is just what Jack says it is."

Yes. Well. That was helpful. But explain that to Tom Dillehay or James Adovasio, both highly competent archaeologists, each of whom had assured us that the pre-Clovis-age occupations at Monte Verde and Meadowcroft were just what they said they were. Why wouldn't we take their word for it?[1]

Why did controversy over antiquity persist in the face of mounting evidence from fields as diverse as linguistics, genetics, and physical anthropology—as well as archaeology? Why did it remain unresolved in the face of a regular crop of purportedly ancient assemblages and archaeological sites, and an almost annual harvest of books and papers on the topic? Why, indeed, all the fuss?

Some blamed this state of affairs on critics either secretly envious or so hopelessly biased they attack even legitimate pre-Clovis claims—Niède Guidon believes that "when you are the first to discover something, people want to kill you because you disturbed the placid waters of the lake". The result, Guidon claims, is a "Kafkaesque" situation in which there are good archaeologists (with Clovis sites), and bad ones (those, like herself, with pre-Clovis sites).[2]

Yet, were I selecting Kafka metaphors, it would not be his dark classic *The Trial*, but his lesser-known *Metamorphosis*, in which a perfectly promising human being turns into a cockroach. Virtually every field season over the last several decades brought a new contender for the pre-Clovis crown. Each was initially heralded with great fanfare, but soon started rolling down the slippery slope of the pre-Clovis credibility decay curve. The problems varied, but the outcome was inevitably the same. Most pre-Clovis claims had a shelf-life of about a decade: better than fish and house guests, but ultimately they, too, began to stink.

Recall that in 1964 Alex Krieger had identified fifty pre-Clovis sites in the New World. By 1976 Scotty MacNeish had tallied just thirty-five; in 1988 Richard Morlan had found only five.[3] The sites on their lists are almost completely different. Supporters of pre-Clovis put a positive spin on these wholesale losses: by 1976, they say, MacNeish had at his disposal a wealth of new and improved sites unavailable to Krieger. Naturally, he would use these weapons in the battle over the antiquity of the first Americans. Maybe so. But then thirty-one of MacNeish's sites failed to make Morlan's list. Critics read a different message: if 90% of what was considered the best pre-Clovis claims in 1964 failed to survive until 1976, and if nearly the identical percentage from 1976 failed to make the 1988 list, why put much faith in any sites on any list?

There is one very good reason: whether the first Americans came here before Clovis times and whether we have detected their traces are independent issues. The fact many sites have been proposed but none accepted only tells us none of those sites is acceptable. It does not prove humans were absent from the Americas in pre-Clovis times. And no matter how many sites fail, that does not reduce the odds of finding one that might pass. Still, and not surprisingly, that decades-long failure meant few archaeologists were willing to accept a pre-Clovis claim at face value. We grew skeptical, cynical, even insulting: Paul Martin (not an archaeologist, but with his own Clovis-first agenda to push[4]) likened the pre-Clovis search to the quest for Bigfoot, as though there were sound reasons why pre-Clovis could not exist. There aren't. But then there's no reason it has to either. Whether it does or not is strictly a question of evidence.

PASSING THE TEST

Although there has been considerable and largely fruitless speculation over what a pre-Clovis site ought to look like, just as there was a century earlier over how an American Paleolithic might appear, one conclusion seems inescapable: we are not looking for—and from what we know of human evolution do not expect to find—Senga-like artifacts here in the Americas. To be sure, in some places "primitive-looking" artifacts were made even in the recent past (it is often a sign of a dearth of suitable toolmaking stone), but overall we should see the relatively complex tools and features that characterize those of other sites produced by anatomically modern humans (*Homo sapiens*), who spanned the globe over the last 100,000 years. Beyond that, however, we cannot be much more specific predicting the look of a pre-Clovis toolkit.

What we can expect—demand, even—of a pre-Clovis site are (1) undeniable traces of humans, either their artifacts or skeletal remains, in (2) undisturbed geological deposits in proper stratigraphic position, which are (3) accompanied by indisputable radiometric ages (speared bison no longer suffice). These are no more than evolved versions of criteria Aleš Hrdlička first proposed in 1907, and no less than what was finally met at Folsom in 1927.[5] On their face, they are not difficult criteria to meet, nor any different than the ones used in Africa to identify the earliest (and often the most ambiguous) traces of human culture.

Still, as the changes between Krieger, MacNeish, and Morlan's lists plainly show, these criteria have proven extraordinarily difficult to meet, so much so that pre-Clovis proponents have cried foul.[6] The criteria, they charge, are unfairly applied and unduly rigid. They insist that if broadly applied, we would be left wondering whether anyone occupied North America in *Clovis* time. Are the criteria so perversely narrow they blind archaeologists from seeing legitimate evidence? A closer look is in order.

The first criterion (and often the most troublesome), requires the presence of artifacts or human skeletal remains. The latter have not figured prominently in the pre-Clovis dispute, simply because few Pleistocene-age human remains have been found in North or South America, and none is significantly older than 11,500 BP (as we'll see in Chapter 5, skeletal remains are central to the most controversial questions about *who* the first Americans were). Artifacts are another story.

Artifacts come in a myriad of materials, but most commonly only those made of stone, and to a lesser degree bone, survive the rough passage of time. Almost certainly the earliest Americans fashioned artifacts of wood, antler, shell, and the like (but not ceramics—those came later), but those materials are quick to decompose, and are seldom preserved or recovered archaeologically.

An artifact is an object made, modified, or otherwise used by humans. Oftentimes, fashioning an object for use renders it easy to spot as the product of human hands: turning a chert cobble into a finely crafted spear point, or carving a piece of mammoth rib into a socket wrench. No recognition problems here.

But at other times, objects are hardly modified, and even after use show little sign of having been made by human hands—like a round river cobble employed briefly to smash open bone for marrow, but which left few traces of that action. And to complicate matters, nature can modify stone and bone in ways that mimic the action of human hands. Grinding glacial ice and roaring rivers, for example, can break stone or bone into forms that resemble rough-hewn artifacts, such as choppers or scrapers. Here it becomes much harder to tell whether an object was made or used by people.

Such ambiguous cases require close scrutiny of *context*: where was the object found and with what? An otherwise nondescript, round river cobble found in a site high above a river, alongside charred and broken animal bones, a charcoal-laden hearth, and a dozen broken spear points all in fine-grained loess, is likely an artifact. Find that cobble in the river bed, amid millions like it, and even if it were an artifact, one would be hard-pressed to prove it. Worse, in that river bed there would be naturally broken cobbles that looked more like artifacts than that round cobble. Where artifact look-alikes grade imperceptibly into broken cobbles bearing no resemblance to artifacts, and then into unbroken cobbles, the context bespeaks a natural origin for all. Seems simple enough.

TROUBLE IN THE HILLS

The Calico Hills site is located in the hills high above Pleistocene pluvial Lake Manix, in what is now the Mojave Desert near Yermo, California. In the 1960s, Ruth "Dee" Simpson found stone artifacts scattered about its surface; some were obviously recent, others looked "Paleolithic" (shades of Charles Abbott).

Louis Leakey, fresh from his triumphant discovery of 2-million-year-old hominids at Olduvai Gorge, in Tanzania, was invited to Calico in May 1963. When he saw the site, and what appeared to be artifacts in place deep in a bulldozer pit, he was instantly convinced this was a spot to find ancient Americans. "Dee, dig here," is what he said.[7] "Here" turned out to be smack in the middle of a pre-Wisconsin-age alluvial fan, a dense mass of gravel and sand (Figure 19) originally deposited by moving water. Its age was initially estimated at more than 100,000 years old; it's now thought to be twice that old. Excavations plunged into the fan, and soon recovered "uniquely different" artifacts. Leakey returned in the spring of 1965, and in front of the anxious excavation crews,

FIGURE 19.

Excavations in Master
Pit 2 at the Calico site, ca.
1968. The untold number
of naturally broken rocks
jutting from the sidewalls
convey why it has proven
impossible to demonstrate
that humans were present at
this site in early Pleistocene
times. (Photograph courtesy
of David J. Wilson.)

examined stones they had cherry-picked from the rubble. He judged at least twenty-five of the specimens to be humanly made. "Leakey's luck," it seemed, had struck again.[8]

That was enough to entice the *National Geographic Society* to fund fieldwork at Calico to the tune of $100,000 (no small change in those days), but with strings attached: they funded Vance Haynes and others to visit and evaluate the site's artifacts and geology. When the visitors saw that "artifacts" were being plucked from amid hundreds of thousands of broken stones occurring naturally in the fan, they urged all specimens be saved, whether thought to be artifacts or not. They were.

Excavations continued at Calico, and as the pile of rocks tossed aside as non-artifacts grew to cover several acres, so did the number of Calico "artifacts" (at last count over 11,400 pieces). No surprise there: where nature breaks millions of rocks (as it can in a fast-flowing river), a small percentage will resemble crude artifacts. Were these found in an otherwise indisputable archaeological context, they might be accepted as artifacts. But found as these were in the Calico alluvial fan—a giant, natural gravel crusher—Haynes suspected none were humanly made, but instead were naturally produced *geofacts,* formed as the rocks tumbled downstream from sources in the Calico mountains.

Haynes's criticisms sparked a lively discussion over the minutiae of flaked stone: did the Calico specimens possess attributes unequivocally made by human hands? Caliper-wielding critics compared known artifacts with naturally broken stones, identified the essential criteria that differentiate the two, and then applied those to the Calico specimens. The specimens proved to be more like naturally fractured stones than humanly fashioned ones.

Proponents insisted that other attributes of the stones bespoke their human origin. Only, they reached that conclusion after discarding all the rocks that did not look like artifacts. Naturally, the ones they kept looked humanly made and had attributes of artifacts. But that only means they were consistent in their sorting criteria, and not that those specimens were flaked by people.

Calico, in the words of Nicholas Toth—who studies Africa's earliest stone tools—represents a worst-case setting for documenting a human presence. Against a backdrop of all those naturally broken cobbles, is there reason to suppose *any* were artificially fractured?[9] And if they were, how could they possibly be reliably identified?

Leakey organized an international conference and site visit at Calico in 1970, hoping it would be the American equivalent of the British geologists' visits to examine Boucher de Perthes' collections and sites in 1859 (naturally, the Cambridge-trained Leakey would think of the landmark European site visit rather than Folsom). Leakey had "no doubt, whatsoever!" about Calico's antiquity, but he failed to carry the day. His wife Mary—an accomplished archaeologist in her own right—later bluntly wrote that no one at the conference "really cared to get up and make a strong outright condemnation of the evidence, because it would have been equivalent to saying that Louis had lost his capacity to think scientifically—which, I fear, was perfectly true."[10] Indeed, there's reason to wonder whether Calico would have been taken seriously by anyone, were it not for his ringing endorsement.

In the end, Calico's best hope for pre-Clovis legitimacy vanished when a semicircular "hearth," found deep in the excavations, proved not to be one: the stones comprising the feature hadn't been subjected to heat in as much as 400,000 years, if ever. It was instead a natural rock scatter that, friend and foe admit, only looked hearthlike after surrounding gravel was removed. A small core of loyalists remain convinced of Calico's antiquity, but otherwise no one considers it an ancient archaeological site.

"Leakey's luck" finally ran out in the California desert, a couple of hours down the road from Las Vegas, where dreams are daily made, and lost, on luck.

STENO'S LEGACY

No matter how sophisticated archaeological techniques have become, we still rely on one developed centuries earlier: stratigraphy. In 1669 Niels Steensen (better known by his Latin pen name, Nicolaus Steno) formulated the first principle of stratigraphy:

namely, that in an undisturbed deposit of strata or layers in the earth, the order of deposition is from bottom to top, and therefore the layers below (and any fossils or artifacts within them) are older than those above. Stratigraphy can be relatively uncomplicated, even where layers were cut through or are cross-bedded, but problems can arise.

In an ideal stratigraphic circumstance artifacts found in, say, Late Wisconsin–age loess, would presumably be Late Wisconsin in age. Yet, in the real world, both natural

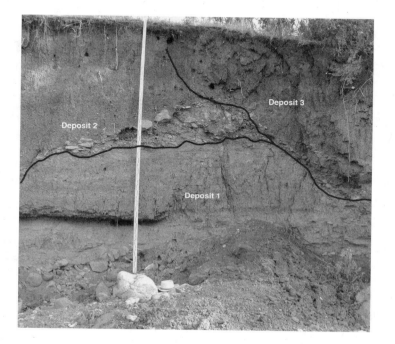

FIGURE 20.

An example of the complications of stratigraphy, Dry Cimarron Valley, New Mexico. Deposit 1 was laid down gradually by river overflow. At times deposition slowed, vegetation grew on the land, and created soil horizons (for example, the horizontal black band under the Deposit 1 label), which were subsequently buried by more overbank deposition. In turn, the top of Deposit 1 was eroded—how much was lost, we cannot say—which was then followed by the accumulation of Deposit 2 sediments. Deposits 1 and 2 were then both eroded and subsequently buried by Deposit 3. If stone artifacts were found on the uppermost surface of Deposit 1, it could mean they were contained in sediments originally part of Deposit 1, which had since eroded; or that they were deposited after erosion cut into Deposit 1 but before Deposit 2 was laid down; or if they were found where Deposit 3 sits atop Deposit 1, they could have been in or atop Deposit 1, or in or atop Deposit 2, or were dropped there just before Deposit 3 began to accumulate. Given what is known of the radiocarbon ages from this site (which range from 7800 to 2200 BP), incorrectly reading the stratigraphy could mean "misplacing" the artifacts in time by as much as 5,000 years. And that assumes the artifacts were dropped at this spot, but because this is a river valley, they could have been washed in from some distance away, and were older (or younger) than their position at this spot would indicate. (Photograph by David J. Meltzer.)

processes (such as burrowing animals, shrinking and swelling of soils, erosion, and redeposition) and human action (our tendency to bury the dead) mix the deposits in which archaeological remains are found. Younger artifacts can easily intrude into older strata, falsely implying they are just as old. Sometimes the reverse occurs (Figure 20). Because of our very-human tendency to pick up curious fossils and even artifacts, older materials can be added to younger sites (fluted points and Paleozoic crinoids have been found in prehistoric Pueblos, but those sites are neither Pleistocene nor Paleozoic in age).

Back in the pre-Folsom days, Holmes and Hrdlička guarded against the possibility of stratigraphic mixing by automatically assuming that *any* artifact or human skeleton found in Pleistocene-aged deposits had arrived there at a later time. A site was stratigraphically guilty until proven innocent. Theirs was an extreme position, but in the sometimes complex stratigraphy of ancient valleys, terraces, and especially caves (recall the worries over Brixham Cave), where claims of great antiquity hang in the balance, stratigraphic caution is appropriate.

A YOUNG CHILD ON OLDMAN RIVER

In the early 1960s, in a modern gully draining into the Oldman River of southwestern Alberta (near Taber), Archie Stalker of Canada's Geological Survey came across the fragmentary skeleton of an infant. Little could be told of the child's heartrending story, but perhaps the tiny presence was a story in itself. The remains were in a block of calcium carbonate–cemented sands, apparently broken away from a layer of ancient stream alluvium exposed on the gully floor.[11]

The alluvium layer was deep below the surface, itself no great sign of age. More significant, it was sandwiched above one glacial till and beneath another. By Steno's law, the child predated the Late Wisconsin till; given how far beneath the till the skeleton was found, Stalker thought it might reach into the Middle Wisconsin.

A geologist by training, Stalker was well aware of the necessity of guarding against possible stratigraphic mixing and intrusion, but he was certain that had not happened at Taber. There was no evidence for erosion older than the modern gully but younger than the alluvium, which might have carried the infant remains down from the surface. The bones were in place, he insisted, their age bracketed by the ages of the enclosing alluvium that had been radiocarbon dated at nearby sites. In round numbers, the remains were likely more than 18,000 years old, and perhaps as much as 60,000 years old. Unfortunately, there was not enough bone to be directly dated using radiocarbon techniques then available, and despite Stalker's almost annual visits to the site, no more bone was found.

Stalker's confidence notwithstanding, few were completely satisfied with the antiquity of the Taber child. In the late 1970s, Michael Wilson and others returned to the site (Figure 21) to look for more remains and clarify its stratigraphic sequence. They had no

FIGURE 21.

Excavations at the Taber Child site, Alberta. The folding and dipping of the deposits (which complicated the stratigraphic interpretation) show up nicely in the pit floor. (Photograph courtesy of Michael Wilson.)

better luck than Stalker finding skeletal remains, but did spot patches of bone-bearing, redeposited Holocene sands on the nearby slopes.

Those recent sands possessed many features thought to be unique to the Middle Wisconsin alluvium, but careful inspection revealed the two sands were quite different in texture, color, and chemistry. Fortunately, the sand adhering to the remains of the Taber child had been saved, and when it was analyzed, it proved to be like the Holocene sands, and not Middle Wisconsin sands. At some time in the recent past, the unfortunate child's bones had been redeposited alongside, but not in, a layer of genuine antiquity.

In the 1980s, after AMS radiocarbon dating was developed, a tiny piece of the Taber skeleton was submitted: it dated to 3,550 ± 500 BP. Stalker rejected that result because of possible contamination of the bone. Perhaps. But if the date is off, the revised stratigraphy shows it is not off by much.

THE DATING GAME

Stratigraphy is a relative dating technique: the age of a layer (or items in it) is only known to be older or younger than some other layer. Knowing the actual age of an artifact or event requires using an absolute dating technique that yields ages in years (usually measured "before present"). There are many such techniques available nowadays, some of which can be used for specimens less than a thousand years old, others on materials more than a million years old.

Many use the clock nature provides in the radioactive decay of elemental isotopes, such as radiocarbon (^{14}C), Argon/Argon ($^{40}Ar/^{39}Ar$), and uranium series dating ($^{230}Th/^{234}U$). Others rely on incremental time markers such as dendrochronology (which uses the annual growth rings in trees), and luminescence dating (which measures the accumulation of electrons in crystal lattices within quartz grains). Still others rely on changes in the position of the earth's magnetic pole (archaeomagnetism and paleomagnetism). In order for a dating technique to be useful, certain conditions must obtain; among them are these: there must be a time-dependent signal (a natural clock) that is relatively constant over time; a means of calibrating the signal in years; a rate of chemical or physical change that matches the time span of interest (some short-lived isotopes disappear in a matter of minutes—those are not useful for dating events thousands of years old); and a means of linking the *sample* being dated, be it charcoal, quartz grains, or magnetized clay, with the *event* of interest—say, the occupation of an archaeological site.[12]

Over the decades since its development by Willard Libby, radiocarbon dating has been the preferred method for dating the sites of the earliest Americans: its range reaches back to around 50,000 BP (Chapter 1), which nicely covers the time period of interest. Its reliability is well attested, since we now know that the amount of radiocarbon in the atmosphere has changed over time but can account for the confounding effects of that through calibration (though as noted, the calibration of ages from the Younger Dryas period is still challenging). And, finally, the technique can be applied to any material that incorporates radiocarbon, such as charcoal and wood from plants, bone, teeth, ivory, keratin, hooves, horns, shell from animals, and even soils, by virtue of the organic acids that seep into them.

Yet, not all these materials, including charcoal and bone from which radiocarbon ages are most often obtained, give results that are equally *reliable* (if we run the sample multiple times will we get the same result?) or *valid* (is the result correct?). Case in point: prior to our reopening excavations at the original Folsom site in the 1990s (Chapter 9), the bison kill was thought to have dated to 10,890 ± 50 BP, based on radiocarbon ages of charcoal found with the bison bones. Yet, our subsequent radiocarbon dating of the bison bones themselves yielded an average age 400 years younger: 10,490 ± 20 BP. Why the discrepancy?

Charcoal has the virtue of being relatively impervious to contamination and thus normally yields a reliable age. But the age may not be valid since the charcoal may not be the same age as the bison kill. If the charcoal came from, say, a lightning-strike forest fire, it will have no bearing on the age of the kill. Even if the charcoal was from wood that stoked the fires over which bison meat was cooked, it could still overestimate the age of the kill if the wood came from a long-lived tree, the inner rings of which could be decades or even centuries older than its outer rings. This is why, as we'll see, there is often a call in contentious cases for the radiocarbon dating of short-lived seeds.

Conversely, radiocarbon dating of animal bones can more precisely pin down the age of a site, since animals live shorter lives than trees, but only if those bones occurred there as a result of human activity (which is certainly the case of the Folsom bison). But dating bone itself can be challenging, since it is made of multiple constituents, some of which are highly susceptible to contamination. The inorganic (apatite) portion, for example, readily absorbs ancient carbon via groundwater or other contaminants. Before that was realized, multiple radiocarbon ages on the same skeleton often gave wildly divergent results—in one case, for example, the oldest and youngest radiocarbon ages on the same mammoth were nearly 3,000 years apart. It was only in the 1980s, after laboratory advances enabled the extraction of protein and even individual amino acids from organic bone collagen, substances less susceptible to contamination, that it became possible to derive ages from bone that were more reliable. Hence, our conclusion the bone amino acid we obtained dates from Folsom better estimated when that kill occurred.

But the handicap of radiocarbon dating, which will be felt if the first Americans prove to be far older than supposed, is that it does not provide reliable ages for any samples that might prove to be significantly more than 50,000 years old. For that, waiting in the wings are a number of dating methods that can go beyond radiocarbon's limits (or can be used where we lack material suitable for radiocarbon dating). Among those tried on pre-Clovis sites are amino acid racemization, cation-ratio dating, fission-track dating, electron spin resonance (ESR), uranium-series dating, and thermoluminescence (TL) and optically stimulated luminescence (OSL) dating (the difference between the TL and OSL techniques is in how the accumulated electrons in the lattice traps are released and measured). Sometimes the results of these applications have been good, sometimes not, and sometimes one can't tell, which is worse than getting bad results or no results at all. Thus far, no dating method has proven to be as tried and true as radiocarbon dating.

FOOTPRINTS IN THE ASHES OF TIME

Every year London's Royal Society puts on a summer exhibition to showcase some of the UK's latest scientific research. One of the exhibits in 2005 was an archaeological show-stopper: a report of more than 150 apparently human footprints in a volcanic ash (the Xalnene ash) at the Toluquilla site in central Mexico, which were possibly more than 40,000 years old. Finding human footprints is unusual, though not unprecedented: 3.6-million-year-old tracks of early hominids were found at the site of Laetoli in northern Tanzania. But the Toluquilla site was rare enough and old enough by New World standards to be noticed, especially for the faint bell it rang in archaeological memory.

Toluquilla, it turned out, was on the edges of the Valsequillo reservoir, which was constructed in the 1940s in the basin of a one-time Pleistocene lake. A 30-meter-thick gravel formation outcropping along the edges of the reservoir had yielded a zoo of Pleistocene fossils, and a slew of archaeological sites including a very controversial one: Hueyatlaco, the lower layer of which had flake tools and possible projectile points associated with the remains of a mastodon. Geologists had dated that lower layer at 250,000–600,000 years old, based on the then-experimental techniques of uranium-series, fission-track, and tephrahydration dating. However, the archaeologist excavating the site put that layer at just 9,000–22,000 years old, based on results from radiocarbon dating of mollusk shells contained within it.[13]

Unfortunately, it proved impossible to close the huge gap separating the two sets of age estimates, and from the 1960s through the 1980s, Hueyatlaco was a poster child for problems that occur when different dating techniques, especially still-unproven ones, cannot be reconciled. Further efforts in the 1990s using the ages of the vertebrate fauna and of the diatoms from Hueyatlaco were equally at odds: the vertebrates pointed to an age less than 30,000 years ago, while diatoms in the deposits appeared to be more than 80,000 years old. "Conflicting and confusing" was the verdict handed down on the site's age.[14]

Worse, many of the original Hueyatlaco artifacts, including mammoth bone ostensibly engraved with serpents and saber-toothed cats, had disappeared under puzzling circumstances. Ugly accusations surfaced that the artifacts at Hueyatlaco had been planted by unscrupulous laborers. The allegations were false. But the odor permeating the site's integrity and age was enough to knock it from the list of pre-Clovis contenders, though its claim to great antiquity lived on in the archaeology conspiracy literature (yes, there is one[15]) as one more site professional archaeologists suppressed ostensibly because it threatened our authority to dictate archaeological knowledge. Alas, the truth is far less sinister: we just didn't know what to make of this site.

In the summer of 2003, a team led by archaeologist Sylvia Gonzalez returned to the Valsequillo region and spotted the footprints. They undertook an ambitious campaign to date the Xalnene ash in which the tracks were imprinted. OSL dating yielded an age of about 40,000 BP, older (as it should be) than radiocarbon and ESR ages from overlying deposits. Better still, since that ash occurs throughout the Valesequillo area, and at Hueyatlaco 2 kilometers away it lies *beneath* the gravels in which the site is situated, that meant (here's Steno again) the 250,000 years and older ages from this controversial locality must be wrong. After forty years, the age of Hueyatlaco was finally known to be younger than 40,000 years old. At least now we had a ballpark age.[16] Or so it seemed.

Within months of the Royal Society's summer exhibition, geochronologist Paul Renne and colleagues announced that $^{40}Ar/^{39}Ar$ and paleomagnetic dating of the Xalnene ash—for those keeping count, we're up to ten separate dating methods brought to bear on this chronologically stubborn region—had yielded startlingly different results: the ash was laid down 1.3 *million* years ago. If those were indeed footprints, they must have been made by an ancestral hominid that was no *Homo sapiens*. Hardly plausible, they argued: more likely, those weren't human footprints at all, or if they were, they were not made when the ash was fresh. An expert on OSL chimed in as well: the Xalnene ash contained negligible amounts of quartz, making it "far from ideal" for luminescence dating, especially since over half the quartz grains analyzed had their electron traps completely emptied, others partially so, making age estimates unreliable. He would not be convinced that ash was 40,000 years old until additional OSL dates replicated the result.

Gonzalez's group had already tried $^{40}Ar/^{39}Ar$ dating, but had no success because they spotted in the ash older mineral fragments that could skew age estimates, so they, in turn, were suspicious of the ages from Renne's group. Besides, if it were 1.3 million years old, why was there no stratigraphic hint of the vast time separating it from the late Pleistocene deposits directly atop it? Clearly, replication of the dating of the ash was in order, not just of the OSL results but of the $^{40}Ar/^{39}Ar$ and paleomagnetic ages as well. Turnabout: it's fair play in science.[17]

So the long-running debate over the age of the Valsequillo gravels is not over. Gonzalez's group has embarked on additional efforts to date the ash, and are optimistic they will eventually be able to nail down its age, and with it the age of the apparent footprints (their human origin is in question), and of the Valsequillo gravels that bury Hueyatlaco. They're optimistic, too, their results will make the Valsequillo Basin one of the oldest and most important regions for the study of the first Americans. On that, they may be correct. And they suggest that even if the Xalnene ash is 1.3 million years old, those could *still* be human footprints. But that's not a rabbit hole most archaeologists will be willing to follow them down, though conspiracy buffs might.

SO MANY SITES, SO MANY PROBLEMS

Ambiguous artifacts, scrambled stratigraphy, dubious dates. The problems at Calico, Taber, and Valsequillo are matched in a roll call of sites from Old Crow in Canada's Yukon Territory, to Lewisville in Texas, to Pikimachay in Peru. Could the search for pre-Clovis be doomed to failure simply because there is no pre-Clovis? Some skeptics suspect as much and point to Australia, where American archaeologists have long looked with interest and more than a touch of envy. We are interested because of similarities in the prehistory of the two continents. Both were colonized about the same time from Asia by highly mobile people possessing Stone Age technology, who found a new landscape teeming with a naïve native fauna, and who—it seems—spread quickly through each continent.

And we covet our neighbor's archaeological record because of our apparently very different archaeological histories. The search for the earliest sites in America has been suffused with ambiguity and controversy. But look at Australia, many a pre-Clovis critic has said: once serious archaeological work began there in the 1960s, its deep antiquity was "established quickly, definitively, and largely without rancor," despite that continent having far fewer archaeologists. Perhaps, Paul Martin suggested, we just need to hire some "Aussies [who] seem to be capable of finding and agreeing on the existence of [ancient] sites . . . unclouded by controversy." Were that only true.[18]

As archaeologists James O'Connell and Jim Allen have recently recounted, in the early 1960s Australian prehistory was thought to extend back a mere 10,000 years. But within two decades, widespread excavations and radiocarbon dating had pushed the arrival of the earliest Australians to just over 40,000 years ago. That antiquity held for a brief time, but then a series of TL and OSL ages suddenly plunged its prehistory back to 65,000 years ago, and then, astoundingly, to almost 176,000 years ago (at the site of Jinmium). That sparked a lively—and to American ears familiar—debate over the antiquity of the first Australians. Is the controversy over Down Under? Hardly. O'Connell and Allen make a compelling case for an antiquity no greater than 45,000 years old, but there are those who continue to push Australia's antiquity earlier still.[19]

Pre-Clovis critics also point to the impoverishment of early American archaeological records compared to those of Europe, Africa, and Asia. Look around before 11,000 BP, Martin suggests, and everywhere but in America artifact quantities are "awesome." If America were occupied at the same time, "why didn't they leave us some similar trophies?"[20] The answer, quite simply, is antiquity and demographics. The Old World was occupied much earlier than the New, no matter how old the New, and by late Pleistocene times, it had relatively large and archaeologically visible populations. Of course there are many more Old World Pleistocene-age sites.

In contrast, as Fekri Hassan and others have shown, a small number of hunter-gatherers could have entered the New World 25,000–20,000 years ago, yet hovered below archaeological radar for a long time thereafter.[21] We'll probe this matter further; for now, bear in mind that the search for pre-Clovis in the Americas may be a search for a very elusive quarry indeed.

That being the case, should the criteria be relaxed? Is it true, as pre-Clovis proponents insist, that by the criteria, Clovis would not pass muster? Hardly. Clovis long ago met these criteria, and the process need not be repeated with each discovery of a new Clovis site, since a site's recognition—assuming it has diagnostic Clovis points—is straightforward (however, what is appropriate to ask are whether claims about, say, Clovis adaptations, can be supported by evidence from those sites, and such questions are being asked—as we'll see). That same privilege cannot be granted pre-Clovis until after a site is found that meets these criteria, and when we have learned how to recognize sites and distinctive artifacts of that age.

That the criteria worked at Folsom, at a time when archaeological methods were less refined than they are today, is also testimony that the criteria are not too rigid. The fact so many pre-Clovis sites since have not measured up speaks more of problems with the sites than with the criteria.

Those problems notwithstanding, there have been payoffs along the way. In the course of evaluating new kinds of evidence, applying new methods and techniques, or trying on new ideas, the field as a whole advances. A prime example: archaeologists working in the Old Crow Basin of the Yukon Territory found mammoth bones that appeared to have been flaked, fractured, and polished in ways that seemed "unnatural." Could they be artifacts? The specimens were sent off for radiocarbon dating, and the results were surprising: some dated to 60,000 years ago. A *paralithic* industry, William Irving called these ostensible artifacts, and he and others showed experimentally they could have been made by humans. Fair enough. But then Irving threw down the gauntlet, insisting only humans could make those fractures. It was "difficult to imagine," he said, nature breaking bones the same way.[22] Upping the ante, he announced some of those broken bones were in 350,000 year old deposits, thereby plunging North American prehistory deeper into the Pleistocene, and even before the appearance of anatomically modern humans.

That did it. The gauntlet was picked up by a gaggle of researchers who soon proved nature could fracture bone in ways identical to the Old Crow specimens with embarrassing ease, whether by carnivores gnawing, large animals trampling, rivers freezing and thawing, or even volcanoes exploding.[23] In the end, it raised the red flag of *equifinality:* if humans and nature can break bones the same way, then one must find independent evidence to prove the bones were artifacts (say, finding them in an undeniable archaeological site context). That hasn't happened. Instead, the paralithic specimens have come from stream deposits that are chock full of naturally broken bones (a sort of "bony" Calico).

Nowadays few believe the Old Crow mammoth bones were humanly modified or attest to an ancient human presence. But credit Old Crow with igniting an explosion of studies that have helped us understand how geological processes, animals, and humans may have altered the bones that we find in archaeological sites. Because of it, we know a great deal more about the mechanics of bone breakage, and along the way, we forever lost our innocent notion that just because something *could* have been broken by people is somehow proof it *was*. Like other knowledge forged in this dispute, lessons learned here have rippled outward to benefit the archaeology of other places and times.

In the end, it comes down to this: no matter how vigorously claims for a pre-Clovis site are promoted, they will never prove compelling unless the site's artifacts, stratigraphy, and ages are unimpeachable. And preaching to the choir will not do: they are already believers. Instead, the site has to convince the skeptics. And it has to be utterly unimpeachable in all respects: one that can withstand withering criticism and meet the criteria with room to spare. Call it the Jackie Robinson Rule, after the first African American to break the color barrier in major league baseball (in 1949 with the then Brooklyn Dodgers), who not only had to be a superb ballplayer, which he was, but also an exemplary human being (he was that, too) who could rise above the inevitable attacks he would bear for the color of his skin. In the last decade or so, archaeology has had a few sites come close, and one that made it over the bar.

Meadowcroft Rockshelter is 50 kilometers west of Pittsburgh, Pennsylvania, on a steep slope 15 meters above Cross Creek, a tributary of the Ohio River. James Adovasio and an interdisciplinary team of geologists, paleontologists, palynologists, and others began work in this large (15 m wide, 6 m deep, and 13 m high), sandstone overhang in 1973. Intensive excavations continued at Meadowcroft for the next five years, and by all accounts—including those of the site's harshest critics—the work was superb. Adovasio has a well-earned reputation for employing cutting-edge field techniques, and for being a meticulous, demanding, highly skilled excavator, even in the stratigraphically challenging setting of a rockshelter.[24]

As a result, the excavations—sometimes done by razor blade to follow the thinnest deposits—went slowly, but ultimately the team dug through 4.6 meters of sediment, divisible into eleven natural strata (Figure 22), in which were artifacts, charcoal, and various features of human occupation, mostly from later (Woodland) occupations (marked by pottery sherds and early cultivated plants such as squash and corn). In the deeper levels of Stratum II was a rich record of Archaic hunter-gatherers, and beneath that, remains from Paleoindian times. The stratigraphic sequence was anchored by fifty-two

FIGURE 22.

Excavations and soil sampling in the deep pit at the Meadowcroft Rockshelter, Pennsylvania, ca. 1976. The white tags on the wall mark individual stratigraphic layers. (Photograph courtesy of James Adovasio.)

radiocarbon dates, analyzed by several different laboratories, which ranged in age from 31,400 BP to the Historic period; the Declaration of Independence was signed on the opposite side of Pennsylvania about the time the last hearth at Meadowcroft was snuffed out. All but four dates are in proper stratigraphic order, and the ones out of order are minor flip-flops in upper, Holocene-age levels. Steno would have been proud.

The pre-Clovis action is down in middle Stratum IIa, bracketed between 11,300 and 12,800 BP and which has produced an unfluted lanceolate projectile point, and lower Stratum IIa, which has six radiocarbon dates from 12,800 to 16,175 BP, along with associated artifacts (Figure 23). These are genuine artifacts, too (about 700 altogether), and include a broken bone punch, flake debris from stone tool making, small blades (long and narrow pieces of stone), unifacial (flaked on one side only) and bifacial knives, and gravers (pointed tools possibly used for bone or wood working). The stone for these artifacts came from distant (or *exotic*) sources in New York, Ohio, West Virginia, and Pennsylvania.

The average of the lower Stratum IIa dates associated with these artifacts indicates a human presence at Meadowcroft around 14,250 years BP. This is Adovasio's "conservative interpretation." It may, of course, be earlier still. But, skeptics say, consider the implications: 14,250 years ago the Laurentide ice sheet was within 150 kilometers of Meadowcroft. At that close range, they say, Meadowcroft ought to have experienced near-glacial climatic and ecological conditions: permafrost and tundra, perhaps, or shy of that, a boreal setting with cold-loving flora and fauna. Yet, there is no evidence for either. Present instead in the Pleistocene levels at Meadowcroft are bits of wood and charcoal from deciduous trees (oak, hickory, and walnut), along with the bones of white-tailed

FIGURE 23.
Miller lanceolate projectile point in situ in Stratum IIa at Meadowcroft Rockshelter, Pennsylvania (Photograph courtesy of James Adovasio)

deer, southern flying squirrel, and passenger pigeon. These speak of a temperate environment, seemingly anomalous for this time and place. Where are the bones of extinct Pleistocene fauna like mastodon or sloth, or the remains of displaced arctic or boreal plant and animal taxa?

Certainly not in the lowest levels at Meadowcroft, but then hardly any organic remains come from there. Nearly 1 million animal bones were recovered from the entire shelter excavations, but only 278 bones came from the lower reaches of Stratum IIa, and of those, just 11 could be identified. Same story with the plants: roughly 1.4 million fragments of wood, charcoal, seeds, and fruits were recovered. The grand total from the glacial-age levels was a pitiful 11.9 grams.

That's a hopelessly inadequate basis to judge whether the anomaly is real, or a freak of sampling that netted only deciduous elements from local, complex boreal-deciduous forests (Chapter 2). Adovasio's team sees no anomaly, attributing the absence of boreal species to Meadowcroft's low-lying (260 m above sea level) and south-facing aspect. Still, without an adequate sample the Meadowcroft paleoenvironmental record cannot well support any interpretation.

Or perhaps, critics add, the radiocarbon dates are simply wrong, and Stratum IIa is actually post-glacial in age? Meadowcroft is, after all, in western Pennsylvania coal country, and skeptics—Vance Haynes foremost among them—wondered whether ancient coal infiltrating the deposits as dust or via groundwater contaminated the Meadowcroft radiocarbon samples and inflated their actual ages.

Adovasio and colleagues counter that the closest coal to the rockshelter is 800 meters away, and neither coal particles nor coal-associated spores were found in microscope scans of the radiocarbon samples. Besides, if there had been contamination, it surely would have been noticed by one of the several radiocarbon laboratories that analyzed the Meadowcroft samples—unless, Adovasio snarled sarcastically, one assumes "a witless incompetence of astronomical . . . dimensions" on their part, which he does not.[25] Moreover, if the samples were contaminated, the fifty-two dates would not have been in near-perfect stratigraphic order.

Maybe, Haynes responded, only the charcoal from lower Stratum IIa sat in contaminated groundwater and got proportionately older as a result. Adovasio calls such a scenario "contorted, even fanciful," since the geologists on his team showed that groundwater never got closer than 10 meters below the deepest occupation surface at Meadowcroft. More important, geoarchaeologist Paul Goldberg conducted an independent microscopic inspection of Stratum IIa sediments searching for traces of groundwater saturation or coal particles, but found no evidence of either.

Haynes, dismissing the Goldberg study (though without explaining why), said the chronological dispute could neatly be resolved if only some short-lived seeds or nuts from lower Stratum IIa were dated, believing they would be less susceptible to contamination. Only, he complains, his "repeated pleas for radiocarbon dating of nuts and seeds . . . have gone unheeded."[26]

Not exactly unheeded, more like emphatically rejected: "I knew," Adovasio writes, "that by Haynes's tortured reasoning, if I took a seed from Stratum IIa at Meadowcroft and the date came back just as old as the 16,000-year-old date from charcoal from a hearth, Haynes would say it had to have been tainted by some soluble contaminant—yet to be identified, of course. And if it came out younger than my older [16,000 BP] dates, Haynes would deem [those older] dates discredited, although it is well known that seeds can migrate between horizons quite easily."[27] Adovasio figures he cannot win for losing. He might be right.

Round and round it goes. So far, Meadowcroft has cheated pre-Clovis actuarial tables, with more written about it than any other potential early site, save Trenton itself, an analogy Adovasio himself has drawn. They are both, arguably, the most controversial sites of their times. And like Abbott, Adovasio has aggressively carried the flag for Meadowcroft, branding his critics "gnats,"[28] and producing to date more than 100 published articles and papers pushing the pre-Clovis antiquity of the site.

But there's one thing Adovasio hasn't done: issue a comprehensive, final report on Meadowcroft. One is needed, as even he admits, in order to lay out all the data in one place, including detailed descriptions of individual artifacts, precise information on their position relative to radiocarbon samples, and the context of those samples relative to the stratigraphy. Then the archaeological community will have the chance to evaluate it in detail. Until that moment, Meadowcroft may remain in archaeological limbo, the exception that proves the rule that if a site is not convincingly older than 11,500 BP before a pre-Clovis presence is generally accepted, then it will not be any more convincing afterward.

EARLY AMERICAN ART?

The prehistoric rock art of the remote, arid *caatinga* ("thorn forest") of northeastern Brazil is rich, colorful, expressive, and in spots, downright racy (some Brazilian customs, it appears, have prehistoric precursors). It was rock art that drew Nième Guidon to Toca do Boqueirão da Pedra Furada ("hole in the wall," after an erosional cavity in a nearby cliff), a rockshelter that covers a roughly semicircular area 70 meters wide and nearly 18 meters deep. Seeking to determine the age of the art, Guidon began excavations in 1978 that continued until bedrock was reached in the late 1980s. Her team dug through nearly 5 meters—and as many strata—of sediment filling the shelter, now dated by more than fifty radiocarbon ages, the earliest of which is greater than 50,000 years old. Two periods of prehistoric occupation were recognized: the Pedra Furada phase, all materials predating 14,300 BP, and the younger Serra Talhada phase, which began around 10,400 BP and lasted through much of the Holocene.

The Serra Talhada phase is marked by distinctive, well-made bifacial and unifacial stone tools fashioned, in many cases, of exotic chert. It has extensive and complicated rock art, abundant ocher (a natural paint), well-defined hearths, and easily detected

FIGURE 24.
A quartzite cobble "chopper" from one of
the deepest levels at the Pedra Furada site,
Brazil. (Photograph by David J. Meltzer.)

living surfaces that preserved animal bones and plant remains. Serra Talhada is unequivocally archaeological.

Less so the older Pedra Furada phase, which lacks bone, wood, or other organic remains, save for charcoal. Nor does it have the rock art, unmistakable hearths, or the obvious living floors of the younger phase. Instead, proof of its human presence rests largely on about 600 crudely fashioned quartz and quartzite cobble choppers (Figure 24), scrapers, and flake cutting tools, along with clusters of rock identified as hearths.

Such wholesale differences between the two phases naturally make skeptics wonder whether the older material is truly archaeological, and not just rocks flaked and arranged by nature. Guidon dismissed such carping, however, and for seemingly good reason: a 1993 doctoral thesis by Fabio Parenti showed that the Pleistocene-age specimens possessed flaking patterns diagnostic of human manufacture (these included the removal of more than three flakes, edge angles less than 90°, and a "logic" to the flake removal pattern). No small triumph that, and even *Nature,* a journal that rarely notices doctoral theses, noticed this one.

Of course, we know the pre-Clovis dispute is littered with sites that testify to nature's uncanny ability to mimic primitive human artifacts, and trap even the expert (think Louis Leakey). Pedra Furada might be such a setting. It is nestled beneath a sandstone cliff more than 100 meters high (see Plate 5). Atop the cliff is a 20-meter-thick rock layer of densely packed quartz and quartzite pebbles and cobbles: a high-energy gravel bar, frozen in time. At both the east and west ends of the shelter are water-stained chutes down which rocks, dislodged from the cobble layer by weathering and erosion, plummeted to the shelter floor, occasionally swept along by "violent torrents of water" (Guidon's words[29]). At the base of the eastern chute there is a large debris cone of shattered quartz and quartzite pieces, flaked by the impact of their 100-meter free-fall; beneath the western chute are potholes scoured from the bedrock, one upwards of 3 meters in diameter, testimony to the erosive pounding of cascading water and debris. Filling the shelter are coarse sediments, pebbles, and hand- to head-sized cobbles, coming from directly overhead. Not surprisingly, the pebbles and cobbles are flaked to varying degrees. Clearly, nature's been dropping rocks into Pedra Furada over tens of thousands of years. But were people?

As *Nature* said of Pedra Furada, seeing may be believing. And believing "meetings of specialists on site, and formal debates, should take place regularly if we are to establish an agreed basis for evaluating evidence,"[30] Guidon generously extended invitations to proponents and critics to visit Pedra Furada in December 1993. We accepted. The meeting was held in Sao Raimundo Nonato, a dot on the map in the northeastern Brazilian state of Piauí and the gateway to a beautiful ecological and archaeological preserve, of which Pedra Furada is the centerpiece. It was a landmark meeting: stimulating, intense, enlightening, and sometimes downright uncomfortable, simultaneously and in several languages.

James Adovasio, Tom Dillehay, and I were among the conferees (Figure 25), and we raised questions (later published), about the site stratigraphy, its supposed hearths, and the 600 Pedra Furada–phase artifacts.[31] We were especially concerned about the artifacts. They had been identified as such during excavations, presumably because they looked the part, and then afterward were sent to Parenti for analysis to see if they could be artifacts. Yet, these were specimens already suspected to be of human manufacture. What was not at all apparent was how much these specimens, plucked from a dense deposit of naturally broken rocks, differed from those left behind.

No matter. The supposed artifacts certainly looked the part, though were decidedly retro: more akin to stone tools hundreds of thousands of years old than to what was being used around the globe 50,000 years ago. Looks aren't everything, of course, and

FIGURE 25.
James Adovasio (left) and Tom Dillehay (to Adovasio's right) examining the witness block left unexcavated at the Pedra Furada site, Brazil, during the international site visit in December 1993. Naturally broken cobbles are visible throughout the witness block. (Photograph by David J. Meltzer.)

their crudeness might easily be explained by the indifference of those who did the flaking, or the mean quality of the raw material. Some of prehistory's most spectacularly ugly stone work was executed in quartz and quartzite.

Even so, many bore a worrisome resemblance to the naturally flaked cobbles that today litter the surface beneath the chutes at the site. Just how guilty nature might be comes clear on a visit to Pedra Furada, as you crane the neck upward to catch a glimpse of the conglomerate layer high overhead. How long would it take for a dislodged cobble to make the Pedra Furada plunge, and how fast would it be traveling when it smashed onto the rock-strewn surface below? The questions aren't entirely academic: one instinctively steps back from the impact zone while calculating the physics of free-falling rocks. From that height, it would take a cobble just 4.51 seconds to hit bottom, impacting at a velocity of roughly 160 kilometers per hour (100 miles per hour). Give that rock a mass of 1–2 kilograms, and there's considerable force involved, far more than a person fashioning a stone tool normally musters. Cobbles would be flaked when they first hit, when they bounced into other rocks after impact, and again when they were struck by later-falling stones. Could such natural impacts produce the patterns seen on the Pedra Furada specimens?

The conference hall in Sao Raimundo Nonato grew warm on this issue, even before an electrical failure knocked out the ceiling fans. A deep division emerged on the question of a Pleistocene human presence at Pedra Furada. The divide was partly, though not entirely, along national lines, but for reasons that have little to do with politics, and much to do with paradigms: the question of antiquity looks very different depending on the intellectual tradition from which one emerges. North Americans tend to be more skeptical than most, but then we brought Calico into the debate and have much for which to atone.

For us, it's the usual rub: it is not enough to show a specimen could be cultural, one must also demonstrate it could not be natural. The Pedra Furada team was ready for that. Parenti had gathered 2,000 cobbles from the modern debris cones. All were obviously flaked by nature, fell in the last couple of decades, and were well above the archaeological levels at the site. He was pleased to report none exhibited the flake scars or patterns observed in the 600 specimens from the Pleistocene levels. Case closed?

Not yet. The issue is one of probability. As stone tool expert Jacques Pelegrin admitted, nature can flake stones in ways that mimic the Pedra Furada specimens, though he thought the odds slim, less than 1 in 100, he supposed. For the sake of discussion, make them even more remote: 1 in 1,000, or 1 in 10,000. Yet, no matter how rare the chances, given sufficient time and raw material (Pedra Furada had plenty of both), nature can magnify even the slimmest odds to the point where such geofacts occur in detectable frequencies. The shelter first opened more than 50,000 years ago, and has been bombarded by cobbles ever since: by conservative estimate, upwards of 10 million cobbles have taken the 100-meter plunge. That's a potential yield, even with odds as slim as 1 in

10,000, of roughly 1,000 geofacts. Nature, it appears, had ample opportunity to produce the 600 "artifacts" that occurred at the site—far better odds, in fact, than waiting for a monkey to type Shakespeare (the analogy Pedra Furada's defenders invoke).

Guidon fired back, fiercely so, denouncing our assessment as "worthless," pointing out several errors of fact in our statements about the site and their excavation methods (errors for which we were guilty), but otherwise largely ignoring the critical question of whether those were truly artifacts, except to say that "only one lithic flaked by man is enough to demonstrate his presence."[32] Alas, were it only so easy in a geological context such as this.

Work has continued intermittently at Pedra Furada, and recently included one analytical surprise: thermoluminescence was used to date forty burned quartz pebbles thought to come from Pleistocene hearths. These results failed to confirm that the burning resulted from human activity. Moreover, the TL ages ranged so widely (to nearly 162,000 years ago) they raised the possibility the deposits had been disturbed and that the stratigraphy was "much more complex than suspected."[33] One can only admire Guidon and colleagues' candor. Clearly, more needs to be known of Pedra Furada, its artifacts/geofacts, features, and stratigraphy. On these, its pre-Clovis status will ultimately stand or fall.

BANKING ON CHINCHIHUAPI CREEK

Monte Verde is on the banks of Chinchihuapi Creek in southern Chile, 50 kilometers up the Rio Maullín from the Pacific Ocean. It was found when local woodsmen, cutting back the creek bank to widen an oxcart trail, dislodged some buried wood, mastodon bone, and a stone artifact, all of which found their way to American archaeologist Tom Dillehay, then teaching at the Universidad Austral de Chile (Valdivia). From 1977 to 1985, Dillehay and an international, interdisciplinary team of nearly eighty collaborators excavated and analyzed the material from the site.

Monte Verde first leapt into archaeological consciousness one afternoon in the spring of 1989 when a couple hundred archaeologists assembled at the University of Maine for several days of wrangling over the origin and antiquity of the first Americans. It was a tough crowd. The evening before, it had sat quietly and largely unimpressed by a presentation on Pedra Furada. But ten minutes into Dillehay's talk on Monte Verde, the fellow sitting next to me (then, a hard-nosed skeptic—now, ironically, the purveyor of his own pre-Clovis site) whistled softly in astonishment and then said, to no one in particular, "What planet is this stuff from?" I was wondering that myself.

The images Dillehay was showing of the well-preserved remains at Monte Verde— wooden artifacts and house planks, fruits, berries, seeds, leaves, and stems, as well as marine algae, crayfish, chunks of animal hide, and what appeared to be several human coprolites found in three small pits—were unlike anything most of us, who long ago had learned to be used to stone tools and grateful for occasional bits of bone, had ever seen. The site's unusual preservation resulted from a geological fluke. Very soon after the people who camped on the sandy banks of Chinchihuapi Creek departed, the creek was dammed, water backed up over the site, and a water-saturated, culturally sterile, grass-matted peat deposit formed atop the 600 m² area they once occupied—what Dillehay calls the MVII surface (there's a much older MVI surface, but let's put that aside for now). In that oxygen-starved setting, the normal decay processes stalled, preserving that astonishing array of organic materials.

Excavating such remains is extremely challenging. They are saturated with water, extremely soft, and easily harmed; worse, once exposed to the air they rapidly dry, crack, and crumble to dust. In order to minimize the damage—especially of fragile wood and animal hide—Dillehay's team had to excavate slowly and without the usual metal tools of the trade, using instead fine-haired brushes, small air pumps, even their fingers. They had to minimize the exposure of these remains, then move quickly to keep them saturated for safe transport to the laboratory where, after 6–8 months in carefully controlled water and chemical baths, the items could be permanently stabilized.

Those painstaking efforts were well worth it, for they revealed a site rich in plant remains: parts of some seventy-five species littered the MVII surface, many more than occur in a comparably sized natural deposit, which Dillehay knew because he had dug where the site was, and where it wasn't: he ventured 2 kilometers up and down Chinchihuapi Creek to examine naturally deposited, non-artifact bearing sediments. That preemptive shoveling helped muffle criticism that the site had formed naturally by highlighting differences between natural deposits and the archaeological ones.[34]

About a third of the plants at Monte Verde were charred from cooking; a quarter were exotic, imported to the site from the Pacific coast some 30 kilometers distant (including seaweed, the residue of which adhered to many tools), from high Andean settings, or from arid grasslands up to 600 kilometers away. Coincidence or not, more than half the plants are still used by the native Mapuche as food, drink, medicine, or construction materials. The archaeological list includes charred and uncharred skins of wild potato, along with burned and unburned juncus and scirpus seeds that were found mashed into the cracks of wooden basins. Among the most unusual of the MVII plant remains were a half-dozen plugs of chewed boldo leaves mixed with seaweed, forming a medicinal tea still used by the Mapuche to relieve stomach ills, colds, and congestion.

Prehistoric chaw, we'd call that, and the plugs were found lying just outside a wishbone-shaped foundation (roughly 3.9 m by 3 m) of sand and gravel that appeared to be glued together by animal fat, and lined with vertical wood stubs and scraps of animal skin, the remnants of a hide-draped frame hut that once stood there. Inside the structure were three small hearths containing a variety of exotic plants and stone artifacts including quartz crystals. Fronting the structure were hearths; a small cache of salt brought from the coast; a litter of leaves and seeds; scraps of mastodon meat and bones; artifacts of stone, wood, and ivory; and a large (30 × 40 cm) cut and scraped block of red ocher. Monte Verdeans worked, ate, and took cures here.

They lived about 35 meters away in a cluster of roughly rectangular huts, the wooden foundation timbers of which survived (Figure 26). The structure was about 20 meters long, arranged like a row house, and framed by log planks anchored by wooden stakes, which supported walls of wooden poles that had bits of animal skin still clinging to them, possibly the remains of mastodon or paleo-llama hide that once draped the structure. String fashioned of juncus reed was knotted around many of the wooden stakes, likely tent pegs for securing the hide cover.

The hut floors were high in nitrogen and phosphate, a sure chemical sign of human waste, and littered with ash, grit, and flecks of animal skin. The huts surrounded two large communal hearths, and two dozen smaller ones, which were clay lined, contained ash and charcoal, and often had been heavily scorched. In the muddy sand and clay floor alongside one of the hearths was a child's footprint (one of three found) (Figure 27). In this area, too, were traces of meals: more plant parts, as well as the bones, tusks, teeth, and fractured ribs of mastodon; paleo-llama bones; and the remains of fish, shellfish, birds, and even seaweed, coastal algae, and egg shells.

Most of the MVII artifacts are wooden (digging sticks, mortars, lances), while others are bone and ivory, and still others stone (some 715 specimens), including bola stones, cores, a polished basalt perforator, grinding slabs, knives, unifacial scrapers, and spear points (Figure 28). Not all the stone pieces, even the majority, are indisputably artifacts. Many are unmodified stones from the stream bed or distant coastal beaches. But

FIGURE 26.
Timbers that served as the base of the hut remains in Area D of the Monte Verde site, Chile. (Photograph courtesy of Tom D. Dillehay.)

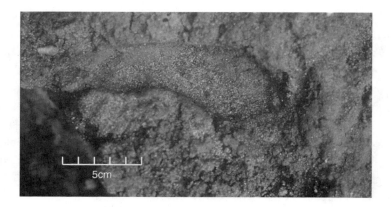

FIGURE 27.
One of the three human footprints left in the muddy sand and clay in Area D of the Monte Verde site. By its size (13 cm long), the print was apparently left by a young person. (Photograph courtesy of Tom D. Dillehay.)

because they were found amid sandy sediments (this was no Pedra Furada) and showed, as Michael Collins demonstrated, an unnatural sphericity, they are accorded human authorship. Most of the stone for the artifacts was locally derived, but over forty were made of material from distant sources. And from the coast came bitumen, a natural tar used to bind several of the stone tools to wooden and bone handles.

The assemblage from Monte Verde is extraordinary, all the more so because of its age. The preservation of organic material made it possible to obtain radiocarbon dates

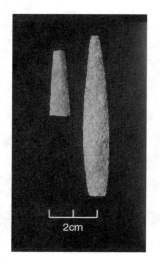

FIGURE 28.
Two of the Monte Verde lanceolate projectile points. (Photograph by David J. Meltzer.)

on wood, charcoal, tusk, and bone. The radiocarbon analyses were conducted by several different laboratories, a handy cross-check of the reliability of the results, and produced ages on the MVII surface and its artifacts that ranged from 11,790 ± 200 to 13,565 ± 250 BP. The average was put at 12,500 BP.

That's certainly old enough to establish Monte Verde's pre-Clovis credentials, but making matters more intriguing, Dillehay's excavations in a different area of the site encountered a couple dozen more stone specimens, along with three small, shallow, circular hearth-like basins, each containing charcoal specks that dated to more than 33,000 BP. These MVI deposits lack the spectacular organic preservation typical of the MVII occupation, and though some of the MVI artifacts are unequivocal, no one, least of all Dillehay, was quite sure what to make of this much older occupation, since it has not been fully excavated. Understandably, he concentrated his efforts and attention on MVII—as I shall here.

FROM RESISTANCE TO RESOLUTION

Monte Verde, like all pre-Clovis contenders, got no free ride: early on, it even failed to make Morlan's 1988 list of the five "best pre-Clovis sites," wary as he was that the wooden timbers were naturally fallen trees and the stakes merely branches. It took Dillehay more than a decade of postexcavation analysis to fully compile the evidence, during which time he published a series of papers and—in 1989—the first of two books on the site. The volume presented the results of the paleoenvironmental and geological studies, and descriptions of the plant, animal, and other organic remains. Yet, it said very little about the site's artifacts and architecture. This was deliberate. Dillehay wanted to prove Monte Verde was an archaeological site by showing nature could not have deposited plant and animal remains in that variety and amount. It was a novel gambit, and one Dillehay admitted imposed "a burden on the reader to understand and accept results not yet presented."[35] He was right, though the effort almost paid off. Almost.

But the burden of proof, as always, had to wait on the archaeological testimony. Skeptics were willing to grant Monte Verde's non-archaeological remains were unusual, but was their "pervasive strangeness" (as skeptic Dena Dincauze put it) necessarily archaeological?[36] That was something only the artifacts—slated for description in the second volume—could resolve. Some were unwilling to wait. Thomas Lynch, who for many years had carefully patrolled the pre-Clovis border in South America, was not giving Monte Verde a pass based on its plant and animal remains, or its promise of artifacts. Indeed, what little he had seen of MVII's artifacts and features failed to convince him that MVII was an archaeological surface. The few artifacts that he deemed genuine might, he suspected, come from later (Archaic) occupations in the area, and even the biface associated with mastodon bones still dated to "essentially Clovis time." Lynch dismissed the older MVI evidence outright: he doubted its specimens were artifacts, and thought it highly unlikely humans would camp twice in this "wet and boggy" creek bottom 20,000 years apart (in point of fact, MVI sits on the high terrace over a lagoon, not on the floor of the creek).[37]

Dillehay and Collins were quick to defend the site: Lynch was biased, ignorant of the Monte Verde artifacts, and misunderstood the site's geological and archaeological context. This was, they argued, an ideal place to camp, and the closest Archaic site was at least half a kilometer away, precluding any mixing of younger artifacts. Lynch didn't back down. The accusation of bias was "baseless," and more to the point, he dismissed their rebuttal as disingenuous. After all, how could he (Lynch) be accused of ignorance or misunderstanding if the evidence itself remained "unpublished or inadequately illustrated"? (To which Dillehay asks, that being so, how could Lynch be so stringent in his judgment?[38]). And to Dillehay's lament that only two archaeologists had responded to his invitation to visit Monte Verde during the excavations and see for themselves, Lynch growled: "If so many of us stayed away, it was in good part because we did not feel free to go and make our own observations."[39]

Clearly, much was riding on the publication of the second Monte Verde volume, and what it would say of the site's artifacts, archaeological features, and architecture. When it finally appeared in early 1997, written by Dillehay and a team of thirty-two co-authors and weighing in at a whopping 1,000 pages, reviewers were awestruck. *Nature* hailed it as "one of the most persuasive and impressive interdisciplinary volumes of any kind of archaeology for a generation."[40] That wasn't hyperbole: the volume bristled with meticulous details on everything from mastodon bones and wooden lances to eggshell fragments and finely made strings; it examined and analyzed evidence that could be seen with the naked eye, and that which could not, such as the sediment chemistry; residues adhering to stone, bone, and wooden artifacts; and the microscopic bits of animal tissue. The effort was analytical overkill. Yet, overkill was necessary, given the great skepticism facing this (or any) potentially early site and the doubts expressed about Monte Verde's antiquity by Lynch and others since the site's discovery. Jackie Robinson could not be baseball player of merely average skill.

But would the second volume be enough to bring about consensus and—if judged positively—establish a pre-Clovis presence in the Americas? A few years before, following the appearance of the first Monte Verde volume, I had put that question to Vance Haynes, the dean of the skeptics: "No," he said.

"But why not?"

"Have you ever visited a site," he replied, "that looked just the way you thought it would, based on what you'd read of it beforehand?"

I had to admit I hadn't. But was a site visit really necessary? Much has changed since the 1920s visit to Folsom. Nowadays, professional competence can usually be assumed, and archaeological fieldwork follows well-defined protocols. More significant, much of the crucial evidence emerges only in postexcavation laboratory analyses of radiocarbon samples, sediment chemistry, artifacts, the isotopic composition of organic remains, and the detailed spatial analyses of the artifact distribution maps. Fortunately, those results can be evaluated without having to look over the shoulders of those doing the analyses, and thank goodness for that: who has time for a six-month watch in a radiocarbon laboratory to insure dating samples are properly analyzed? A site visit might lead one to reject a claim of great antiquity, but cannot demonstrate one. Folsom was the first, and almost assuredly the last site in which a visit fully and without ambiguity resolved a claim of great antiquity. In the 1990s, a site visit could reasonably be viewed as an anachronism, even irrelevant.

So why bother? One good reason: as Haynes said, what one reads of a site and what one sees are often very different. Unlike researchers in the experimental sciences, archaeologists cannot replicate a critical study in our own labs, nor fully recreate in words and pictures a site's setting and surroundings, the complexity of the sediments and stratigraphy, the artifacts in the ground (if you arrive in time), a measure of how difficult the sediment was to excavate, how easy it might have been to miss potentially mixed deposits, or what the back dirt piles look like (one sometimes learns an embarrassing amount about an excavation by looking through the backdirt at what was missed).

"The next book's important," Haynes said, "but one day on the site of Monte Verde would be worth all the words they could write." Overstatement, perhaps, but no matter.

Dillehay's response? "Fine. Let's go look at it."

And so Dillehay, Haynes, and I began organizing a visit to Monte Verde, which took place in January 1997. A group of pre-Clovis proponents and skeptics—including Dincauze and Haynes—armed with the proof pages to the forthcoming second Monte Verde volume, traveled to the site via the University of Kentucky (where Dillehay was then teaching) and the Universidad Austral de Chile (Valdivia), both of which housed the artifact collections from Monte Verde. Over a week-long, nearly continuous conversation and debate reaching from North to far South America, the Monte Verde evidence—detailed in the pages of Dillehay's book; visible in the specimens laid out on tables in Lexington and Valdivia; and seen in the geology, setting, and stratigraphy of the site itself—was thoroughly vetted.[41]

The trip to Monte Verde was extremely interesting, educational, and mostly enjoyable, though it also had its share of tense and downright unpleasant moments. No one ever promised breaking the Clovis barrier was going to be easy. Awkward as it occasionally was, that was a sure sign the visit was doing what it was supposed to do: allow a careful inspection of the evidence and a frank exchange of views. The one thing the Monte Verde trip was not was a counterpart to the 1927 Folsom site visit, but then it wasn't intended to be. The excavations were long-since completed, and much of the site was gone. There were no artifacts to be witnessed in place.

On the final evening of the trip at a quiet bar in the seaside town of Puerto Montt, Chile, we had a long and sometimes rancorous discussion that finally culminated in a consensus: all agreed Monte Verde was archaeological and 12,500 years old (Figure 29). Everyone around the table knew *exactly* what that meant.

FIGURE 29.

The calm before the storm: at Monte Verde late in the afternoon, January 1997, as the site visit was wrapping up (the author is on the far left). (Photograph courtesy of Alex Barker.)

A couple months after our return, and after we'd had a chance to mull what had been seen and said, Donald Grayson and I drafted a summary report of our visit and assessment that was circulated for comment and signature by the participants—if they accepted its conclusions. Everyone had a chance to examine and reassess the Monte Verde case in the cold light of day and well after any bruised egos or perceived twisted arms had fully healed. To be sure, some of us had lingering questions about the character of the Monte Verde artifact assemblages and their interpretation (for example, does the MVII occupation represent, as Dillehay suspects, a nearly year-round encampment?), but that did not change our central conclusion about the site's age and archaeological status. Everyone signed.[47]

The statement was published in *American Antiquity*, the premier professional journal of American archaeology.[48] In it, we urged others to judge Monte Verde for themselves based on their own evaluation of the evidence in Dillehay's two volumes on the site. Many did just that, and the fact that a few (but only a few) reached different conclusions than we did is testimony our say-so did not carry the weight that Lyell's or Kidder's had 140 and 70 years earlier.

So, too, is the fact that Haynes has since had second thoughts, claiming the site has only six unequivocal artifacts; those six, naturally, are the ones that most resemble the North American Paleoindian projectile points with which he (and I, for that matter) is

A VISIT TO MONTE VERDE

We came, we saw, we were convinced. And a few weeks after our return from Monte Verde in January 1997, the Dallas Museum of Natural History and the National Geographic Society (which co-sponsored our trip) organized a press conference to announce our conclusion that Tom Dillehay's work at Monte Verde had finally broken the Clovis barrier. It was a moment the *New York Times* ranked with "aviation's breaking of the sound barrier."[42] Heady stuff this—but *The Right Stuff*? I wasn't convinced. We'd long been open to the possibility of pre-Clovis; most thought the sound barrier unbreakable. However, part of the comparison rang true: as Chuck Yeager had discovered fifty years before, when approaching the sound barrier, a plane is badly buffeted. Now *that* I'd experienced.

The visit to Monte Verde began in calm skies, when from across the United States and distant points in South America participants assembled at the University of Kentucky. That first evening Dillehay briefed us on South America late Pleistocene archaeology. To those of us who work in North America, it was an eye-opener: an archaeology with a scant bifacial technology, few formal tools, little in the way of "big-game" kills, and yet a gaggle of sites as old as Clovis—and a few possibly even older.

The next morning we assembled in Dillehay's lab where many of Monte Verde's stone, bone, ivory, wood, and soft tissue specimens were neatly laid out. Dillehay and his colleagues explained their analyses and results, but the talks were no match for seeing Pleistocene string tied in overhand knots, a wooden lance with mastodon flesh adhering to it, or a log mortar with wild potato skins pounded into it (dinner was meat and potatoes?). There was give and take: in a room full of pros, there was no avoiding it—or so I thought. Deep into what I judged a riveting, statistics-studded debate over whether some rounded stones were artifacts, I glanced over at the *National Geographic* writer traveling with us. He was dozing. Maybe it wasn't so riveting.

After several long plane rides, we repeated the process in Valdivia, Chile, with the Monte Verde collections housed there: more stunning specimens, more exchanges. Finally, to the site itself. It was a spectacular austral summer day. We examined the stratigraphy in the walls of Chinchihuapi Creek, studied the different MVII occupation areas, and visited off-site spots where Dillehay had tested the natural deposits. We peered into a trench in the MVI area, and wondered if people had also been on this spot 33,000 years ago. But no one wanted to go there just yet: we stayed focused on MVII. Vance Haynes, as is his custom, collected charcoal samples; Jim Adovasio, as is his custom, wove string from juncus reed.

Since it was the last day of the trip and participants would scatter the next morning, we headed back to Puerto Montt looking for a restaurant where we could chew over dinner and all we'd seen. None were open that early, so we headed instead for a local bar: La Caverna. I had a sinking feeling. There'd been tense moments already. Dillehay was dealing as best he could with a nonstop barrage of questions from a dozen of his peers about a project to which he'd devoted twenty years of his professional life. None of us had ever experienced that kind of intense grilling. I thought it best if everyone was feeling satisfied from a meal before turning to what I suspected would be a hard conversation. It wasn't to be. Drinks were ordered, and the talk began.

And it was tough talk, lasting over two hours. We probed Monte Verde for flaws, and hashed and rehashed alternative explanations. Maybe the dates were contaminated by local volcanic eruptions,[43] perhaps those wooden "lances" were natural, where exactly were the artifacts relative to the dated remains, were the deposits at all mixed, and on, and on. Dillehay, who'd spent years thinking about such matters and was unquestionably more familiar with the evidence than we ever could be, grew increasingly short-tempered, even insulting (which is *not* his personality). Tempers flared, and at a moment when the words grew especially heated, Alex Barker, the Dallas Museum's representative on the trip, gave an ear-piercing whistle. Stunned silence. The museum, he announced, had sponsored the trip with the proviso we'd seek some degree of closure over Monte Verde. He didn't want to have to report that we only learned "archaeologists don't play well together." The tension broke, the discussion continued. Finally, Barker called for a vote on whether we believed Monte Verde was a pre-Clovis site. It was a theatrical ploy, but I was not entirely surprised by how it turned out—though I was intensely curious. Nor was I the only one. As I was furtively looking around the table to see whose hands were up, so was everyone else. But all hands were raised: everyone agreed MVII was archaeological and 12,500 years old.

Barker called for a toast, and then a dreadful slip of the tongue: Dennis Stanford, perhaps looking at Jim Adovasio sitting across the table, raised his glass and inadvertently toasted *Meadowcroft*, not Monte Verde (I *knew* we should have had food first). Stanford recovered as best as he could, saying we should all do this next year at Meadowcroft. Too late. Haynes innocently chimed in, suggesting that if Adovasio would date just one more seed from the deepest levels at Meadowcroft, he might believe in the antiquity of the site.

By his own admission, Adovasio lost it: "I burst out in derisive laughter," and he then spat that "never would I accede to any request [Haynes] made for further testing of the Meadowcroft site because if I did he would simply ask for something else in a never-ending spiral of problems. . . . Meadowcroft's antiquity was settled as far as most other professionals and I were concerned, and that if any remaining skeptics did not believe it, I could not care less. I then stormed out of the bar."[44]

It was painful. Haynes looked like he'd been slapped. A pall settled over the room. We'd been so close to getting through without an explosion. The *National Geographic* writer sat intently witnessing it all. I wondered if he was trying to memorize everyone's lines. I headed for the restroom to wash my hands (it seemed the right thing to do). The restaurant was now open, so we drifted out of *La Caverna*, much more subdued than we'd been just five minutes earlier. The moment of triumph had passed.

Ironically, Adovasio and Haynes were sharing a hotel room, and I hoped they would have the chance to smooth over matters (I can't help it: both are good friends of mine). But Adovasio got in late that night, and Haynes left early the next morning. He was not flying back to the States with us, but was headed to a conference in Cairo. He was gone before any of us saw him.

In the weeks after we returned, Donald Grayson and I circulated to our colleagues drafts of the statement for *American Antiquity*; a press conference was held; we began to ponder the full implications of Monte Verde. There was much to ponder, but the one thing I thought was beyond pondering was the site's antiquity and archaeological status. I was wrong.

After our *American Antiquity* statement appeared, archaeologist Stuart Fiedel took our advice and examined Dillehay's Monte Verde volume. He found errors, which mostly boiled down to discrepancies in artifact location, numbers, and mapping. I was not surprised: errors are inevitable in complex, multiyear, multidisciplinary research programs and the massive books they yield. All of us who do such projects live with that unfortunate reality, however much we attempt to prevent errors from happening. But were the errors in the Monte Verde volume trivial or profound? It was hard to say, and Fiedel's critique hard to swallow since he gave all his criticisms equal weight, and blanketed them in a patina of conspiratorial mistrust and accusations, layered with snide remarks. He did his critique no favors.

Worse, instead of sending it to Dillehay prior to publication—it's professional courtesy, and would allow Dillehay to correct errors of fact—then submitting the piece to a rigorously peer-reviewed academic journal (the usual procedure), Fiedel instead sent it to *Discovering Archaeology*, a "popular" archaeology magazine (but not too popular: it's now defunct). Dillehay learned of Fiedel's attack only after returning from a long season of fieldwork. *Discovering Archaeology*'s editor told him Fiedel's piece would be published, and if Dillehay wanted to respond, he had two weeks: the magazine was being hurried for distribution at the Clovis and Beyond conference that fall of 1999. As David Hurst Thomas observes, it looked like an ambush.[45] To those of us also asked to comment under this looming deadline, it certainly felt like an ambush. Fiedel claims this was not his doing: the editor assured him the piece would not be published if Dillehay chose not to respond. Right. The editor—knowing full well controversy sells magazines—failed to mention that option to Dillehay, who responded as best he could in the time allowed, then later produced a much lengthier reply for posting on the Web.[46]

So was this much ado about nothing? Fiedel's critique focused on the "unambiguous" artifacts (projectile points, mostly): Where were they found? How were they associated with dated samples? Why were there no photographs of them in place, as at Folsom? Dillehay and colleagues clarified their provenience and provided photographs. I could only shake my head. Could a picture of a point in place make that much difference? Adovasio had published pictures of a projectile point in pre-Clovis age levels at Meadowcroft (see Figure 23 in this book), but where was the groundswell of support for that site's claims to great antiquity?

The Clovis and Beyond conference buzzed with talk of the Monte Verde critique, but it was clear what the 1,200 people in the hall listening to Dillehay's response to Fiedel thought of the whole matter: when he finished speaking, Dillehay was given a long, standing ovation. I was in the back of the hall with one of the conference organizers when he was approached by a furious Fiedel asking for "ten minutes on stage to respond to this crap." "No, Stuart," came the reply, "you can write a letter to the editor of *Discovering Archaeology*."

most familiar.[49] Frankly, I think that's a red herring, ignoring as it does the wooden artifacts, modified bones, ivory pieces, chunks of mastodon meat, the strings tied in overhand knots around the wooden stakes, the exotic plant remains, and so on. We can agree to disagree about those other remains, but it doesn't matter: even with just six artifacts, we are still left with an archaeological site that is 12,500 years old. Haynes's second thoughts have not reversed the archaeological tide of opinion on Monte Verde.

And, indeed, the tide keeps rolling in for this site: in early 2008 Dillehay and colleagues reported further analysis of hearth sediment from Monte Verde yielded half a dozen previously unrecorded species of algae and seaweed (the latter still adhering to a stone tool), and radiocarbon dates neatly affirming the site's great age. These new results, as long-time skeptic Daniel Sandweiss graciously admitted, "remove any lingering doubts about the antiquity of human presence at that site."[50] And as Dillehay observed, here was further testimony to the importance of coastal resources in the diet and health of early Americans.

Although Monte Verde is just a thousand years older than Clovis, what a difference a millennium makes: to have reached southern South America by 12,500 years ago, the first Americans must have left northeast Asia and crossed Beringia—16,000 kilometers distant—much earlier. But when, from where, and how? And why is the oldest acceptable site in the Americas about as far from Beringia as one can reach, with no sites in North America demonstrably as old or older? Monte Verde has forced a sea change in our thinking of the possible entry route(s) into the Americas, the timing and mode of that entry, how and where we look for ancient archaeological sites, and just who the first Americans were and from whence they came.

COASTING TO AMERICA

Current geological evidence has the Laurentide and Cordilleran obstructing the interior route into North America from before about 28,000 until sometime after 12,000 years ago (Chapter 2). That being the case, no matter how fast one supposes the first Americans traveled south from Alaska, they could not have gone fast enough to arrive at Monte Verde 500 years before the ice-free corridor opened. This leaves a couple of interesting possibilities: either they departed Alaska *before* the interior corridor opened—that is, in pre-LGM times—or they traveled south ice via a coastal route.

So far, no pre-Clovis sites have been found in the interior corridor region. A very few are Clovis age, but most are younger than 10,500 BP. And no sites here or, arguably, anywhere else in the New World, precede the LGM. That would seem to preclude entry through an interior route—for the moment (we have learned the hard way to be circumspect about such pronouncements, since new discoveries often surprise and contradict our received wisdom).

In the meantime, and in the wake of Monte Verde, archaeological attention has turned to the Pacific Northwest coast to explore the timing and viability of a coastal entry route. Opinion is divided over whether in-migrating groups coming down the coast would have traveled on foot or by boat. Since we've not found any late Pleistocene watercraft, much of this debate rests on speculation and circumstantial evidence—both for and against the possibility.

Pleistocene Australia, for example, even though joined during times of lowest sea level to New Guinea and Tasmania in a land mass known as Sahul, was throughout the Pleistocene separated from mainland Southeast Asia by at least 90 kilometers of open ocean. That the first Australians crossed open water over 40,000 years ago is taken as indirect testimony that the first Americans could also have paddled the coastal waters of Beringia and northwest North America. But the North and South Pacific are very different oceans, especially during glacial episodes. No early Australian on a raft had to contend with a northern ocean's dark skies and frozen seas of winter, the treacherous ice floes and squalls of summer, and year-round frigid water, where death by hypothermia would come quickly when capsizing or falling into those inhospitable waters (in any case, regular, planned, and repeated voyaging around the South Pacific began only 3,500 years ago).

The first evidence watercraft were used in the North Pacific—the occupation of islands off the Pacific Northwest coast and the hunting of marine mammals—mostly postdates 10,000 BP, too late for initial colonizers skirting south around the ice sheets (Chapter 6). Early Americans made it onto a couple of California's Channel Islands, at the Arlington Springs site on Santa Rosa Island and Daisy Cave on San Miguel Island, as early as 10,960 BP.[51] Even at the LGM, these islands were separated from the mainland by at least 10 kilometers and had to be reached by watercraft (see Plate 6). That said, they are too far south to be relevant to the question of whether the first Americans got from Alaska to unglaciated North America by boat.[52] Nonetheless, based on such evidence, scenarios are conjured of early American seafarers who hunted and fished saltwater, and gathered shellfish and other intertidal resources, and thus could have made it down the coast in short order. That they may have, but whether they had boats to travel in remains unresolved.

Walking (rather than rafting) down the Pleistocene coast became a viable option once the outer coast was free of glacial ice (Chapter 2). Although the terrain of the present-day Pacific coast is highly dissected and inhibits movement, the now-drowned Pleistocene coastline would not have presented such a challenge. For the moment, Occam's razor points to groups walking down the coast—if they came via that route.

Forestalling the resolution of the by-foot vs. by-boat debate, and indeed the larger question of whether colonization proceeded via the coast, is the fact that much of that route is presently under water, drowned by rising post-glacial seas. Yet, as noted in Chapter 2, not all the Pleistocene coast was drowned: rapid isostatic rebound and tectonic activity has uplifted segments of the Pleistocene coastline in Alaska and British Columbia. Unfortunately for archaeology, dense temperate rainforest hinders surface visibility along much of that stretch. These challenges notwithstanding, in the last decade, Daryl Fedje and colleagues (among others) have sought early coastal sites on land and under sea, an effort enhanced by sophisticated geological and bathymetric mapping and seafloor coring, along with efforts to predict the most likely places to locate sites.[53] To date, none older than 10,300 BP have been found, but it's still early in the search.

IT'S ABOUT TIME

Monte Verde implies an arrival in the New World much earlier than 12,500 years ago, given how long it must have taken to travel from Beringia to southern Chile. The initial round of speculation, and it was little more than that, put the arrival at perhaps 20,000 years ago. It could have been later, of course, or even far earlier, if one had to reach Monte Verde in time to make the MVI occupation. But estimates of the timing depend on how long it would take for colonizers to traverse the continent. Chapter 7 looks into what we know or might infer about how fast the first Americans dispersed, in light of obstacles or challenges they may have met along the way, how readily they adapted to the diverse and increasingly exotic New World in which they found themselves, how easily they may have coped with novel pathogens and diseases, and how they might have maintained their population size and reproductive viability while living in relatively small numbers spread thinly over a vast and otherwise unpopulated continent.

For now, consider this: hunter-gatherers can travel quickly within familiar environments (especially ones that are relatively homogenous) in which they know the plants that are edible and when they are available, how animal prey behave, where rock outcrops to replenish their stone supplies are located, where (or how) to find water, and so on. Such knowledge is usually hard won over lifetimes and generations. Yet, to the first Americans, this was a new, exotic, and utterly unknown landscape, which became ever more alien to their Beringian experience as they entered different habitats on their journey south. With every ecological boundary and geographic barrier they crossed, and there were many across late Pleistocene North and South America, the process of finding water, food, and critical materials began anew. This was not the Oklahoma Land Rush, with a known destination, wagons bulging with homesteading provisions, and a strict timetable to meet. The process took time: just how much, we don't know (more on this in Chapter 7).

SEARCHING FOR THE VERY FIRST AMERICANS

What we do know is that if people were in southern South America 12,500 years ago, sites of equal or greater age should be found between there and the Beringian gateway. Some argue they already have been. In Venezuela, projectile points similar to those at Monte Verde were found with the skeleton of a young mastodon. Skeptics suspect the points sank through overlying wet sediment, and came to rest next to the bones. If so, nature is wonderfully perverse: one of the points rests snugly in the cavity of the mastodon's right innominate (pelvic) bone. The radiocarbon ages on the bone do not much help the case, ranging as they do from 12,980 to 14,200 BP, with some suggestion the site may be much younger. Taima-Taima's antiquity and association must stand on its own merit, but having artifacts similar to those from Monte Verde raises the possibility it is an expression of the same early archaeological presence. However, until the age of Taima-Taima is pinned down, it will remain in archaeological limbo.[54]

As for pre-Clovis age sites in North America, there's Meadowcroft, of course, though we still await its final report. In the wake of Monte Verde, a flurry of additional pre-Clovis contenders have appeared—almost twenty, by Adovasio's count—including the sites of Big Eddy (Missouri), Cactus Hill and Saltville (both in Virginia), the localities of the Chesrow complex (Wisconsin), Paisley Cave (Oregon), Pendejo Cave (New Mexico), and Topper (South Carolina). Most of these have only recently been discovered and/or reported, some are still undergoing excavation and analysis, but a few—like Paisley Cave, which has yielded ancient DNA from human coprolites that date to more than 12,000 BP—are nonetheless extremely promising: the ages of the coprolites were determined by two different radiocarbon laboratories, the DNA analysis was likewise cross-checked by independent laboratories, the investigators carefully guarded against the possibility of contamination (in the field and the laboratory), and the mitochondrial DNA haplogroups matched those seen in American Indians.[55]

So far, however, none of these sites has been fully accepted by the archaeological community in North America that, I think rightly, still maintains a healthy skepticism toward pre-Clovis claims. It is a point worth emphasizing: seventy years earlier when

Folsom broke the Holmes-Hrdličkian log jam, Abbott's paleoliths got no older as a result. Merely because a barrier is breached does not mean it should be removed. Put another way, reason to *expect* a pre-Clovis site should never be mistaken for *evidence* a site is pre-Clovis. Each must stand on its own merits.

All of which begs a question: do we need to find more than one site to prove a pre-Clovis presence? A few critics insist on it. But that insistence logically confounds the process of establishing antiquity—finding a site that meets the necessary criteria—with the process of discerning the broader cultural patterns of that period (which does require multiple sites). The initial establishment of a Pleistocene human presence in America required only one site (Folsom), and there is no reason why that should be any different today. Indeed, in 2004 V. V. Pitulko and colleagues announced the discovery of the Yana RHS site in northeastern Siberia, which at 71°N and dated to 27,000 BP was nearly twice as old as any previously recorded far northern site, and had artifacts startlingly like those found in Clovis assemblages (Chapter 6).[56] No one batted an eye, or insisted that Yana would be unacceptable until replicated.

So why hasn't a pre-Clovis presence been better established in the decade since Monte Verde? Within a decade of the discoveries at Brixham Cave and Folsom, dozens of Paleolithic and Paleoindian sites (respectively) were found. But the circumstances then and now are very different. Both Brixham Cave and Folsom lent themselves to ready generalizations about how or where more sites like them would be found (Chapter 3). Equally important, the occupation at Brixham Cave was preceded by many hundreds of thousands of years of Paleolithic prehistory. Folsom, at the very least, came on the heels of a millennium or more of a well-established human presence in the Americas. Neither Brixham nor Folsom were the very oldest sites ever found in England or America, and that's no surprise: the very oldest archaeological sites are extraordinarily hard to find, and almost never the first ones discovered.

Here's why: the number of sites produced in a given period or region is the sum of many things, not least the size of the contemporary population. Lots of people produce lots of debris and ample evidence of their presence; if there are fewer people on the landscape, and they are highly mobile and widely scattered, their traces are much rarer. And certainly in the earliest centuries and millennia of colonization, large areas of the American continent were thinly occupied, or simply unoccupied. This is why we know so much less about the first peoples of a continent than about a continent's first farmers: the latter were numerous, more firmly rooted in place, and produced heavier artifact "rains."

We do not know how large the first group(s) of colonizers in the Americas was, but it was almost certainly not large. We assume, in the absence of any evidence but in keeping with studies of modern hunter-gatherers, that minimally viable numbers for an initial population might have ranged between several score and several hundred. So one part of the answer to the question of why there were so few early sites is that there were few early people.

The other part of the answer is that the odds of site preservation and discovery can be slim. The number of sites recovered archaeologically from a given period is determined by, among other factors, erosional and depositional processes that help erase or bury ancient sites and make them less visible; the antiquity of the period, older sites being rarer because time has had longer to erase them; the search techniques used to find sites, because if sites are deeply buried or, say, underwater on the now-drowned continental shelf, they will be more difficult to locate (of which, more below); and, finally, the number of sites relative to the size of the area being searched by archaeologists.

The last point perhaps requires a bit of explanation, easily done by comparing the colonization of North America with the colonization of, say, any one of the scores of islands that dot the Pacific Ocean and were settled in just the last few thousand years. In those instances, the number of colonizers might well have been no larger than the group that came to the Americas. Yet, when these seafaring peoples arrived on an island, they had an immediate and devastating ecological impact: the native, flightless birds were driven to extinction, trees were felled for fuel and construction materials, and exotic plants and animals (such as yams and stowaway rats) were introduced. Ecosystems changed radically on the heels of human arrival, and given the small size of these islands, those changes were sweeping and are usually highly visible archaeologically (sediment cores, for example, show a sharp drop in the pollen of native trees and a sudden rise in the amount of charcoal, while the fossil record is marked by a precipitous decline and disappearance in the number of bird species). Moreover, the size of the area to be searched on islands is smaller than a continent by many orders of magnitude, and thus the likelihood of finding the archaeological "footprints" of the first colonists correspondingly greater.

Not so in the Americas, where the area to be searched is a haystack of many millions of square kilometers, and the needle a tiny number of sites, left by people that could not have had a visible impact on the hemisphere's ecosystems (though as we'll see in Chapter 8, some have claimed as much[57]).

Archaeological geologist Karl Butzer has observed, further, that because of the converging circumstances of low population and destructive geological processes, "recoverable" sites dating to, say, 30,000 years ago, will be about ten to fifteen times less common than those dating to 11,000 years ago.[58] All of which implies that

- people were likely present in the Americas long before their traces appear on our archaeological radar screen;

- the oldest archaeological site in the Americas that we currently know of—Monte Verde—represents an early presence in the New World, but assuredly not the earliest; and

- earlier or the earliest sites of the first peoples will be rare, hard to find, and may never be found.

If Monte Verde is pushing the limits of antiquity in this hemisphere, then the likelihood of finding dozens or scores more as old or older diminishes.

For that matter, and again unlike the circumstances of Brixham Cave and Folsom, the unusual geology and archaeology of Monte Verde do not readily lend themselves to easy generalizations about how or where more sites like it will be found. The absence of confirmed North American pre-Clovis sites raises the possibility that archaeologists haven't looked in the right places or in the right way for sites of this age (Albert Goodyear even admits he never used to excavate below the Clovis layer; once he started doing that, he discovered the Topper site). It is, for example, a curious but well-known fact of the archaeology of North America that Paleoindian traces are rarely found in rockshelters or caves, Meadowcroft being (once again) the obvious exception. Yet around the world, from Arago (France) to Zhuoukoudian (China), caves were used throughout prehistory.

Various scenarios are conjured to explain why Paleoindians failed to inhabit North American caves, notably that they were moving through the Americas at such a fast clip they hadn't had time to find them (though, as we'll see later, they managed to find small and isolated stone sources). Archaeologist Michael Collins has another idea: North American archaeologists only excavate the younger (generally Holocene) deposits in caves, and routinely fail to peer beneath massive roof spalls—as Adovasio did at Meadowcroft—to see what traces may be found in older, Pleistocene sediments (more on this in Chapter 9).

Indeed, there are many regions and settings yet to be searched systematically for evidence of the first Americans: deeply buried late Pleistocene surfaces, especially at the intersection of major rivers (which were likely corridors of travel); bogs and wetlands; now-drowned or uplifted coastlines; resource-rich forests; and especially the 60-kilometer-wide Panama bottleneck connecting North and South America, through which the earliest South Americans must have passed. Searching those settings will require more than the usual archaeological field techniques: geological and geophysical remote sensing techniques will become increasingly in demand.[59]

WHO WERE THE FIRST AMERICANS?

Whatever may ultimately prove to be the antiquity of the first Americans, Monte Verde has renewed discussion of who they were, and where they originated. The archaeological "footprints" of those who occupied Monte Verde cannot be followed back to North America, let alone across Beringia or into northeast Asia (Chapter 6). The absence of an obvious Siberian precursor has prompted some breathtaking speculations, among them that the first Americans came across the Pleistocene Atlantic from Europe, or sailed across the Pacific, possibly first setting foot in South America. Those scenarios are highly doubtful (of which, more later).

Monte Verde's decidedly non-Clovis look raises further questions of the historical relationship between the people who occupied this site and later (and northern) Clovis groups. Were they related, and if so, how? Do they represent the same or different colonizing pulses? More broadly, how many early migrations were there to the Americas; was it a single event from which all subsequent groups were descended, or (as Hrdlička supposed) did it involve a series of separate migratory pulses? And what might we say from this of the population history of contemporary Native American populations: are they descendants of the first Americans?

These are not strictly archaeological questions, but ones to which geneticists, linguists, and physical anthropologists have offered answers. Indeed, few archaeological questions lend themselves more readily to an interdisciplinary solution than that of the peopling of the Americas.

NON-ARCHAEOLOGICAL ANSWERS TO ARCHAEOLOGICAL QUESTIONS

"The hour has at last arrived," Pliny Earle Goddard crowed the spring of 1927, "for an extensive reorganization of our conception of the peopling of America." About time, too. Goddard, a linguist, had just read Harold Cook's proclamation that the Lone Wolf Creek site (Chapter 3), with its artifacts and extinct bison, proved beyond "reasonable doubt" humans were here in the Pleistocene. Goddard had long suspected as much and was delighted by the news, for it meant Holmes and Hrdlička, whom he despised (the feeling was mutual), were finally proven wrong on their terms. Goddard scolded them for refusing to admit a deep human antiquity on this continent. People inclined to be critical and conservative, he sneered, should doubt that humans were recent, not the other way around. How else could one account for the time required for New World peoples to domesticate maize, develop the "higher civilizations of Peru, Mexico, and our own Southwest," and allow for the divergence of some "fifty languages [families] so distinct from one another in their vocabularies, that no relationship can be traced"?[1]

Assuming the first Americans came as a single group, they must have spoken one language "as pure biology would require," and had to have arrived "many millenniums"[2] ago in order for that single language to evolve into the hundreds now spoken. Goddard guessed 100,000 years or more was needed. Not a bad number, Cook supposed. He thought Lone Wolf Creek might be even three times older.

Holmes and Hrdlička weren't buying: not Cook's Lone Wolf Creek claims (Chapter 3), nor the idea that the diversity of American Indians, whether cultural, linguistic, or physical, implied a lengthy period of divergence from a common ancestor. Not only were

Native Americans not as diverse as they appeared—Hrdlička had long argued American Indians and Asians were so similar physically they could not be long separated—but such diversity as existed could be readily explained by processes other than divergence over time from a single ancestor. The many Native American languages, for example, could reflect successive migrations from Asia of different groups speaking "radically different tongues."[3]

It was a reasonable counterargument, so far as it went: cultural, linguistic, or physical diversity could result from either cause, so by itself was no measure of antiquity. That required independent evidence, such as a well-dated archaeological site. But the archaeological evidence in those days was equally unsettled. Even when Folsom came along a few months later, its antiquity was far from what Goddard deemed necessary. He wasn't alone. The giants of his day—like linguist Edward Sapir, who thought it "intrinsically highly probable" there had been multiple migrations, and even anthropologist Franz Boas who did not—nonetheless supposed that 10,000 years was "a hopelessly inadequate span of time" to account for American Indian language diversity.[4]

Archaeologists, physical anthropologists, and linguists haven't stopped bickering over the origin, antiquity, and number of migrations of the first Americans. Yet, for a brief period in the 1980s, interdisciplinary peace threatened to break out. It began with a visit by Stanford University linguist Joseph Greenberg to the University of Arizona, where he presented his idea that American Indian languages could be grouped into three families, each a descendant of three independent groups that had separately migrated to America. In the audience was geneticist Stephen Zegura, and he was astonished. He'd heard that punch line before, only not from Greenberg but also from physical anthropologist Christy Turner, who studies teeth. Zegura wondered if the pattern evident in languages and teeth was also true of American Indian genes. He took a look: with a bit of prodding, the genetics might also support a model of three migrations to America.[5]

In December 1986, Greenberg, Turner, and Zegura jointly announced the sweeping conclusion that independent linguistic, dental, and genetic evidence was converging on a single story of the identity, number, relative order, and antiquity of migrations to the Americas. Yet, with a speed and ferocity unusual even for the long history of controversy over the peopling of the Americas, their coalition was attacked. And unlike in years gone by, where disputes mostly raged between disciplines, strong opposition also came from behind their own disciplinary lines. The convergence was short-lived.

So we're back to bickering. Although this debate has an almost NASCAR-like appeal (watching for the crashes), much is being learned from the non-archaeological evidence being brought to bear on these archaeological questions. This chapter explores what language, skeletal biology, and genetics can say and have said about the origin and antiquity of the first Americans. We'll postpone for Chapter 6 the more challenging question of how (or whether) these various lines—along with archaeology—converge on answers to the central questions of the peopling of the Americas.

When Europeans arrived, American Indians spoke a bewildering number of languages. Just how many, no one knows. Linguists recognize over 1,000 languages (out of about 6,000 spoken worldwide), but only about 600 survived as of 1950 (see Table 3). The loss of approximately 40% of America's native languages resulted from several factors, the most important being the arrival of infectious epidemic diseases on the heels of European contact, which wiped out many groups of speakers, or forced survivors to amalgamate and adopt a common tongue. That cause, coupled with centuries of acculturation and the steady decline of native speakers (which has been precipitous since the 1950s—only about forty-five of North America's languages are still spoken), has meant that many languages are poorly known, some by no more than a single phrase or word. South America has suffered the greatest losses, where only 15% of surviving languages have been adequately documented.[6] The losses, linguist Ives Goddard (no relation to Pliny Earle) grimly observes, are "a cultural tragedy of almost unimaginable dimensions."[7]

Given the loss of so many American Indian languages, what might we say about the historical relationships among them? The answer lies in classifying languages, though the aim is not just to bring order to Babel. Languages evolve. That process, as Charles Darwin long ago observed, parallels biological evolution: over time, languages change and diversify into new forms (try tackling Shakespeare as it was originally written just 400 years ago). There are, of course, language rules, but those are not strictly enforced by the community of speakers, though they might have been by your grade-school English teacher, nor are they even enforceable. As new words or pronunciations are

TABLE 3   Counts of languages and language families in the Americas

	North America	Central America	South America	Total
Estimated number of languages in 1492	~330	~125	~550	~1005
Number of languages surviving in 1950	~200	~100	~300	~600
Established language families (traditional)	33	7	~50	~90
Isolates	27	4	~50	~81
Families + isolates	60	11	~100	~170
Established language families (radical)	3	1	1	3

NOTE: Data from Goddard 1996, Golla 2000, Greenberg 1987, and Kaufman and Golla 2000; given language losses, other estimates of American languages and language families naturally vary, but are within the same ballpark.

invented by an individual (or individuals), new variants of semantics, syntax, and sound spread through the community. Most members adopt the new forms: speaking the same as your peers is a way of being part of a group.[8]

But then, some of the speakers moved away—to America, say. Months pass, then years since these individuals (or their ancestors) were part of that original community. They may have maintained sporadic contact over that time, rendezvousing at intervals to exchange information, resources, or mates. But because contacts are limited, the inevitable changes taking place within each group's speech have led to the emergence of different *dialects*, or regional varieties of a language. Although no longer identical, the dialects are to a degree mutually intelligible. Were the splinter group to return to the ancestral homeland or meet their now-distant cousins, they would be able to make themselves understood.

However, if the descendants are separated long enough so as to lose all contact, their speech will become mutually unintelligible, and their dialects different *languages*. At that point, you can't go home again—or more properly, make yourself understood if you do, for whether you remember the old language or not, the speech of those who stayed behind would have since changed as well.

The time it takes for a new language to develop is not known precisely, largely because rates of language change vary owing to the size of the group, the degree of geographic and social isolation or contact, and as Daniel Nettle adds, circumstances such as speakers dispersing across a new and empty landscape (obviously relevant in this instance), which often leads to a proliferation of new languages.[9] But as a very rough estimate, linguists suggest a new language can emerge after 600–1,000 years.

Repeat this process over time, such that many groups splinter off and move to distant regions, or otherwise become linguistically isolated, and a Babel of different tongues emerges. The descendant languages are considered part of the same *language family*, such as the Romance languages of Europe (including French, Spanish, Italian, Portuguese, and Rumanian, descended from Latin). In turn, the Romance and Germanic families, along with ancient Greek, Sanskrit, and dozens of others, are more distantly evolved from Indo-European, a mother tongue spoken sometime after about 9,000 years ago (since words for wheeled transport and domesticated animals are common) but before about 6,000 years ago (since words for metallurgy are not).

Some languages, however, are not easily linked to any family. Euskara, spoken by the Basque inhabitants of the Pyrenees region between France and Spain, is an *isolate*, completely unrelated to its neighbor languages. Isolates in North America include Tonkawa, Tunica, Zuni, and (possibly) Haida. Assuming their uniqueness is not a function of inadequate documentation, any linguistic ancestors these share with other American languages may be very deep in time.

In principle, then, separate languages sharing an ancestor retain vestiges of their common inheritance in words (meaning and form), syntax (grammatical structure), and sound (pronunciation). But how distant are the ancestral echoes we can hear? Can languages descended from a common form spoken in, say, Pleistocene times, retain traces of their shared ancestry? There are two camps on the issue. One argues the rate of language change is so rapid that within 8,000–10,000 years, all vestiges of once-common elements will be erased. The other believes they can detect in the world's languages subtle reverberations of words that were once spoken by the first modern humans 40,000–60,000 or more years ago. A few among them think they hear faint sounds of the first words spoken 100,000 years ago (apparently one of those was *tik* [finger], according to Merritt Ruhlen).[10] Not surprisingly, the two camps take very different routes in arriving at such divergent estimates (of which more below), but in both cases, finding like languages is the vital first step in recognizing their shared ancestry.

The process of classifying American languages began in earnest in the late nineteenth century, when the BAE's John Wesley Powell—examining similarities across word lists—grouped North America's languages into fifty-eight broad families. Although he believed that these families could "not have sprung from a common source," he supposed "future and more critical study will result in the fusion of some of these families."[11] He was right. Linguists progressively lumped languages together until, by the 1920s, Sapir—who considered Powell's idea of nearly sixty genetically independent families "tantamount to a historical absurdity"—had whittled that number down to just 6 families, each presumably sharing elements and a common ancestor.[12] Powell was long dead by then, but Sapir still faced formidable opposition: his mentor, Franz Boas, who chided Sapir for assuming like languages had to be related. After all, Boas reminded him, languages (like other elements of culture) can be similar for other reasons, borrowing and chance, chief among them.

Sapir understood that, but he possessed remarkable linguistic intuition and the cool nerve of a gambler. He was playing a hunch with his 6-family scheme, sensing the similarities among these many languages meant they were related. As even his critics granted, Sapir's hunches were often better than most others' evidence. Still, Sapir expected amendments to his scheme, friendly or otherwise, and he got them. One camp, the great majority, splintered his 6 families into 150, believing no fewer could be justified. The other camp, led by Joseph Greenberg, out-Sapired Sapir himself, grouping all American languages into just 3 families. It was, Greenberg acknowledged, "a very bold thesis indeed."[13] Could it also be right? It depends on the method one uses.

THE TRADITIONAL METHOD

Traditional historical linguists—and Greenberg is not one—identify language families from the bottom up, comparing two languages side by side, starting with a search for *cognates* (words alike in form and meaning). They recognize that shared words themselves may not indicate common ancestry. When speakers of different languages come into contact, words often jump from one language to another: the Norman Conquest in 1066, for example, injected a heavy dose of Norman French (a Romance language) into Old English (a Germanic language), resulting in Middle English, which combined elements of both.[14] Likewise, some words are similar across languages because of deep-rooted patterns in the evolution of human speech (the reason why certain consonant-vowel combinations such as "mama" recur), and some are similar just by chance. How then, to sort the historically meaningful cognates from ones that are merely accidental?

A key answer for historical linguists lies in finding sound correspondences. As languages evolve, the words change, but so, too, do the sounds. And they do so, as Donald Ringe puts it, with "statistically overwhelming regularity."[15] Thus, Latin words with the consonant pairing *ct* (for example, *octo* and *noctem*) consistently shifted in Spanish to a *ch* sound pairing (as in *ocho* and *noche*), and so on (see Table 4). The phonetic expression of the sound change, Ringe explains, is irrelevant: what matters is the regularity of the pattern and the number of those changes. Where a sound change is broadly consistent, it provides strong and, importantly, statistically reliable proof the cognates result from common descent; borrowing, because it is more haphazard, will not be accompanied by regular, widespread sound correspondences.

Traditional historical linguists take the analysis a step farther. If a number of different languages are part of the same family, they should be able to reconstruct the *proto-language* by identifying the core elements shared by the related languages, including words and grammatical forms. The hypothesized proto-language can then be put to the test, and this was done dramatically in the case of Indo-European, when archaeologists in the 1920s excavated a stone tablet in Turkey on which was carved the predicted consonant sets of the now-extinct Indo-European language of Anatolian.

It took more than a century to work out the details of Indo-European, despite the fact that the number of languages under study was relatively few and well attested: approximately 100 Indo-European languages are still spoken by half the world's population. Understanding the history of American Indian languages is far more challenging, given

TABLE 4   Cognates and sound correspondences among Romance languages

Latin (English)	French	Italian	Portuguese	Spanish	Rumanian
directum (straight)	directement	Diritto	direito	derecho	Drept
factum (made)	fait	Fatto	feito	hecho	—
lact (milk)	lait	Latte	leite	leche	Lapte
lectum (bed)	lit	Letto	leito	lecho	—
noctem (night)	nuit	note	noite	noche	Noapte
octo (eight)	huit	Otto	oito	ocho	Opt
Sound change from Latin	ct→it	ct→tt	ct→it	ct→ch	ct→pt

NOTE: Data from Ringe 2000:144.

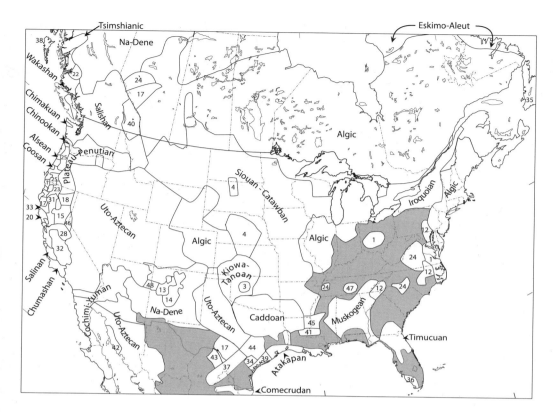

FIGURE 30.

Map of Native North American language families and languages, based on traditional historical linguistics. The northern portions of the Eskimo-Aleut and Na-Dene regions are not shown (see Figure 31). Areas in gray are regions where data are inadequate or languages are unclassified.

The 33 North American language families (keyed to map numbers): (1) Algic, (2) Alsean, (3) Atakapan, (4) Caddoan, (5) Chimakuan, (6) Chinookan, (7) Chumashan, (8) Cochimi-Yuman, (9) Comecrudan, (10) Coosan, (11) Eskimo-Aleut, (12) Iroquoian, (13) Keresan, (14) Kiowa-Tanoan, (15) Maiduan, (16) Muskogean, (17) Na-Dene, (18) Palaihnihan, (19) Salishan, (20) Salinan, (21) Pomoan, (22) Plateau-Penutian, (23) Shastan, (24) Siouan-Catawba, (25) Takelman, (26) Timucuan, (27) Tsimshianic, (28) Utian, (29) Uto-Aztecan, (30) Wakashan, (31) Wintuan, (32) Yokutsan, and (33) Yukian.

There are as well some individual languages and language isolates identified, as follows: (34) Aranama, (35) Beothuk, (36) Calusa, (37) Coahuilteco, (38) Haida, (39) Karankawa, (40) Kootenai, (41) Natchez, (42) Seri, (43) Solano, (44) Tonkawa, (45) Tunica, (46) Washoe, (47) Yuchi, and (48) Zuni. (Adapted from Goddard 1996.)

the severity of language loss and the attendant difficulty of fully identifying cognates, sound correspondences, and the like. Nonetheless, based on the data at hand, traditional historical linguists sort North American languages (Figure 30) into about 33 separate families, and about as many isolates (for a hemisphere-wide total of approximately 170 families and isolates).

We are a long way from ascertaining precise historical relationships among them. Consider the Algic language family. It includes a couple dozen languages scattered from California (e.g., Yurok and Wiyot) to the Plains (Arapaho, Cheyenne), around the Great Lakes (Fox, Menominee), and along the eastern seaboard (Delaware, Micmac).[16] Though demonstrably related, there are considerable differences among these languages, bespeaking a common ancestor deep in the distant past. How deep is a matter of speculation. So too, Lyle Campbell adds, is the question of where the ancestral language was first spoken, and how or why the languages became so geographically far-flung.

It is assumed, and reasonably so, that additional and still deeper historical relationships exist between Algic and other American Indian languages and families. As Jane Hill says, "We can never prove that two languages are *not* related to each other."[17] However, the approximately thirty-three families currently identified are as far as traditional historical linguists care to go with the available data, which in many cases is the most data we'll ever have, given the continued, relentless loss of American languages.

Traditional historical linguistics yields robust groupings. Yet, robust as the method might be, obtaining those groupings is also extraordinarily labor-intensive and, worse, "discouragingly slow."[18] Others searched for an easier way.

SEEKING A SHORTCUT

Sapir, the godfather of the radical historical linguists, certainly thought there was; Joseph Greenberg is convinced he found it. Greenberg rejects the traditionalists' approach and results, and thinks North American language families are not nearly so difficult to group as traditionalists portray them. He uses a technique he dubs multilateral or *mass comparison,* which works not from the bottom up but from the top down, by simultaneously comparing a relatively few words across as many different languages as possible.

Doing so sacrifices linguistic details, but Greenberg thinks the cost is worth it. With so many American languages to examine, comparing only two or three at a time would make the process impossibly long. Besides, he argues, two distantly related languages may actually look different when compared only with each other, but their similarities would emerge when compared against many others simultaneously. As Greenberg explains, English and German, though related, have only vaguely similar words for *tooth* (*zahn* in German). Yet, alongside Dutch (also a Germanic language) the connections become more apparent *(tand),* as do their farther-reaching similarities to Italian *(dente),* Greek *(dhondi),* and the even more linguistically distant Lithuanian *(dant-is)* and Hindi *(dāːt).*[19] Hindi and English share an Indo-European ancestor, but perch on distant branches of the tree of its descendants.

Of course, just because many languages have a similar word for *tooth* does not a shared ancestor make. Any single match can easily result from borrowing, onomatopoeia, or chance. To show common origin, Greenberg argues, play the odds: Compare a list of

FIGURE 31.

Map of Joseph Greenberg's
three New World
language families, based
on his method of mass
comparison. The Eskimo-
Aleut and Na-Dene
families are relatively
uncontroversial; Amerind
is not. (Adapted from
Greenberg 1987.)

words. The more that are shared, the more likely they result from common ancestry and the closer their historical relationship. Family trees, even linguistic ones, are like that.

In applying his method of mass comparison to American languages, Greenberg checked all the languages against a word list of about 300 personal pronouns, body parts, and common natural objects (e.g., *I, you, mouth, tooth, leg, foot, hand, dead, blood, water*). He selected these because they ought to be enduring, non-culture-specific words. From that investigation, he concluded, as presented in his 1987 book, *Language in the Americas,* that all American languages fell into just three families: Eskimo-Aleut, Na-Dene, and one vast family called Amerind (Figure 31).

Eskimo-Aleut comprises ten languages, including Aleut, spoken on the Aleutian Islands; Yupik or western Eskimo, spoken on the Siberian and Alaskan coasts; and eastern Eskimo, Inuit-Inupiaq, spoken in the northern Arctic from Alaska to Greenland. These languages are striking in their homogeneity, Inuit-Inupiaq especially so: early European travelers learning a dialect at one end of the range were able, albeit with increasing difficulty, to converse with other Inuit-Inupiaq speakers as they made their way across their territory. The lack of more pronounced differences suggest all spoke a common language in the not-too-distant past, and dispersed across the far north from an Alaskan homeland relatively rapidly. The Eskimo-Aleut languages, Greenberg believes, show close affinities to Asian Pacific Rim languages such as Ainu and Chukchi-Kamchatkan.

The thirty-eight Na-Dene or Athabascan languages are today spoken by three widely separated peoples: the majority, Northern Athabascans (including Koyukon, Dogrib, Chipewyan, Sarcee) inhabit a large area of interior subarctic Alaska and Canada; Pacific Coast Athabascans live in the large river valleys of Oregon and northern California; and Apachean, including the Navajo and Apache, inhabit the southwestern United States (the Apachean apparently separated from Northern Athabascan in just the last millennium). According to Greenberg, Na-Dene languages have distant links with a few Old World language families, notably Sino-Tibetan and Yeniseian.[20]

Greenberg's Amerind family embraces the remaining approximately 900 indigenous languages of the New World, strikingly different tongues spoken across a vast region from Hudson's Bay to Tierra del Fuego by peoples surely separated by thousands of years of prehistory. Yet, by Greenberg's accounting, all share some words and a distinctive first person/second person pronoun pairing of *n/m* (in many Indo-European languages, the pattern is *m/t* as in me/thee, *moi/tu*, etc.). He believes the *n/m* pronoun pattern is so pervasive it should lead any "historically minded" linguist to conclude it marks a common ancestry.[21]

Profound differences between Eskimo-Aleut, Na-Dene, and Amerind convinced Greenberg these were distinct families in Asia before they migrated to America, each at different times. But in what order? The greater the differentiation within a language family, and the greater its geographic extent, Sapir argued, the older the family (the underlying assumption being that over time, language families become more heterogeneous). And the older the family, Greenberg added, the fewer the links to Old World languages and the farther south its location: they left Siberia earliest, and traveled the farthest.

By this reasoning, Amerind, the most heterogeneous and widespread, which lacks all but the slightest affinities with Old World languages, is the oldest. That Amerinds inhabit nearly the entire hemisphere and are far from Beringia suggests to Greenberg their ancestors arrived when no other humans were here to impede their travel. They were the first Americans.

That was apparently not the case with the Na-Dene or Eskimo-Aleut, who today occupy lands north of Amerind speakers, and whose southward progress was presumably blocked by them. Ancestral Na-Dene speakers, whose language family is less internally differentiated (relative to Amerind), and still has attenuated ties to a few Old World languages, must have followed the ancestral Amerinds. The ancestral Eskimo-Aleuts must have come last, since their languages are the least differentiated and have the most conspicuous links to Siberian and Asian languages.

Languages cannot be directly dated. But Greenberg hazarded a guess based on presumed rates of language change: the Eskimo-Aleut family was about 4,000 years old, Na-Dene about 9,000 years old, and Amerind greater than 11,000 years old. About as old as, say, Clovis.

HARSH WORDS

In the later decades of the twentieth century, Sapir-inspired lumping fell from grace in American linguistics, and hard-nosed, traditional historical linguists busily dismantled language families they believed had been sloppily pasted together. *Languages in the Americas* was bucking that trend, and Greenberg knew it, predicting 80%–90% of American linguists would receive his ideas with "something akin to outrage."[22] He was right. His claim that the vast majority of all New World languages were historically related within Amerind met with scorching criticism. The attack began even before *Languages in the Americas* appeared, when Lyle Campbell, one of the most outraged—and who was aware of Greenberg's forthcoming work—proclaimed it "should be shouted down."[23]

Campbell and other critics did not dispute Greenberg's goals: they, too, sought historical relationships among languages. They also believed many of the current American language families and isolates might be historically related. And they accepted the Eskimo-Aleut family (but then this grouping had long been uncontroversial). Beyond that, common ground vanished. Greenberg's critics absolutely rejected the Amerind grouping, and were at best lukewarm about Na-Dene. They accused Greenberg of going too far, too fast, using bad evidence and a method that Ringe denounced as "antiscientific" and "crap." Most critics privately shared this opinion; few were quite so publicly blunt about it.[24]

Cynics might dismiss the dispute as no more than an academic border war: Greenberg had made his reputation studying African languages, and only later came to work on American ones. Naturally, those who worked in the Americas would resent his presence. Maybe. But Greenberg's critics—experts in American languages—had legitimate complaints. They readily showed he had badly mishandled the American word lists, failing to recognize borrowed words (including Spanish ones), attributing words to one language (e.g., Quapaw) that actually came from another unrelated one (Biloxi), and including so many spurious words that one reviewer declared Greenberg's word matches were wrong more often than they were right. Nor were these harmless mistakes, since his criteria for spotting similarities between languages (never consistent, the critics snarled) sometimes involved a single consonant, making it all too easy for chance resemblances "masquerading as historical evidence" to occur.[25]

In tackling all the languages of America, Greenberg admitted there would be mistakes: "It would be miraculous," he said, if there weren't.[26] But he also insisted errors were inconsequential. Comparing a few words broadly across hundreds of languages, he argued, reduced the statistical likelihood of having a match by chance alone. Yet, as Ringe and others quickly countered, the situation is just the opposite: Greenberg's method *compounds* the problem by multiplying opportunities for chance resemblances (since there are a limited number of consonant possibilities appearing in a vast number of languages).

Errors in the word lists were a serious problem, given how little attention Greenberg paid to other aspects of language such as grammar, sound, and meaning. Greenberg insisted he was not neglecting those elements, pointing to the key role the "enormously widespread" *n/m* pattern played in grouping Amerind. That defense gained no traction either: the frequency of *n/m* in America, critics groused, was "grossly overstated" (oddly, neither side actually tallied *n/m* frequencies, nor could agree on whether counts at the level of languages or language families mattered more). More telling, critics argued this supposedly unique American form occurs elsewhere, while the ostensibly non-American pronoun pattern (*m/t*) occurs here.[27]

The indifference Greenberg showed to testing his language families by establishing sound correspondences, at times belittling their importance (even Sapir never did that), made his critics all the more dismissive of his results and, indeed, of mass comparison itself. In reply, Greenberg and his supporters played the Africa card: mass comparison worked there —despite, they added, initial hostility—without sound correspondences, etymologies, and reconstructions of proto-languages. Why shouldn't it ultimately triumph in America as well? The "African Fallacy," Campbell shot back: not only were Greenberg's African language families still in dispute, but "posturing for the American Indian classification on the basis of an African classification is irrelevant."[28]

Greenberg wanted *Languages in the Americas* judged as a whole. His critics did, and rejected it wholeheartedly. Realizing he would never convince his fellow linguists, Greenberg branded them timid souls unwilling to seek bold, broad patterns, prejudiced against his efforts to do so, and in the unkindest cut, not even well acquainted with their own data.

"Megalocomparison," James Matisoff dubbed Greenberg's method, contrasting it with the more traditional microcomparison of languages. Those who practiced microcomparison Matisoff called "micromaniacs." He didn't need to fill in the blank about who he thought practiced megalocomparison. He was right. Greenberg had no need of linguists' approval. Not when he saw vindication in an "almost exact match" of his language families with independently grouped teeth and genes, which he proclaimed "virtually confirms the validity of the Amerind family" and his three-migration model.[29]

TALES FROM TEETH

In the 1920s, just as Sapir was reducing all American languages to six families, Hrdlička was identifying physical traits common to all American Indians. One of the most significant was a distinctive "shovel shape" to the upper incisor (where extra ridges on the tongue side give the crown a shovel look). This inherited trait satisfied Hrdlička that all Native

Americans shared a common ancestor. That was important to him because their physical unity meant they hadn't been here long enough for evolutionary divergence to occur, proof that his staunch opposition to their Pleistocene arrival was correct. Several decades later, Christy Turner's extensive analysis of teeth both fortified and falsified that conclusion.

Why study teeth? They are hard, dense parts of the skeleton, more resistant to weathering, and not an especially attractive target for carnivores or scavengers, or easily broken down by insects, fungi, or other agents. As a result, they invariably outlast the body's soft parts and often bones as well. Although all humans share the same general dentition (thirty-two adult teeth: incisors, canines, premolars, molars), we differ in secondary dental traits, such as tooth shape and size, the number of roots, crown geometry, and the like. Importantly, Turner notes, tooth anatomy is inherited and little modified by environment, use, health, or diet. If two groups have the same dental traits, he argues, they are almost certainly related to one another. Even better, because tooth structure changes slowly, it is possible to trace a dental lineage across space and through time. And teeth can be directly radiocarbon dated. We may never know what language prehistoric Americans spoke, but we can examine their teeth and group them by their similarities. *Ethnic odontology* Turner calls it.

Turner examined over 200,000 teeth from more than 9,000 individuals in sites throughout the New World, and for comparison across many areas of the Old World, especially China, Siberia, and other regions of eastern Asia. He tallied the occurrence of twenty-eight crown and root traits, such as the occurrence of shoveling or winging of the upper incisors, the presence of Caribelli's trait or hypocone on the upper molars, groove and cusp patterns on lower molars, and the number of roots (either two or three) on some premolars and molars (some examples are shown in Table 5).

These are traits that generally occur in all peoples, though in varying frequency in different populations. Incisor shoveling, as Hrdlička first realized, is common among American Indians but rare elsewhere. On the other hand, a Y groove on the lower first molar occurs in nearly the same percentage (about 10%–20%) in populations around the globe. Hence, doing ethnic odontology is a statistical exercise in ascertaining how one sample of teeth differs from another based on a constellation of traits.

Using such analyses, Turner distinguished Asian teeth from those of Africa and Europe, then broadly divided the Asian sample into two groups: *Sundadonts* and *Sinodonts*. He separates the two based on the higher incidence of shoveling and three-rooted molars in Sinodonts, but the division is not altogether a clean one, since on certain traits (such as winging on the upper incisors), Sinodonts and Sundadonts are indistinguishable. One can be hard-pressed to assign jaw fragments or individual teeth to the correct group if all that remain are teeth with traits common to many.

Turner suggests the Sundadont pattern, widespread in Southeast Asia, is the older of the two and the root from which Sinodonty, characteristic of northern Asian and *all* American populations, evolved. Sundadonts, he believes, "had nothing to do with the peopling of the Americas."[30]

TABLE 5 Dental morphology of Native Americans and other groups

Sample group	Winging UI 1	Shoveling UI 1	Hypocone UM 2	Carabelli UM 1	Y groove LM1	3-rooted LM1
Europe	4.2	0.1	75.9	58.3	19.2	0.9
SE Asia	23.8	50.6	92.0	43.9	17.8	10.8
NE Asia	23.4	80.7	87.6	27.8	10.3	30.1
Eskimo	8.3	78.1	82.7	14.3	20.2	19.8
Aleut	36.5	71.8	58.4	6.1	19.8	42.2
Na-Dene/Athapaskan	40.9	75.0	68.2	22.6	9.8	10.4
Eastern U.S. & Canada	48.7	91.5	93.2	33.8	11.0	7.0
California	42.6	97.7	92.9	44.9	12.0	8.2
Southwest	45.1	90.8	82.5	33.0	9.5	6.0
Mesoamerica	66.2	93.5	83.9	47.5	13.2	5.7
South America	55.4	92.5	89.6	41.9	7.7	4.9
Paleoindian	25.0	83.3	100.0	100.0	28.6	25.0

NOTE: From Turner 1985: Table 1 and Turner 1986: Table 1. Values are expressed as percentage within the sample; sample sizes vary.

Turner puts the earliest Asian appearance of Sinodonty in the jaws of early modern humans from the Upper Cave site at Zhoukoudian, China, although the ages of these specimens are uncertain (the dates on them range from 10,000 to 29,000 BP[31]). In America, the Sinodont pattern appears in teeth from the Cerro Sota and Palli Aike caves of southernmost Chile, dated to 9000–10,000 BP. As might be expected given the time elapsed since Northeast Asian and American Sinodonts diverged, Turner sees dental differences between the two. American Sinodonts have, for example, greater frequencies of Carabelli's trait, shoveling and double shoveling, and lower frequencies of three-rooted lower first molars and Y-groove patterns than their Northeast Asian counterparts (let's ignore for the moment that many of the trait frequencies for the populations in Table 5 are at odds with these general statements).

Nor are all American Sinodonts alike. Turner subdivided them into three groups: Eskimo-Aleut, Greater Northwest Coast (which includes southwest Athabascan speakers), and all other American Indians. Eskimo-Aleut and American Indians are least like each other, the first having a lower incidence of shoveling and Carabelli's trait, but a higher frequency of three-rooted molars. Between these extremes the Greater Northwest Coast cluster perches, its dental traits betwixt the other two, yet showing less internal differentiation than either. Turner interprets that as evidence the ancestors of this group were present in Siberia, and were not more recently hybridized between Eskimo-Aleuts and American Indians.

Like Greenberg, Turner explained his three groups as the result of three separate migrations from the Old World (he even uses Greenberg's group labels). All three are traceable back to the same northern Chinese late Pleistocene Sinodonts, but none have distributions that correspond to "natural areas" of the Americas, and so Turner believed each must have formed separately in China or Siberia sometime after the first appearance of Sinodonty, and then gone their separate ways, arriving in the New World at different times via different routes. He supposes ancestral Amerinds drifted initially to the west then north, through the Lena River basin and then along the Arctic shelf across northern Beringia. The Eskimo-Aleuts went east and north, down the Amur River to its mouth, then sidled over to Hokkaido, up the Sea of Okhotsk, and then skipped across the Aleutian Islands. The first Na-Dene traveled the gap between the two. However, no teeth have been found along these routes.

Turner believes the Amerind dental group was the first to reach the New World (manifest archaeologically as Clovis). By his dental reckoning, Eskimo-Aleuts came second, and the Na-Dene were a distant third. That relative order differs, of course, from the one Greenberg arrived at based on languages, as they candidly admitted in their 1986 announcement of the convergence of their views.

Their estimates of an absolute chronology for the migrations naturally differed as well. For want of sufficient radiocarbon dated teeth, Turner invented a quasi-absolute dating technique he called *dentochronology* (not to be confused with *dendro*chronology, or tree-ring dating). On the assumption teeth change at a constant rate, dental differences between

TABLE 6 Mean Measure of Divergence (MMD) between sample groups

Sample group	Europe	SE Asia	NE Asia	Eskimo	Aleut	Na-Dene	Eastern U.S. & Canada	California	U.S. SW	Meso-america	South America
Europe	.000										
SE Asia	.361	.000									
NE Asia	.789	.135	.000								
Eskimo	.831	.281	.082	.000							
Aleut	.882	.293	.085	.034	.000						
Na-Dene	.754	.184	.040	.042	.029	.000					
Eastern U.S./CAN	.967	.252	.128	.203	.219	.083	.000				
California	1.11	.336	.160	.236	.271	.136	.013	.000			
Southwest	.944	.248	.093	.152	.150	.021	.036	.052	.000		
Mesoamerica	1.147	.400	.214	.240	.255	.128	.039	.029	.072	.000	
South America	1.108	.361	.167	.212	.236	.107	.034	.026	.051	.008	.000
Paleoindian	.836	.293	.204	.237	.364	.265	.206	.120	.218	.177	.193

NOTE: Data from Turner 1986: Table 2. Significance values not provided in original, but assume differences are significant if MMD > ~.040. Smaller values indicate more similar groups.

groups—quantified using the mean measure of divergence (MMD) statistic—become a measure of the time elapsed since the groups split from one another (see Table 6).

To calibrate the rate of change, Turner calculated the MMD between groups whose divergence times were independently dated, then divided the MMD by the total elapsed time, to derive and average global dental evolution rate of 0.01003 ± .004 MMD per 1,000 years.[32] Multiplying that rate by the MMD between various Amerind and Northeast Asian populations suggested to Turner they split about 14,000 years ago, a time he thought nicely corresponded to when Clovis ancestors may have first emerged in Siberia.[33] Calculations for Eskimo-Aleut and Northeast Asian populations yielded an age at divergence of approximately 11,000 years ago. The numbers for Na-Dene were more elusive, but nearly all postdated the Eskimo divergence. Altogether, not too bad a fit with the archaeological evidence, Turner supposed. Given the vagaries of estimating antiquity from language, Greenberg deemed it not a bad fit with the linguistic evidence either.

CAVITIES

Indeed, much was made of the "independent" confirmation the teeth provided of Greenberg's language groups. It was "strong presumption," Merritt Ruhlen argued, that Greenberg's groupings and the three-migration model were correct.[34] But as Greenberg, Turner, and Zegura candidly admitted, when an investigator in one field is aware of the conclusions proposed in another, "he or she may be influenced by this knowledge in developing a theory." Thus, the "ultimate test"—as they saw it—was whether the data from the field itself could justify the conclusions drawn.[35] Did it in this case? And just how separate were the two lines of evidence?

Ideally, analysis of the teeth ought to proceed independently of any preconceptions about the language group to which they might belong. Turner's analysis certainly began there, but soon thereafter, he spoke with Greenberg about his linguistic groupings. After that, Turner explicitly linked his dental clusters with Greenberg's language families, and the change was in more than name only: for analytical purposes, he pooled his tooth samples into "regional sets" by dental similarity and by their "known or presumed linguistic affiliation."[36]

Now, teeth may chatter, but they don't talk. Not by themselves, anyway. How, then, can dental groups be assigned to "common culture and closely related languages"?[37] Turner did so by where the teeth were found: a sample of teeth from sites in regions *now* occupied by Eskimos or Aleuts are assigned to the Eskimo or Aleut group. It can be as simple as that, and can cause trouble just as easily. For doing so requires seeing what Goldilocks saw: that each group, like the Three Bears, has its own—and only its own—language, which co-occurs over time with a particular set of genes, artifacts, and teeth in a territory that does not change hands (or teeth).

That might be true at certain times and places. Teeth and artifacts from a 900-year-old site on the northwestern shore of Hudson Bay, for example, are likely those of ancestral

Eskimo speakers: there is good archaeological reason to think no one else was there in later prehistoric times. But 500 years earlier in the same place, the archaeological record looks very different. Whose teeth are those? What language did they help voice? The Goldilocks Standard, if it might be called that, is harder to achieve as one goes deeper into the past, and also in places that have long trafficked in human beings—like Alaska, the corridor for all American migrations (however many there were). What language was associated with 5,000-year-old teeth found there? Can one presume a linguistic affiliation of undated teeth? Not surprisingly, teeth from Alaska and the far north often stubbornly resist falling neatly into corresponding language groups.

The dilemma is obvious: places and times where we might hope to come onto the dental trail of the first peoples are the very times and places where we have the least chance of figuring out who those people were or what language they spoke. Of course, if Turner found three dental groups in Alaska, one dated to 12,000 years ago, the other never occurring before 10,000 years ago, and the third always postdating 4,500 years ago, it would be powerful evidence of three migrations.

But that's not happened. Nearly all the teeth in Turner's sample are less than a couple thousand years old, and their young age, tied to the undemonstrated (if not false) hope that inhabitants haven't changed over that time, makes it hard to do just that. To be sure, Turner's sample includes all known Paleoindian and Early Archaic specimens, but that amounts to teeth from no more than twenty sites, few of which are actually dated to the Pleistocene.

The consequences of this complication are plain enough to see in MMD values: teeth that ought to cluster together are dissimilar, and those from unrelated groups overlap. For example, Aleut teeth, which ought to be most similar to Eskimo teeth by their linguistic affiliation, are statistically more like the Na-Dene teeth. For that matter, some Amerind groups are dentally closer to Na-Dene than to other Amerinds. These complications spill over into the dentochronological estimates, which have Eskimo-Aleuts departing Northeast Asia 8,000 years ago and Na-Dene groups 4,000 years ago, and yet by Turner's reckoning, the Na-Dene arrival is marked by the Paleoarctic archaeological tradition, which is radiocarbon dated to 10,000 BP (thus predating both the dental divergence age and the earliest Eskimo-Aleut sites) (see Table 7).

Likewise, the time separating Sinodonts from Sundadonts (a dentochronological date of about 13,460 years ago) ought to be much greater than that separating Asian from American Sinodonts (which by dentochronology occurred about 20,400 years ago, making it curious, too, that Turner insists there is nothing in the dental evidence to suggest "entry much before Clovis"). Coincidentally, a recent study of eastern Asian teeth failed to support the validity of the Sinodont and Sundadont groups.[38]

In light of the messy overlap of the dental groups, physical anthropologist and geneticist Emoke Szathmary wondered whether Turner was forcing his results and interpretation to fit Greenberg's language families and three-migration model. A harsh accusation, but one easily put to the test by seeing how the teeth cluster when examined

TABLE 7 Elapsed Time from Divergence with Northeast Asian Sinodonts

	MMD from NE Asia	Elapsed time from NE Asia split
		(dentochronological years)
Eskimo	0.082	8,175
Aleut	0.085	8,475
Na-Dene	0.040	3,988
Eastern U.S./CAN	0.128	12,762
California	0.16	15,952
Southwest	0.093	9,272
Mesoamerica	0.214	21,336
South America	0.167	16,650
Paleoindian	0.204	20,339

NOTE: Based on Mean Measure of Divergence values (from Table 6) and the global dentochronological rate (.01003 ± 004).

strictly by their dental attributes, without presuming what language they once voiced. Unfortunately, Turner has never done such a test nor published his raw data, making it impossible for others to do so.

How, then does the genetic evidence fit?

A COMPLEX INHERITANCE

When Zegura joined forces with Greenberg and Turner, he knew the genetic evidence was not as well developed. The basics were certainly known: packaged within the nucleus of cells are chromosomes, of which there are twenty-three pairs per cell. Within each pair are long strands of deoxyribonucleic acid (DNA) made of a sugar molecule (deoxyribose or D) attached to one of four chemical bases: guanine (G), cytosine (C), adenine (A), and thymine (T). These, in turn, are linked to phosphoric acid (P), the combination forming a nucleotide. A DNA molecule is a very long sequence of paired nucleotides joining A to T, and C to G in a double-stranded helical structure. There are approximately 100 million nucleotide pairs (or *base pairs*) per chromosome, and some 3 billion base pairs in the human genome. Segments along either DNA strand with specific functions (not all have them) are genes, and together these are responsible for the biology of an individual.

The nucleus at fertilization joins twenty-three chromosomes from the mother and father: twenty-two of these chromosome pairs make up the autosomes, and the twenty-third pair are the sex chromosomes (either XX [female] or XY [male]). The vast majority of the DNA in the genome is inherited from both parents, joined in a process much like the shuffling of cards, making the offspring about a 50/50 mix of parental DNA. Since autosomes undergo recombination each generation, autosomal DNA is a highly complex skein derived from a geometrically expanding pool of ancestors (two parents, four grandparents, eight great-grandparents, and so on). Go back more than a few generations, and it becomes impossible to tell which specific ancestor contributed a particular piece of DNA to an individual, at least not without comparing the DNA of all.[39]

However, one can spot genetic differences between groups by measuring the frequency with which certain genes and mutations occur in populations that have been isolated for periods of time. The study of genetic distance between human populations began with serological markers—blood group antigens (A, B, and Rh blood groups), serum proteins, immunoglobulin (antibody molecules), and various enzymes. These were the classical genetic markers.

Not all New World populations are genetically alike, but at the time Zegura was working with Greenberg and Turner, he faced a challenge in showing genetic differences among New World peoples. The classical markers then available were known to vary on the scale of continents: useful for separating, say, African, American, Asian, and Europeans.[40] They were stubbornly irresolute, however, on whether there were identifiable genetic groups within Native Americans, let alone which matched Greenberg's Amerind, Na-Dene, and Eskimo-Aleut. The closest Zegura came were patterns in immunoglobulin Gm allotypes, but only because their distribution was "analyzed with respect to the three-migration hypothesis."[41] Had he not been looking for those three groups, he might not have seen them.

The genetic evidence simply didn't match up with the languages and teeth, as Zegura admitted. Mostly, it produced a picture of "discordant variation," the result of thousands of years of gene exchange, recombination, random flutters in gene frequencies, and mutation, all conspiring with natural selection to obscure what once may have been discernable genetic genealogies. Unraveling millennia of history from classical genetic markers was going to be a "challenging and highly speculative enterprise"—or so it appeared in 1986 (it is still mostly that way, though recently inroads have been made using genome-wide autosomal markers to detect signals of the peopling process[42]). And so Greenberg, Turner, and Zegura relegated the genetic evidence to secondary status, unable to confirm their tripartite migration model.

Ironically, within a decade, rapid advances in genetics in two separate corners of the human genome made it possible to reject that model and replace it with a very different picture of the peopling of the Americas. When that happened, Zegura "fully opted out" of the tripartite model, much to the dismay of his co-authors.[43]

ANCESTRAL GENES

Those advances occurred when geneticists shifted their attention away from classical markers and began looking at DNA in mitochondria, the thousands of small, bean-shaped organelles in the cell responsible for the cell's energy metabolism (think of them as cell batteries). Mitochondrial DNA (mtDNA) and nuclear DNA fundamentally differ: both occur in males and females, but through a strange quirk of nature, the mtDNA from the sperm is discarded when an egg is fertilized, and hence mtDNA is inherited strictly from mother to child (male or female). The bane of human genetics—the recombination of genes in mating—does not occur in the passage of mtDNA down the maternal line. Conveniently enough, geneticists peering into another corner of the genome soon realized that uni-parental inheritance also took place down the paternal line, via the non-recombining portion of the Y chromosome (or NRY), which is inherited directly from father to son to son (Figure 32).[44]

With mtDNA and NRY, it is possible to identify the female and male descendants (respectively) of a particular lineage, for they will directly inherit the markers possessed by their common ancestor. But what makes these two inheritance pathways even more valuable for tracing ancestry is the fact that the DNA undergoes *mutations,* or changes over time.

At reproduction the nucleotides in DNA are copied almost—but not entirely—without error, which is quite astonishing, given that about 3 billion must be copied and in the same order in which they are found on the original chromosomes. Although this takes place with amazing accuracy each cell generation, mistakes or mutations occasionally occur.[45] These might include the replacement of one nucleotide by another (C→T, called a single nucleotide polymorphism, or SNP), or perhaps involve the insertion or deletion of a base pair(s) at a particular site. These become variants *(alleles)* of those nucleotides.

Although we tend to think of mutations as trouble, particularly the ones that directly impact the function of a particular gene, many are trivial and indeed neutral with respect to our genes or our phenotype (outward appearance) and are not subject to the forces of natural selection. These neutral traits are sometimes uncharitably referred to as "junk DNA" because they do not encode proteins or regulatory information. Each of us carries about 100 new neutral mutations our parents did not have.

These mutations in our junk DNA can be remarkably useful for tracing population history, since similarities among them can help trace common ancestry. Put simply, individuals or groups that share mtDNA and NRY markers are related; the more they share, the closer they are related, and the closer in time their common ancestor. Consider a very simplified example (Figure 33): the mtDNA molecule has 16,569 nucleotide base pairs. Let's say a SNP occurs in one of the women of a population at nucleotide pair *(np)* 16327 in the control region of the mtDNA molecule. Her several daughters inherit that marker. In one of them, another SNP occurs at *np* 16325. This woman and all her daughters (but neither her sisters nor their children) now have the same markers at *np* 16327 and 16325. In a subsequent generation, another mutation occurs, this time in the coding region at *np* 493, so all members of this lineage now have identical markers at *np* 16327 and 16325 and 493. Through the generations, the SNP list will grow, including all the mutations that occurred previously—unless a new one happens to occur at the same

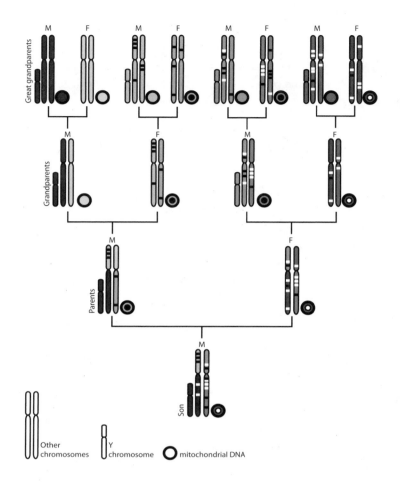

FIGURE 32.

Schematic illustration of the inheritance of mitochondrial, Y chromosome, and autosomal DNA through four generations of males (M) and females (F). The DNA on the twenty-two pairs of autosomes, represented by the two tall bars, combines each generation, and by the fourth generation, the son's autosomes are a complex mix of maternal and paternal DNA from all eight great grandparents (as illustrated by the different patterns). The Y chromosome (the short bar) and the mtDNA (the circle) are inherited uni-parentally, so each son carries the Y chromosome of his father, and each child the mtDNA of his or her mother—but only the daughter passes along mtDNA to the next generation. Hence, the son in the fourth generation traces his Y chromosome solely to his paternal great grandfather, and his mtDNA solely to his maternal great grandmother. (Adapted from Hammer and Zegura 1996.)

site as a previous one, erasing an ancestral marker. Meanwhile, in the other branch of the family, the marker at *np* 16327 is also being passed down, but in those other lines of descent, mutations are also taking place, say at *np* 16356. All branches carry the original mutation at *np* 16327, but each will have added its own shared by daughters down those lineages, but not in other more distantly related lines.[46]

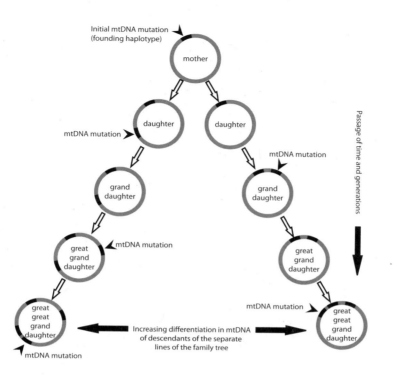

FIGURE 33.

Schematic (and highly simplified) illustration of the descent with mutation of mtDNA in two maternal lineages. The mutations are not as rapid as depicted here, but the schematic is intended merely to convey the larger point that over time, mutations occur in the separate lines of maternal descendants, such that subsequent generations along those separate lines grow increasingly different from one another in their mtDNA. If both lines are descendants of a common founding lineage, they will share a set of markers in their mtDNA (unless later mutations have changed those as well). The analogous process also takes place down the paternal line.

A brief vocabulary time out: all individuals who share an ancestral nucleotide signature (*np* 16327 in the example) are part of the same *haplogroup*, or genetic lineage, while the more specific markers within the branches of the haplogroup (*np* 16325 and *np* 16356) are known as *haplotypes*. Though one doesn't want to put too fine a point on the analogy, think of haplogroups as language families and haplotypes as specific languages.[47] A haplogroup is considered a *founding lineage* if it occurs in both Asian and Siberian populations, but is also widespread in America and one from which New World–specific haplotypes have diverged.[48]

As one might surmise, mtDNA and NRY founding lineages may not trace back to the same actual population, since male and female movements can differ significantly. In some cultures, males travel long distances to find geographically far-flung mates; in others, it is females who leave behind the group into which they were born.[49] By all indications, the first Americans were highly mobile, but it remains uncertain whether, once their descendants began to settle down, differences in male and female migration patterns emerged. Given the adaptive challenges of being in a vast new landscape, there's reason to think differences might emerge (Chapter 7). But whether these, or other gender-specific trends (higher male mortality, or more widespread polygyny) have confused the resulting picture of mtDNA and NRY diversity, is yet unknown.

CLOCKING THE FIRST AMERICANS

Changes in mtDNA and NRY accumulate rapidly (though not quite as rapidly as in the example above); by one estimate, mtDNA mutation rates are 6–17 times faster than those in nuclear DNA. Thus, even after the roughly 400 generations people have been in the Americas, we can expect significant differences to have accumulated between American Indians and their Northeast Asian relatives, as well as between American Indian lineages that went their separate ways once here.

In principle, mutations in mtDNA and the NRY can serve as a molecular clock timing the splitting of genetic lineages. This is so because mutations also occur at a relatively steady rate: about 2.81–6.62 mutations per 10,000 years on the mtDNA side (depending on where in the mtDNA molecule those changes are taking place), and over on the NRY, a slower 1.45 changes or so per 10,000 years (that's the SNP mutation rate; other mutations on the NRY, as noted below, are faster). By measuring the sequence diversity of mtDNA (or NRY) in a population, and knowing the mutation rate at which those changes accumulated, we can in principle convert genetic distance into the time necessary to have generated that level of diversity (that's one method; another estimates time in a different manner).[50]

But molecular clocks must be used with caution, for they can run fast or slow owing to variation in mutation rates (which differ by mutation type and region in the mtDNA and NRY molecules), depending on whether mutations occur uniformly over time (for clock purposes, it is assumed they do), whether the mutations get "fixed" with the population or disappear (what geneticists refer to as the difference between mutation and substitution rates), how reliably the mutation rate has been calibrated,[51] and whether the number of founding haplotypes within a haplogroup is known. A bit of elaboration on this last point: if all the diversity within a haplogroup accumulated in a single haplotype, then mutations must have occurred in a more or less serial fashion one after another, and that will yield an older age for the haplogroup. But if the diversity within a haplogroup is the result of mutations within multiple haplotypes, those could have arisen concurrently, and the molecular age for the haplogroup will be correspondingly younger. Without knowing the number of haplotypes in a founding haplogroup, one cannot know whether the molecular clock is telling the right time. Geneticists are aware of these problems and are rightly wary of ages based solely on molecular clocks.[52] We should be, too.

Those matters aside, what time is told by the ticking of the molecular clock? Unfortunately, it is not necessarily the moment when a group split from their Northeast Asian relatives and headed for America—though it could be. Rather, the clocks measure the time elapsed since two or more molecular lineages began to diverge. Molecular divergence can occur (and often does occur) in a population long before a part of them goes off to discover a new homeland, in which case the molecular age says nothing about when that divergence occurred. But if those genetic lineage and population splits coincide, genes become a handy clock indeed.

MITOCHONDRIAL MEMENTOS

All the nucleotide positions on the mtDNA molecule have been mapped, and it is known where SNPs are likely to occur. Accordingly, geneticists seeking clues to mtDNA history pay particular attention to the *coding* region, which directs the biochemical processes of the mtDNA (and which comprises approximately 94% of the mtDNA and contains thirty-seven genes), and especially to the smaller *control* region (CR), where mtDNA is initiated and regulated, and where changes occur at a faster rate (though mutations here also have the unfortunate tendency to be recurrent, and thus can partially obscure earlier, founding markers).[53]

In the mid-1980s, Douglas Wallace began looking at mtDNA in Arizona's Pima peoples. Their genes bespoke an ancestry of only a few mtDNA lineages. Later, casting the sampling net wider on the chance that the small number of haplogroups he identified were Pima specific, Wallace and his team examined mtDNA from the Yucatan Maya and South American Ticuna. To his surprise, these geographically separate groups showed high frequencies, but not the same high frequencies, of the same marker (which became haplogroup C), again bespeaking common ancestry. That all three were in Greenberg's Amerind group seemed no coincidence. And something else: this marker occurred among Native Americans in frequencies twenty times higher than in Asians. Here was the first indication that the ancestral population of these widely separate groups had been very small, likely having passed through a *population bottleneck,* as geneticists call it, and left Asia with only a tiny fraction of the genetic markers present in the parent population (this is the geneticists' *founder effect,* which recognizes that a new population founded by a small number of individuals necessarily represents only a fraction of the diversity of the source population).[54] Once here, those markers naturally came to dominate in the descendant populations.

Within a few years of Wallace's study, three more haplogroups were identified among American Indians (haplogroups A, B, and D). Together, these were found in 98% of all Native Americans.[55] That's a large percentage, but the absolute number of individuals who have participated in such studies was small, and small samples (statistically speaking) are more likely to encounter the most common haplogroups first, and only later detect the rarer ones (a complicating factor to which we shall return).

These four haplogroups vary in their frequency, and coincidentally do so in alphabetical order: haplogroups A, B, C, and D are present in (respectively) 40%, 32%, 17%, and 7% of American Indians.[56] Because they all occur widely (if unevenly) among Native Americans, they are thought to be founding lineages. Originally, geneticists sorted their results by Greenberg's language families (geneticist Antonio Torroni admits it was hard to avoid using that scheme[57]), and doing so revealed certain patterns. All four haplogroups occurred in peoples identified by Greenberg as Amerind. Yet, Na-Dene speakers generally harbored only haplogroup A; it appeared they shared a deep Siberian ancestor with Amerinds, but because haplogroup A showed less internal variation in the Na-Dene, it seemed they split off from their common Siberian ancestor sometime after the Amerind split. Haplogroup B was virtually absent from both Na-Dene and Eskimo-Aleuts, and was not as old as haplogroups A, C, and D, suggesting it, too, was derived from a separate migration, but one sandwiched between an early (haplogroups A [in Amerinds], C, and D) and a later migration (dominated by haplogroup A [in Na-Dene]). For a time, it seemed the three-migration model might just be saved from scientific purgatory and Greenberg's language families vindicated.[58]

The moment didn't last. Once additional mtDNA samples were analyzed, Na-Dene and Eskimo-Aleuts proved to have all four haplogroups. Andrew Merriwether and others then began lobbying for a simpler scenario: a single migration to the Americas. The rationale was straightforward: given there were only four founding haplogroups found in all Native Americans, their ancestors must have arrived as part of the same migration. It seemed unlikely that different groups of people leaving Asia at different times from different places would all be members of the same genetic club.[59]

Most geneticists accepted the idea of a single migration, even after they realized there weren't just four mtDNA haplogroups. In the late 1990s, a fifth, more-minor haplogroup (X) was spotted in very low frequencies (about 3%) among Native populations in the northeastern United States. Because X was known to occur in Europeans, its appearance among American Indians was initially written off as a case of post-1492 admixture. But then a closer look revealed X was scattered across the Americas (even in pre–European contact burials from North and perhaps South America), never appeared alongside other European markers, and was sufficiently distinct from European X their shared ancestor must be deep in time. Clearly, X was a founding lineage that came to America. But from where?[60]

Haplogroup X was rare in Siberia and eastern Asia, and its presence in western Asia and Europe prompted a team of geneticists to make the breathtaking speculation that *Pleistocene* Europeans had crossed the Atlantic to colonize the Americas. But almost immediately the geneticists recanted, after X also turned up among the Siberian Altai and proved to be an extremely old, possibly central Asian lineage on a different branch of the haplogroup X tree, whose descendants went both west (to Europe) and east (to America). It was too late: by then several archaeologists and skeletal biologists had jumped on the bandwagon and were trumpeting the idea Europeans colonized the Americas (more on that below, and Chapter 6).[61]

Five founding lineages,[62] all from a single migration. Curiously, despite denying the three-migration model, geneticists continued to use Greenberg's three language families to classify their samples—testimony to the time it takes for knowledge from one discipline to fully penetrate another. And even though it was noticed by the late 1990s by Tatiana Karafet and colleagues that the linguistic-genetic fit was problematic,[63] it was not until 2004 that robust statistical tests were conducted to assess whether genetic haplogroups actually coincided with Amerind, Na-Dene, and Eskimo-Aleut.[64] They didn't. In fact, they missed badly, not just among groups assigned to Amerind, but even with Eskimo-Aleut, the one language family on which Greenberg and his linguist critics actually agreed. Aleuts did not cluster genetically with Eskimos, but instead were closer to several of Greenberg's Amerind tribes, as well as to the Navajo, a Na-Dene group.

The mismatch between languages and genes testified there had been, as geneticists Keith Hunley and Jeffrey Long put it, a long "history of pervasive genetic exchange across linguistic boundaries."[65] Or perhaps language sharing across genetic boundaries. Either way, the last hope for independent support of Greenberg's language families vanished. Lyle Campbell and Ives Goddard, Greenberg's longtime adversaries, took a victory lap: joining a pair of geneticists, they pronounced Amerind, abandoned in linguistics, could no longer be used in genetic analysis either.[66]

Although the distribution of the major mtDNA haplogroups did not neatly align with languages, they were not randomly scattered among tribes across the hemisphere either (see Table 8)—and that is a good thing, for it indicates the subsequent millennia of gene flow and genetic drift have not obscured the molecular signal. Haplogroup A is especially frequent in Arctic populations, common among Pacific Coast groups but otherwise rare in western North America, present in eastern North America and Central America, and infrequent throughout much of South America (and there are hints the northern groups possess a different, possibly later haplotype of A). Haplogroup B, on the other hand, is virtually absent in the far north (Siberia, too), though it, too, increases in abundance in more southern regions, especially in the American Southwest and Central America, only to diminish altogether in far South America. Haplogroup C is particularly abundant in South American populations, less so in Central and North American groups. The distribution of haplogroup D is a mix: common in far northern populations (especially among the Aleut), present but rare in North American populations, and quite frequent among South American groups (like C). Finally, haplogroup X is widely scattered but in very low frequencies across North America.[67]

What might these broad patterns reveal? One possibility: the original colonizers carried the full complement of founding lineages, but as they dispersed found themselves isolated for long periods of time. Strongly affected by *genetic drift*—random changes in gene frequency, which are often pronounced in small populations—certain markers diminished in frequency or were lost altogether. Alternatively, there was more than one migratory pulse, and groups with especially distinctive haplogroup and haplotype profiles (like many Eskimo-Aleuts, for example), represent separate migrations.[68]

Molecular ages may shed light on the matter (if one haplogroup, say, was much younger than the others). Early molecular ages ranged from 20,000 to 41,000 years for haplogroups A, C, D, and X, and from 6,000 to 12,000 years for haplogroup B.[69] These provided the first molecular indication ancestral American Indians left Siberia far earlier than the generally accepted Paleoindian ages, and hinted that haplogroup B might represent a later migration.

In subsequent years, the molecular clock seemed to spin wildly around the dial, with estimated ages growing younger in some haplogroups, older in others, and in some in both directions simultaneously. The spinning has since slowed, and in general the molecular ages are younger, but there is still considerable scatter in them (and in the accompanying error estimates), the result, Jason Eshelman and colleagues explained, of varying haplotypes used to estimate diversity, and the methods employed to calculate molecular divergence.

Change may be on the horizon, however: recent work by Erika Tamm and colleagues examining whole mtDNA sequences for haplotypes within haplogroups A through D put their ages generally younger than 20,000 BP, with haplogroup B now aligning in age with the others (indeed, it's now estimated to be slighter *older*).[70] This makes good sense, since haplogroup B is present in coprolite samples from Paisley Cave alongside haplogroup A (Chapter 4). Still, it remains uncertain just when the mtDNA clock started ticking relative to the population dispersal of the first Americans, and (naturally) whether that occurred at the same time for all haplogroups.[71]

As to *where* they originated, haplogroups A, C, and D—or at least haplotypes within them—are present among eastern Siberians, though haplogroups B and X are virtually absent. Yet, all five markers occur elsewhere in Asian populations, from as far west as the Altai Mountains of Siberia, to as far east as Japan and Korea. If we simply use present-day occurrence to map Pleistocene distributions, then the potential home range was extensive.

One region all haplogroups occur is Mongolia, which consequently has been identified as the source of ancestral American Indians.[72] Of course, this claim is based on little more than the observation that haplogroups present in Native Americas are also present among Native Mongolians. Yet, people living in Mongolia today may not have been living there more than 20,000 years ago. They could have arrived there in recent times, just as their modern Native American counterparts arrived at their residences relatively recently. The most we can reliably say is that Native Mongolians and American Indians are descended from the same population; Mongolia cannot be confirmed as a

TABLE 8 Founding mtDNA lineages in the Americas

Founding haplogroup (haplotypes)	Approximate geographic occurrence	Molecular age in years BP (control region)[a]	Molecular age in years BP (complete mtDNA)[b]	Earliest ¹⁴C dated aDNA occurrence[c]
A (A2)	Frequent in far northern groups (Na-Dene and Eskimo-Aleut), common in eastern North America and Central America, and infrequent in South America	27,241–35,909	13,900 ± 2000 (A2)	4975 ± 40 BP (Big Bar Lake, British Columbia)
B (B2)	Virtually absent in northern North America, increases in Central and South America	13,448–17,727	16,500 ± 2700 (B2)	8000 BP (Hourglass Cave, Colorado)
C (C1b, C1c, C1d, C4c)	Present in North and Central America, abundant in South America	42,069–55,545	14,700 ± 4700 (C1b) 15,800 ± 4700 (C1c) 9500 ± 3400 (C1d)	9225 ± 60 BP (average) (Wizard's Beach, Nevada)
D (D1, D2, D4h3)	Common among far northern groups (Na-Dene and Eskimo-Aleut), present but not abundant in North and Central American populations, and frequent among South American groups	19,655–25,909	10,800 ± 2000 (D1) 10,300 ± 6000 (D2)	9820 ± 40 (average) (49-PET-408/On Your Knees Cave, Alaska)
M	Only presently known North American occurrence from aDNA recovered from specimens in British Columbia			4950 ± 170 BP (China Lake, British Columbia)
X (X2a)	Trace occurrences in North America	13,000–17,000 or 23,000–36,000		7400 BP (average) (Windover, Florida)

NOTE: 2007 and Smith et al. 2006. Haplotype markers are ones widespread in the Americas but generally absent in Asia, suggesting a post-arrival appearance. Molecular ages, as noted in the text, vary widely depending on method and region; only two sets of estimates are provided here. Ancient DNA haplogroup assignments are unconfirmed: not all defining markers for all haplogroups were sequenced.

[a]Data from Schur 2004a. [b]Data from Tamm et al. 2007. [c]Data from Malhi et al.

molecular homeland until we find Pleistocene-age DNA in Mongolian skeletal remains that matches those founding lineages.[73]

AMERICAN ADAM(S)

The Y is one of the smallest of the chromosomes, comprising approximately 51 *megabases* of DNA (a megabase is 1 million base pairs).[74] Although relatively small, most of it is non-recombining, making it the largest such block of the human genome. NRY mutations can occur in one of several forms, which include SNPs, insertions/deletions, and Short Tandem Repeats (STRs). STRs differ from the other mutations in that instead of being binary (e.g., a C→T change), they are characterized by multiple repeat states (e.g., CG→CGCGCG change), and their mutation rate is higher, making them more variable in human populations and more useful than SNPs for tracking more recent events in human history. In the NRY, haplogroups and founding lineages are generally defined by SNPs (the broader groups), and the haplotypes by STRs.

Investigations of the NRY began in the mid-1990s, and owing to the small number of samples examined and the large regions to be mapped, there are fewer studies than of mtDNA. Complicating matters, different laboratories were not looking at the same segments of the NRY initially, and when they did, they were not using the same labels. Fortunately, such confusion has been straightened out, and there is a now standardized nomenclature for the eighteen NRY worldwide haplogroups (labeled A through R).[75]

Zegura and others have detected three major NRY lineages in America: C, Q, and R (see Table 9). Together, these comprise over 95% of the Native American NRYs

TABLE 9   NRY lineages in the Americas

NRY haplogroup	Diagnostic SNPs	Geographic occurrence	Molecular age range, in years BP
C	M217—ancestral form P39	Asia (Altai); Apache, Navajo, Tanana, Cheyenne, Inuit, Sioux	13,900 ± 3200 to 13,600 ± 4100
Q	P36 (23.8%)— ancestral form M3 (52.6%)	Asia (Altai, Ket, Selkup); all New World groups	17,200 ± 4600 to 14,700 ± 5700 (M3 split at 14.9–10.1 ka)
R	M173, M207, SRY10831b, P25	Europe (British, Greek, Italian, Russian); Sioux, Cheyenne	historic period (post-1492) admixture

NOTE: Data from Zegura et al. 2004.

examined thus far. The remaining 5% are haplogroups E through G, and I through J, all of which are European markers and are attributed to recent genetic admixture.[76] Haplogroup R might also have arrived on these shores after 1492. Most American Indians who are members of haplogroup R are in lineage R-P25, one widely shared by Europeans and Native Americans, but not Asians. Either Europeans and American Indians have a common ancestor deep in time that left no descendants in Asia (akin to the situation with mtDNA haplogroup X), or—more likely given the strong similarity in the STRs between Europeans and American Indians—that shared ancestor is relatively recent (post 1492).

We are left then with NRY haplogroups C and Q, both of which appear to be founding lineages. Haplogroup Q occurs in 65%–75% of all individuals and in all American Indian tribes. Haplogroup C is less abundant and patchier in its distribution, being present primarily in Na-Dene groups, and sporadically in Eskimo and ostensibly Amerind groups. Both of these haplogroups thus crosscut Greenberg's three linguistic families (Eskimo-Aleut, Na-Dene, and Amerind), indicating those language families are once again "a poor predictor of paternal genetic affinities among Native American populations."[77]

As was the case with mtDNA, Native American Y chromosome diversity is much lower than in Asia, showing that here, too, the ancestors left Asia with only a fraction of the genetic variability of the parent population. But did they come as one migration or two? Back to the molecular clock. The age of Q-P36, the marker defining the Q lineage, is estimated at 17,200–14,300 years ago but with a large margin of error. The estimated age for the C lineage is 13,900 years ago (± 4,000 years). These generally put the molecular split between 18,000 and 14,000 years ago; population expansion must have come later. It hasn't escaped notice these ages are generally younger than those based on mtDNA.[78]

Given that the molecular ages for haplogroups C and Q broadly overlap, they could be part of the same, single migration to the Americas. If so, they ought to have come from the same place and show similarities to the same Asian populations. Both, in fact, are present in relatively high frequencies in the Kets and Selkups of western Siberia, though this is where populations with those particular markers happen to live now. It is supposed by some historians that the "ancient homeland" of the Kets and Selkups was farther south in the Altai region (where haplogroups are also found in modern populations), though it is not clear on what actual evidence that is based.

As it happens, mtDNA haplogroups A-D and X are also found among some Altai populations, but not the same populations that yielded the NRY haplogroups C and Q. Of course, that difference might mean little: mtDNA and NRY haplogroups could have been present in a single population living here in late Pleistocene times, and the fact they are scattered today only means their ancestors dispersed over time. We know that happens: after all, at least some of them went off to colonize the New World. Nor will we know, without ancient DNA from Pleistocene skeletal remains from this area, whether their ancestors lived in the Altai region at the time of migration to the Americas. It is reasonable to suppose present-day Altaians are descended from the same population as Native Americans, and that's why they have complementary mtDNA and NRY. Assuming, that is, a single population harbored all those haplogroups.

DOES 5 + 2 = 1?

Certainly, that's been the view for much of the last decade: one relatively small, presumably closely related founding population harbored five mtDNA and two NRY founding haplogroups. How small? Jody Hey has shown DNA diversity in the Americas could result from an initial *effective* population of around 70 individuals (the effective population is the number of individuals likely to contribute genes to the next generation; the actual on-the-ground size of the population is routinely about three to five times larger). Andrew Kitchen and colleagues have more recently pegged the initial effective population at about 1,000–5,400 individuals (~5,000–27,500 people altogether). Hey's estimate seems an especially small number with which to found a New World (~200–350 individuals), but there's no anthropological or genetic reason to preclude the possibility. There are signs in all genetic systems (mtDNA, NRY, autosomal) of a population bottleneck, so whatever the absolute number, those who made it to the New World were some fraction of the Pleistocene population of East Asia and Siberia.[79]

But was it a single migration? Much of the support for that idea is the difficulty of imagining, as Merriwether and colleagues put it, that "these [founding] lineages were drawn from separate populations . . . or from the same population thousands of years apart."[80] In part, our inability to conjure that image is rooted in the idea there are just a handful of founding lineages in the Americas, and all of them have been accounted for in studies of American Indian mtDNA and NRY. That may be the case.

Yet, many New World groups have not participated in mtDNA or NRY studies, and among groups that have, the number of participants is often relatively small, on average fewer than thirty individuals. Small samples are the bane of scientific conclusions since (as noted) they typically detect only the most common haplogroups and haplotypes in a population. Recall that it took more than a decade before geneticists came across haplogroup X, which occurs in only about 3% of Native Americans. Such gaps in sampling, Ripan Malhi and colleagues argue, "make it likely that significant undocumented genetic structure still exists in the Americas."[81]

The National Geographic Society and IBM's five-year Genographic Project aims to rectify that situation by assembling the world's largest collection of mtDNA and NRY, with the lofty goal of mapping human genetic diversity and history.[82] Shifting analytical attention from the mtDNA control region to the whole mtDNA molecule will help.

So, too, will greater attention to genome-wide autosomal patterns. After all, mtDNA and NRY markers represent uni-parental descent in small corners of the human genome. Yet, each of us has two parents, four grandparents, eight great-grandparents, and so on; go back just twenty generations, for example, and each of us has more than 1 million ancestors. All of them were potential contributors of some piece, however tiny, of our autosomal DNA.[83] Thanks to ongoing advances in genetics and the statistical analysis of the genome, the "discordant variation" that stymied Zegura in the mid-1980s is being parsed, and as computational biologist Nick Patterson reports, the study of autosomal (nuclear) DNA is likely to lead soon to a much richer genetic understanding.

All of these efforts will show whether our current picture of Native American genetic diversity and the conclusions drawn from it about the number of founding lineages and the timing of their divergence are correct (I'll return to these issues but from another angle in the next chapter).

But it still might not be enough. The genetic studies to date are primarily based on modern populations, which represent only the most recent time slice of the more than 12,500 years (and about 400 generations) Native Americans have occupied the New World. There's long been reason to suspect that over that span, there were mtDNA and NRY lineages that once existed but have since vanished, and thus would never be detected in modern populations, even if those populations were adequately sampled. Recently, that suspicion was heightened.

In January 2007, Malhi and colleagues announced that mtDNA haplogroup M, known from contemporary populations on other continents but heretofore unreported in the Americas, had been found in ancient DNA recovered from two skeletons (possibly brothers) from the China Lake site in British Columbia dating to about 5000 BP. Coincidentally, that same month saw the announcement by Brian Kemp and others that a bone sample from On Your Knees Cave, Alaska, dated to 10,300 BP had yielded a new founding haplotype within haplogroup D.[84]

Neither the discovery of a new haplotype, nor even the discovery of a previously undetected haplogroup, is surprising. But if these discoveries are any indication—and they need to be fully sequenced and confirmed[85]—the recovery of ancient DNA (aDNA) has the potential to revolutionize our understanding of population history, since genetic data from contemporary populations may not be fully representative of the genetics of the earliest populations in the New World. It may not even fully represent pre-1492 populations. This is not to say, however, that modern groups are not descended from those early populations, only that certain specific markers in two corners of the vast human genome may have disappeared along the way (a matter taken up more fully in Chapter 6).

Studies of aDNA are in their infancy, and present formidable analytical challenges.[86] aDNA recovered from teeth or bones routinely occurs in badly degraded fragments, and when amplified in the laboratory is highly prone to contamination, even from a researcher's stifled sneeze or lost split end. Laboratories working with aDNA must be extraordinarily vigilant, and maintain a file of the DNA of all who come through the lab, to ensure an unexpected haplogroup does not belong to anyone other than the ancient individual whose sample is under study.

In the relatively few studies of *a*DNA (less than 1,000) to date, most on archaeological specimens fewer than 4,000 years old, traces of the mtDNA founding lineages seen in modern American Indians are present in varying frequency. In some cases, *a*DNA and modern mtDNA from the same region are alike, but in others not. Present and past mtDNA signatures tend to match if the *a*DNA is not very old: *a*DNA from the high Arctic that's fewer than 1,000 years old matches that of Eskimo populations, but *a*DNA from Windover, Florida (dated to 7400 BP), is mostly haplogroup X, in a part of the country where X has not been documented among American Indians. It is too soon to conclude that DNA continuity occurs only where there is a relatively shallow time depth, and that the further back in time, the more likely the DNA signal will flicker and then disappear. But this possibility has far-reaching implications for scientific and legal efforts to document the modern identity of ancient skeletal remains (as explored in Chapter 10).[87]

Of more immediate consequence, the newly discovered haplogroups and haplotypes in *a*DNA may imply that the diversity of the founding Native America population was larger than previously proposed (and will lead to different estimates of the effective population size of the founding group); that molecular estimates for the timing of the peopling of the Americas will have to be recalculated to reflect the greater number of haplotypes; and that we're still open for business on the possibility of multiple migrations, or at least the possibility of multiple *sources* contributing to a single founding population.[88] In this regard, it's better to think of Beringia as a funnel that was open to the west, with many genetic streams flowing in (from Siberia) but just one coming out (in Alaska), rather than as a pipeline (a single stream from Siberia to Alaska).

That funnel might have been open for some time, for there are four haplotypes now known within haplogroup C. Of those, haplotype C1a is restricted to Asian populations, while haplotypes C1b, C1c, and C1d are widespread in the Americas—and only the Americas. Those haplotypes must have originally arisen in Asia, perhaps in the region of Beringia, apart from other populations harboring haplogroup C, but before heading to the New World. A Beringian "standstill" Tamm and colleagues call it,[89] one that possibly lasted 10,000–20,000 years (the time necessary to give rise to the separate haplotypes C1b–C1d). It is an intriguing, though as yet untested, hypothesis, not to mention one that sees little archaeological support of a long occupancy of Siberia and Beringia (though admittedly that might be the fault of the archaeological record).[90]

There's one more thing *a*DNA can do: help resolve a particularly contentious subplot over whether ancient and modern skeletal remains are related.

TALKING HEADS

Skeletal remains and particularly crania have long figured in this debate, though they carry a seedy legacy since the size and shape of the head—where, of course, the brain is housed—was typed and often stereotyped to fit prevailing notions of race, ethnicity, and intelligence. A soft-core racism persisted in physical anthropology well into the twentieth century. Harvard University's Earnest Hooton distinguished among a sample of the nearly 1,800 crania A. V. Kidder had excavated from Pecos Pueblo (New Mexico) seven distinct craniofacial types, including "Pseudo-Negroid," "Pseudo-Australoid," "Plains Indian," and "Long-faced European." Although Pecos had been occupied in Spanish colonial times, Hooton insisted all of his cranial types were found in prehistoric deposits. Like other "heriditarians" (his term), Hooton believed these affinities to Africans and Europeans were no coincidence, but rather physical testimony of a prehistoric melting pot, one that enabled him to shamelessly credit the American Indians' development of agriculture and civilization to their partial European heritage. Hooton believed cranial form was permanent, and had no use for "fanatical environmentalists" who refused to accept that a skull could "withstand the moulding [sic] and modifying influence of physical environment and . . . endure through generations as a racial earmark."[91]

One of those so-called fanatics was Franz Boas, who (in addition to having a low opinion of Hooton) steadfastly challenged claims of cranial immutability, showing that the *cephalic index,* the ratio of maximum width to maximum length of the skull that was often used to sort races, varied widely among adults of a single group and even within the lifetime of a single individual. He later argued, based on a massive study he conducted of immigrants who passed through Ellis Island in the early twentieth century, that American-born children of immigrants had significantly different cephalic indices than their parents—a result of the healthier American environment in which they were raised. Partly as a result of Boas's efforts, and also of the development of intelligence testing (which moved the discussion of race onto even more slippery ground, craniometry had lost much its luster by mid-twentieth century.

Now it's back, having been revived in a far more sophisticated form in the last decades of the twentieth century by practitioners armed with complex statistics and powerful computers for analyzing cranial variation. They take multiple measurements of a skull (as many as 60–75 on a complete cranium, far fewer on fragmentary archaeological specimens), including the width of the skull at its maximum point, its distance from front to back, and the width and height of the eye sockets. The data are then analyzed using a variety of statistical techniques—principal components and discriminant function analysis are the tools of choice—that attempt to discern combinations of variables that best describe skull shape and size, and can reliably cluster like crania. They then compare specimens from other times and places to see how they resemble or differ from one another, and how much variation each group displays.[92]

The approach is advertised as "exciting and dynamic," but in reality much of it is just old wine in new skins. Many steadfastly hold to Hooton's assumption that cranial form faithfully reveals group identity (some, in fact, still haven't forgiven Boas for claiming otherwise). In so doing, they hold to a deeper belief as well: that race is a natural category, and individuals can be assigned to one or another race on the basis of their skulls.[93] We'll come back to this point.

TABLE 10 Late Pleistocene and Early Holocene (~pre-8000 BP) skeletal remains from North America

	Site	Age (^{14}C years BP)	Description	References
1	49-PET-408 (On Your Knees Cave (AK)	9820 ± 40 (average)	mandible, pelvis, vertebrae	Dixon 1999
2	Anzick (MT)	10,680 ± 50	cranial fragments, 1–2 yr old	Owsley and Hunt 2001
3	Arch Lake (NM)	10,220 ± 50; 8870 ± 40	skeleton, female	Montgomery et al. (in preparation)
4	Arlington Springs (CA)	10,000 ± 310 10,080 ± 810 10,960 ± 80	femora and humerus	Orr 1962; R. Berger and Protsch 1989; J. Johnson et al. 2002
5	Brown's Valley (MN)	8700 ± 110	skeleton, male, 20–30 yrs old	Jenks 1937
6	Buhl (Idaho)	10,675 ± 95	skeleton, female, 17–20 yrs old	Green et al. 1998
7	Fishbone Cave (NV)	8370 ± 50 8220 ± 50	postcranial fragments	Dansie and Jerrems 2006
8	Gordon Creek (CO)	9455 ± 110 (average)	skeleton, female, 25–30 yrs old	Breternitz, Swedlund, and Anderson 1971; Swedlund and Anderson 1999
9	Gore Creek (BC)	8250 ± 115	cranium, male, adult (skeleton missing)	Cybulski et al. 1981

10	Grimes Point Burial Shelter (NV)	9470 ± 60	cranium, mandible, postcranial material, female, ~10 yrs old	Touhy and Dansie 1997
11	Horn Shelter (TX)	9875 ± 110 (average)	two skeletons: male, 35–45 yrs old, adolescent (female?)	Redder and Fox 1998; Young 1998
12	Kennewick (WA)	8410 ± 35 (average)	skeleton, male, 35–45 yrs old	Chatters 2000, 2001; J. Powell and Rose 1999
13	LaBrea (CA)	9000 ± 80	skeleton, female, ~25 yrs old	R. Berger 1975, Kroeber 1962
14	Spirit Cave (NV)	9415 ± 25 (average)	mummy, male, 40–44 yrs old	Touhy and Dansie 1997
15	Warm Mineral Springs (FL)	10,260 ± 190	three remains: skeleton, male, 30–40 yrs old; partial cranium, female; vertebra and pelvis, child, ~6 yrs old	Clausen, Brooks, and Wesolowsky 1975
16	White Water Draw (AZ)	8000–10,000	two skeletons, female, both in 30–50 year old range	Waters 1986
17	Wilson-Leonard (TX)	10,000–10,500	skeleton, female, 20–25 yrs old	D.G. Steele 1998
18	Wizards Beach (NV)	9225 ± 60 (average)	skeleton, male, 30–40 yrs old	Dansie and Jerrems 2006

NOTE: Data from Chatters 2001, D. Steele 1998, and other sources. These are primarily specimens for which a significant portion of the skeleton has been recovered. A comparable number of far more fragmentary human fossils have been found, but are not included here since the information that can be derived from them is limited.

They have set their analytical sights on early New World crania, or at least the few there are. Highly mobile peoples spread thinly on a vast landscape, as the first Americans were, rarely used cemeteries, and very few skeletal remains have survived the passage of time. Skulls are especially thin walled and easily broken (unnervingly so, given what they protect), and almost inevitably disintegrate over time, occasionally leaving behind the hard and dense teeth they once packed. In the whole of the New World, only a few crania predating 10,000 years ago survive. To bolster that tiny sample, osteologists often include Early Holocene remains (ones more than 8,000 years old) in their analyses (see Table 10). Even then, it's a total sample of just a few dozen.

We will never know how many people were in America from 12,500 to 8,000 years ago, but it was assuredly more than three dozen. That's a pitifully small number on which to draw sweeping conclusions about what these people looked like, how cranial size and shape varied, and why. Were that not enough of a challenge, the remains are rarely complete, making it difficult to collect the full suite of measurements. That complicates analyses since, as Gentry Steele and Joseph Powell explain, the form of a skull and its similarity to others can vary depending on how and how many measurements are used in the analysis.[94]

So if the remains are few, fragmentary, and unrepresentative, it's fair to ask, Why bother? One good reason: with these bones we come face to face—literally—with what early Americans actually looked like. But beyond being able to look them in the eye (sockets), what else? It gets complicated. Like teeth, skulls vary over time, but unlike teeth, skulls change easily and more readily. Cranial plasticity, it's called, and it happens (1) in response to changing environments, where natural selection confers an advantage to a particular facial shape or structure; (2) by virtue of isolation—genetic drift again, this time expressed phenotypically; (3) as a result of cultural practices, such as from diet or cooking methods; or (4) as a consequence of growth and development (including the effects of disease). Morphological change happens, and can even be more pronounced within a group than between groups.[95]

There are fierce debates over just how rapidly craniofacial form changes on an evolutionary time scale. A century ago Aleš Hrdlička, believing evolutionary change was gradual and unhappily facing the prospect of having Pleistocene peoples in America, emphasized the morphological similarity between the ostensibly ancient crania and those of modern American Indians. His was not a terribly sophisticated view: if a cranium was truly ancient, it ought to resemble a Neanderthal. If it was younger, but still thousands of years old, it should look modern. As far as he was concerned, all of them did. It was a conclusion for which he, like Boas, has not been forgiven.[96] Most specimens Hrdlička deemed modern were not very old, so it's no surprise he came to the conclusion he did. Yet, he failed to appreciate that even at 10,000 years remove, archaeological crania can differ strikingly from modern ones. That has only become obvious in the last couple of decades as radiocarbon dated skeletal remains appeared.

But can a 10,000-year-old cranium look different from one just a few thousand years younger, and yet both be members of the same historically related population? Physical anthropologists Charles Roseman and Timothy Weaver certainly think so, especially where population size was small and the likelihood of genetic drift was increased, or where populations migrated into regions that posed different environmental challenges than the one they left, a circumstance faced, for example, by the first Americans (this is akin to Nettle's argument for early language proliferation).[97]

Under such circumstances, would differences (or similarities) in skull form say anything about who's related to whom? Perhaps, but the challenge lies in determining whether those differences (or similarities) are historically significant, and can reliably distinguish members of different populations who do not share a common ancestor, or who may have come to America in different migrations (or, if similar, show they are a part of the same lineage); or result from *structural* changes wrought by natural selection, genetic drift, and/or genetic change over time acting on related groups (or whether their similarity results from adaptive convergence); or are a complex mix of history and structure.[98]

Here's what's been found. Loring Brace and colleagues see evidence of two major craniofacial "types." The first links recent and archaeological American Indian specimens from across the hemisphere with ones from as far away as Europe (and Paleolithic Europe, at that), as well as from Bronze Age Mongolia and the Japanese archipelago (those of the Ainu and Jomon). The second group includes Eskimo-Aleut and Athabascan crania, along with those of East and Southeast Asians. The first group—of "Eurasian" origin—is thought to represent an earlier Pleistocene migration, and the second a later, Holocene-age expansion. By this analysis, East Asians, including recent Mongolians, made only "partial contributions" to the peopling of the Americas.[99]

Walter Neves likewise detects two broad craniofacial groups, which he attributes to two migrations, only they are not the same two groups Brace spotted. Instead, Neves sees in Late Pleistocene and Early Holocene crania in South America what he calls the "Paleoamerican" type, one strongly similar to present-day Australians, Melanesians, and sub-Saharan Africans. That type is very distinct, he says, from recent American Indians, whose crania are more akin to the "typical Mongoloid morphology" of northern Asians. The only modern Native American group that fell into Neves's Paleoamerican type was the Pericú, who live at the southern tip of California's Baja Peninsula. Neves suspects they are survivors of the ancient Paleoamerican migration, somehow having avoided mixing with other groups over their entire history in the New World.[100]

Like Neves (but unlike Brace), Richard Jantz and Douglas Owsley agree that early American crania are distinct from those of recent American Indians, who are said to be more typically "Mongoloid." Yet, unlike the South American specimens Neves studied, the Late Pleistocene and Early Holocene North American crania Jantz and Owsley examined, ostensibly Paleoamericans as well, show no particular resemblance to southwest Pacific or African populations. In fact, they appear highly variable, half falling within the range of recent American Indians, the other half "reluctant members of any recent group."[101]

Finally, Steele and Powell report that the crania of South American Paleoindians are, as Neves suggested, closer in their size and shape to those of Australians and Africans than to those of any other group. Yet, unlike Jantz and Owsley, they find that some North American Paleoindian crania are also similar to those of Australians, Ainu, and Polynesians, while others are more like those of Africans and Southeast Asians. In general, however, Steele and Powell believe all Late Pleistocene and Early Holocene skeletal remains are highly variable (as Jantz and Owsley, but not Neves found), and are consistently more like those of southern Asians and Pacific Rim populations than northern Asians, and are like skeletons of Archaic age here in the Americas, but otherwise quite distinguishable from more recent Native American populations (they do not, however, automatically assume that the Americas were peopled by Polynesians, say, or the Ainu, only that these are forms with general, and perhaps historically meaningless, resemblances).[102]

Overall, it's a very confused picture. Depending on which analysis one accepts, the first American crania are either alike, or they are not. They either look like recent American Indian crania, or they do not. They either resemble African, Ainu, Australian, European, Melanesian, or Polynesian crania, or not. They originated in eastern Asia, or not. There's even disagreement about individual specimens: Jantz and Owsley liken the Wizards Beach cranium to those of modern Native Americans; Steele and Powell think it more similar to Polynesians and Australians.[103] All this is to say nothing of the vast gulf between the cranial results and the groups derived from teeth, language, and genes (a thorny problem to be confronted in the next chapter).

What's going on here?

COMPLICATING CRANIA

There are several possibilities. For one, this could all be a prank of sampling error: with so few cranial remains, it may be impossible to detect any meaningful patterns, however powerful the statistics we apply. Steele and Powell—and Neves, too—doubt this, since they see widespread patterning in their analyses. But since their separate analyses often are looking at the very same specimens yet seeing very different patterns, their rebuttal loses much of its potency.

Alternatively, put the craniofacial variability to use, as Jantz and Owsley do, and argue it testifies to separate and multiple migrations. Replace (or assimilate) more ancient immigrants with more recent ones, bring some from southern Asia and the Pacific Rim and others from Northeast Asia, and perhaps this craniofacial cacophony makes sense. Only, that assumes the specimens are representative of the populations from which they were derived (doubtful, given the time and space over which they are scattered), and that craniofacial morphology is slow to change, especially given the brief time the Americas

Human skulls are—as no stone tool can be—a tangible connection of a people to their past, a connection that has powerful resonance. But which people, and whose past? And who makes the call?

Traditionally, anthropologists in the United States rarely gave those questions much thought. Now, they can't afford to ignore them. It's the law: the Native American Graves Protection and Repatriation Act (NAGPRA), signed into law by President George Bush in 1990, stipulates that American Indians are entitled to the return, for burying, of the remains of their ancestors housed in museums and universities. This well-intentioned legislation was aimed at righting the wrongs of earlier generations of anthropologists, who too often indiscriminately collected the bones of American Indians for study and display. It was often little more than genteel grave robbing. Although it was not only Native American remains that received such treatment, theirs were gathered in lopsided numbers. To American Indians and many others, this was one more instance of mistreatment at the hands of Euroamericans. NAGPRA required institutions having Native American skeletons to inventory and repatriate (return) those remains and any associated artifacts at the request of a known "lineal descendant" or the tribe with the "closest cultural affiliation."

In NAGPRA's wake, thousands of skeletons and associated artifacts have been (or are being) repatriated. Many of those remains were only a few hundred years old or younger, and there was rarely much debate over who the lineal descendants or cultural affiliates were.

Then Kennewick came along. Its story began in July 1996, when two twenty-somethings were trying to sneak into the annual hydroplane races on the Columbia River near Kennewick, Washington. They skirted the perimeter fence and were wading along the edge of the Columbia River when one of them spotted a skull in the shallow water. Thinking it might be a murder victim, they called the police, who collected the skull and turned it over to the county coroner. He, in turn, called James Chatters, a local archaeologist who'd also done forensic work. With a police escort, the coroner and Chatters returned to the site and collected what bones they could in the muddy bottom as it grew dark. Near the bones were some late nineteenth-century artifacts, which were also gathered.[104]

Chatters's initial diagnosis of Kennewick, based on characteristics of the skull and the associated artifacts, was that he was a middle-aged European. But then he noticed what appeared to be a projectile point embedded in the pelvis and was curious about "a white guy with a stone point in him,"[105] particularly since it was a point thought to have gone out of style thousands of years ago. Chatters sent a few bone fragments off for radiocarbon dating. Over the next few weeks, he revisited the site and collected more fragments; then in late August, he got the radiocarbon results: 8410 BP.

A news conference was called a few days later, and the date was announced. Here is where everything started to go very wrong. Chatters described the skeleton as a "Caucasoid" similar to a "pre-modern European."[106] Whether or not he inadvertently used the term "Caucasian" is disputed, but no matter. That's what some of the reporters heard, and that's what went out over the wires: 8,400 years ago a white man was stabbed on the banks of the Columbia River. That racial impression was certainly fueled by the reconstruction of Kennewick's face which, by Chatters' own admission, took shape after watching *Star Trek: The Next Generation* one evening, and reckoning Captain Picard (actor Patrick Stewart) was a dead ringer—or, more properly, a live one—for Kennewick[107] (ironically, the fictional Picard was said to have had academic training in archaeology). After that, one didn't even have to say 'Caucasian.' It was there for all to see. Naturally, no one read the fine print that putting flesh back on a skull is more art than science.

The local tribes were enraged. They called the U.S. Army Corps of Engineers (the Columbia River and its shoreline are in their jurisdiction) and insisted on a meeting with Chatters and the Corps. The meeting by all accounts was extraordinarily tense. Chatters felt physically threatened: one of the tribal members accused him of calling Kennewick European just to dodge the law (NAGPRA). The tribes left to draft a legal document demanding Chatters immediately turn over all human remains in his possession, while Chatters hustled home to record as much as he could of the skeleton, fearful that marshals would arrive at any moment to collect it. He also began rounding up legal, scientific, and congressional support to gain more time to study the remains.[108]

It didn't work. On orders from Corps attorneys, the sheriff soon came to collect the bones. In mid-September the Corps gave thirty days' notice of its intent to turn the remains over to a coalition of Northwest tribes for reburial under NAGPRA (five called Kennewick their own: the Colville, Nez Perce, Umatilla, Wanapum, and Yakima). They later buried the site beneath 600 tons of rock and sediment and planted trees. They called it site protection. Most everyone else called it a destructive and heavy-handed act to prevent further investigation. The Corps works on hundreds of projects across the country, many on tribal lands, and it was not unnoticed that it was in their best interest to get along with the tribes. If that meant fast-tracking the transfer of the Kennewick bones and shutting down the site, they would.[109] Unless there was a challenge to the validity of the tribes' claim.

Chatters intended there would be. He frantically worked the phones and e-mail, and found a receptive audience: skeletons of this age and completeness (about 90% of Kennewick's bones survive) are rare, and the study opportunity would be lost if the Corps followed through on their plan to repatriate. The Corps was bombarded with messages insisting Kennewick's very un-American Indian appearance demanded further investigation to determine who he was, and maintaining it was impossible for any tribe to assert biological or cultural affiliation with a skeleton of that antiquity. Soon, members of Washington's congressional delegation—some with seats on the Appropriations Committee (powerful adversaries for any federal agency)—weighed in on behalf of a study of the remains.

Inevitably the media picked up the scent of battle, and the story hit the airwaves as nothing less than a conflict between science and the religious beliefs of Native Americans. *Newsweek* and the *New York Times,* among others, questioned the Corps' intent to repatriate: after all, if it was a Caucasoid, how could it be American Indian? Kennewick catapulted from Early Holocene anonymity to high-profile appearances on *60 Minutes* and *People* magazine.

Fearing that the Corps would hunker down, ignore the media firestorm and congressional probes, and proceed with the repatriation, eight archaeologists and physical anthropologists filed suit in federal court to block the transfer.[110] They argued the Corps' decision was arbitrary and capricious, that NAGPRA may not apply to remains this old, and that repatriation denied their First Amendment rights of access to knowledge and freedom from establishment of religion. Their suit lacerated the anthropological community (there was hardly unanimity behind the action), badly damaged relations with Native American groups—even ones who had historically welcomed the work of archaeologists—and triggered a divisive, long-running, and expensive legal tug-of-war, complete with carnival sideshow: members of the Asatru Folk Assembly, who worship Odin the Norse sky god, filed their own suit asking the remains be turned over to them since, as a white man, Kennewick must be one of their own. There's sometimes a fine line between tragedy and farce.

The Department of the Interior took over for the Corps and brought in a team to examine Kennewick. They confirmed its age, and on the basis of Joseph Powell and Jerome Rose's analysis, deemed Kennewick no Caucasoid. Nor was he like any modern American Indian, but instead was similar morphologically to East Asian and Polynesian populations. Powell was well aware that "such a finding means very little, unless one presumes that all skeletal features are fixed and immutable over time." He, for one, does not, so was hardly surprised Kennewick's cranium looked different.[111] In January 2000, the secretary of the interior ruled Kennewick was indeed Native American under NAGPRA, and in September of that year, decided to turn over the remains to the five tribes, based on their oral traditions of ancient times and on the fact that Kennewick was found on their ancestral land.

The plaintiffs scrambled to have the federal judge (John Jelderks) stay any transfer; he did just that, and then followed with an Opinion and Order (in August 2002) that accused the Corps of Engineers of violating federal law and judged the Department of the Interior in error for determining Kennewick belonged to the tribes. He granted the plaintiffs the right to study the remains. Appeals were quickly filed by both the federal government and the tribes. But two years later, the Ninth Circuit Court of Appeals upheld Jelderks's ruling and ordered the study to proceed (more on this in Chapter 10).

It has, yet in the end the affair proves a nonstarter. Kennewick is no European. Rare, interesting, and important he may be, but like many of his skeletal brethren, all we can securely say is that he *appears* distinctive. However, that may only be because he is so rare and was a member of a once-larger population poorly represented in the fossil record. His apparently unusual cranium could be readily explained by evolutionary processes of isolation and genetic drift, migration, population splits, and recurrent episodes of gene flow. Such evolutionary processes, of course, fail to account for why he would bear an uncanny resemblance to a Starfleet Commander.

The myth of Kennewick as the "Great White Hope" for Caucasians colonizing America has died, at least in scientific circles, and a well-deserved death it was, though news of the specimen can still command the cover of *Time* magazine (as it did in March 2006). Yet, the claim the Americas were colonized by Caucasians has not gone away. Instead, it has seeped into the only place such foolishness passes as fact: the poisonous corners of the Internet where white supremacists continue to claim Kennewick as one of their own, to promote the idea that members of the Aryan race—whatever *that* is—were the real discoverers of America.

were occupied, and thus is a meaningful indicator of historical (and historically unrelated) groups. If crania look alike, it is because they are related; if they look different, it is because they are unrelated, Boas be damned.

Another possibility: there was but a single migration (perhaps from a fairly heterogeneous source), and the morphological differences are a result of generations of isolation, mutation, and genetic drift, mixed in with recurring episodes of gene flow. The variation in the early crania of the Americas is not beyond what one might expect of evolutionary change over that span. Indeed, change and variation, as Roseman and Weaver said, would be especially pronounced during the early period of colonization when human populations were small and spread thinly across the vast continent. There is a strong likelihood colonizers dispersed and were geographically isolated from one another for long periods of time, and only later, as populations increased and isolation broke down, was gene flow reestablished: hence the differences between ancient and modern crania.[112]

Underpinning Roseman and Weaver's argument is the principle that we must understand crania not in terms of differences between individual specimens, but as part of larger populations that extended and varied over space and time. Physical anthropology's in-house critic, Jonathan Marks, makes the point more broadly and bluntly: "Today the anthropological community en masse rejects as pseudoscientific the notion of race as a natural category, and as equally bogus the assignment of individuals to such transcendent categories on the basis of their skulls. What exist are features and populations; sometimes one can map them onto one another, and sometimes one cannot."[113]

How, then, might we track whether there was a single or multiple migration, or whether such were drawn from the same or different anatomical well(s)? Not from crania, or at least not without a much larger sample of the populations of interest. That's unlikely to happen (it's never wise to bet against something turning up in the ground, but your money's probably safe on this one).

Geneticists are only too happy to jump into the fray, confident their data provide a far more reliable indication of population histories. If *a*DNA is obtained from ancient skeletal remains, it might reveal whether an individual harbored any of the modern haplogroups among Native American populations. That would certainly establish continuity within that single locus marker. Yet, suppose it's not one of the haplogroups present in modern populations (think haplogroup M here). That may not say much either, since it need not represent a separate population (or migration), but instead just an mtDNA or NRY lineage that went extinct, or one that simply has yet to be detected among more recent American Indians. And, of course, autosomal DNA could well give a very different picture.

Even so, osteologists aren't ceding any ground. Jantz and Owsley have, instead, announced we may be entering a period when cranial morphology will emerge "as *more* informative than genetic data in elucidating the history of our species."[114] Hooton couldn't have said it better himself.

But it's not going to be an easy battle for them to win. Even some of their fellow travelers, like Gerrit van Vark, a onetime believer in using craniofacial morphology to determine ancestor-descendant relationships, are rethinking their views in light of genetic evidence (in this case, from Europe) that indicates continuity where on craniofacial grounds none was previously thought to exist.[115] And it is sure to be an intensely personal battle, for if cranial morphology does not prove to be a useful indicator of population relationships and history, it would make much of their work questionable and, far worse, irrelevant.[116]

6

AMERICAN ORIGINS
The Search for Consensus

In March of 1774, on its second circumnavigation of the globe, Captain James Cook's *Resolution* anchored off Easter Island (Rapa Nui), a tiny island and one of the most isolated on the planet. Some of the islanders canoed out to greet his ship, and the moment they hove into view, Cook was astonished by the natives' tattoos, tapa (bark-based) clothing, weapons, and especially—when they began speaking—their language. He instantly recognized unmistakable variants of words he'd heard spoken by Tahitians and Maoris (of New Zealand), who lived thousands of miles away across vast stretches of open ocean. That the inhabitants on remote Rapa Nui shared bits of distant languages was testimony, he realized, the "same nation" must have spread themselves across the South Pacific. Their descendants may have adopted different "custom and habits," but "the Affinity each has to the other" was still obvious.

Fast forward to the summer of 1778, during Cook's third (and ill-fated) voyage. The *Resolution* had sailed north to chart the coast of Alaska, explore the Bering Strait, and probe the possibility of a sea passage to the North Atlantic. Optimistic he just might be able to sail across the top of the world, Cook was packing a *History of Greenland*: the descriptions of its Natives' manners and customs would prove a handy guide when they met. Alas, Cook found no passage to Greenland. But once again, he found connections. The dress, tattooing, and fishing and hunting gear of the "Esquemaux" in Alaska matched what a *History of Greenland* described of its Natives. So too, did many of their words. Even with the limited Greenlander and "Esquemaux" vocabularies he possessed, it seemed to Cook there was "great reason to believe that all these nations are of the same extraction."[1]

Could the similarities between the Natives of New Zealand and Rapa Nui, or between the inhabitants of Alaska and Greenland, be mere coincidences? Cook didn't think so. He appreciated the possibility unrelated peoples might "accidentally" produce similar clothes or tools in similar environments. We call that *convergence* nowadays, and it would explain why fishhooks, say, were alike on remote islands of the South Pacific (or even like those used in coastal Alaska). But could convergence explain a similarity in words across such vast distances? Cook doubted it: those languages must have had a common origin, however much they had changed since. *Divergence,* we call that. Cook concluded that the peoples who were scattered across thousands of miles of Pacific Ocean must be related; likewise, those who inhabited the top of America, from Alaska to Greenland (he did not think, nor do we, that Pacific and Arctic peoples were related, though we, of course, recognize they must share a common ancestor at some point in the distant past).

In the two centuries since Cook connected those distant dots, his insight has been proven correct. Analyses of their respective genes, languages, and material cultures have amplified the links among these peoples, and archaeologists, geneticists, and linguists have come to broad agreement on the central questions of the origin, antiquity, and timing of the colonization of the Arctic and the Pacific.

Might the same be said of the peopling of the Americas? At the outset of Chapter 5, I postponed discussion of how or whether archaeological, dental, genetic, linguistic, and osteological evidence agreed or disagreed about who the first Americans were and where they came from, when, how they got here, and how many migrations there were. It's now time to pay that note.

FINDING THE HOMELAND

On one point, as few others, there is broad consensus: the first Americans came from Asia. There is nothing to suggest otherwise in the artifacts, teeth, languages, or genes (that includes autosomal DNA, as well as mtDNA and NRY). It's true even in regard to skeletal remains—Kennewick and other like-headed crania having now been rescued from their dubious assignment as "Caucasoids" and all that wrongly implied of their ancestry.[2] The near-universal consensus that America's first peoples were bygone Asians occurs no matter the differences of evidence and opinion regarding exactly where in Asia they originated.

Pinning down a more specific place of origin(s) is no easy task, least of all for archaeology. Sites in northeast Asia are sparse, and there are few that can be unequivocally linked to the first Americans. But then we don't expect too close a correspondence in the Pleistocene archaeology of Asia and America. The migration from one continent to the other, adapting and settling along the way, assuredly took centuries, possibly even millennia. By the time the first Americans arrived in southern South America, their toolkits had changed in significant and subtle ways. These were not colonists who clambered aboard a ship in Old England and disembarked in New England just a few months later with material culture and ideas intact.

Archaeologist Frank Hibben was charming, gregarious, a gifted lecturer and writer, and though born to modest means, had married extremely well (running off with the wife of his patron). With degrees from Princeton and Harvard, he was destined for the good life, even in Depression-era America. But he aspired to much more, and got it. As a graduate student, he excavated in Sandia Cave, New Mexico, where—as Loren Eiseley described it (straining hard to conceal his envy)—Hibben had the "rare and longed-for good fortune, to be the discoverer of a new and early cultural manifestation in the New World."[3]

The Sandia culture, Hibben proclaimed in 1941, was North America's oldest, and more intriguing, its projectile points bore an uncanny resemblance to Solutrean artifacts of Paleolithic Europe. When radiocarbon dating was invented, Hibben got into the queue to obtain an age on a mammoth tusk from Sandia Cave: the result was announced in no less prominent a place than *Time* magazine. The Sandia culture was 20,000 years old—the same age as Solutrean. The possibility the Americas had been peopled directly from Europe, suspected since Charles Abbott's time, now seemed close to proof.

But the archaeological community had grown suspicious. No matter how good a story Hibben told, and he told them awfully well, Sandia was suspect. The fieldwork was scarcely complete when Wesley Bliss, a fellow graduate student who'd excavated there, reported the cave deposits were badly mixed, and hence any results utterly unreliable. Worse, he announced that very publicly in *American Antiquity*. The chairman of the Department of Anthropology at the University of New Mexico, who'd just hired Hibben to his faculty, reacted furiously. Hibben did, too.[4] Bliss's association with the university was terminated.

That hardly stopped the sniping. Hibben's student who had shipped the mammoth tusk to the radiocarbon lab noticed it had suddenly and mysteriously acquired a coating of the yellow ocher typical of Sandia Cave's deposits. Was this a ham-handed effort to make remains found elsewhere look as though they'd come from Sandia? He wasn't sure, but he quietly washed off the ocher before boxing up the piece. Even the radiocarbon laboratory director was skeptical, taking the unusual step of publicly questioning whether the 20,000 BP age was actually associated with the human occupation of the cave.[5]

And what of Sandia points? Archaeologist Bruce Bradley later examined them and, seeing how much they varied in technology and raw material, and realizing some had been shaped on a grinding wheel to more closely approximate European forms, could only conclude Sandia was a hoax. But why? Bradley supposed the hoaxer "needed an assemblage of artifacts that looked as Solutrean as you can get, and I think the motivation was to try to bolster the theory that Clovis originated as

Solutrean, and the thing that's most distinctive about Solutrean is the shouldered point. If you could prove that Clovis originated in southwestern Europe and came across the Atlantic—my goodness!"[6]

"My goodness," indeed. By the 1990s, Sandia had slipped into archaeological oblivion (though not before Hibben's shenanigans inspired a murder mystery).[7] But the effort to link Paleolithic Europe and Paleoindian America had not, and it was revived once more by none other than Bruce Bradley, along with archaeologist Dennis Stanford. Their claim, not surprisingly, emerged in the wake of the apparent discoveries of "Caucasoids" (Kennewick) and supposed European mtDNA (haplogroup X) in America, and gained traction in the post–Monte Verde ferment, when it seemed all bets were off about the peopling of the Americas. Why not shake up the archaeological establishment with a radical new idea? (Never mind the idea was hardly new or radical). Kennewick and haplogroup X have since been declared inadmissible evidence of Europeans in America.[8] Bradley and Stanford shrug off the losses: they see plenty of archaeological proof. They even have a catchy jingle: "Iberia, not Siberia."[9]

As they envision it, a band of Solutreans rafted to America in skin boats across an Eden-like Pleistocene North Atlantic, thereby successfully completing a several-thousand-mile voyage that routed the *Titanic* tens of thousands of years later. Only now, instead of heading to Sandia, Solutreans became Clovis. To be sure, Clovis and Solutrean groups were both Ice Age hunters, who occasionally took down big game with large, bifacial, lanceolate spear points. But if they are related, we face a significant problem: Solutreans vanished from Europe at least 17,000 years ago, 6,000 radiocarbon years before Clovis ever appeared in North America (it must have been an *awfully* long journey).

To the dilemma posed by the great gaps in time and space between Solutrean and Clovis, Bradley and Stanford offer the half-hearted assertion that Solutrean and *pre*-Clovis look alike. Given how little we know of pre-Clovis, and that what's known looks even *less* Solutrean-like than Clovis, the assertion rings hollow. But then, so does the link between Solutrean and Clovis.

Nowadays, our knowledge of Pleistocene prehistory is far richer than in Hibben's—or Abbott's. It's no longer enough to daisy-pick superficial similarities between American and European artifacts to show a hypothesis *could* be right (indeed, the usual scientific procedure is to try to prove a hypothesis wrong). Yet, Bradley and Stanford offer little more than that. The shared Clovis and Solutrean attributes they identify are the use of red ocher (an iron-rich mineral paint), the production of bifacial stone tools, a reliance on exotic stone (sometimes, not always), and especially the presence of a specific technique of thinning stone bifaces. *Outre passé* (overshot) flaking, aficionados call it, and

Bradley and Stanford insist only Clovis and Solutrean knappers did this "intentionally" and so must be related (others might strike overshot flakes, but only by accident).

Bradley and Stanford are undisputed experts in stone tool technology and accomplished flintknappers. If they say *outre passé* flaking proves historical divergence, we cannot dismiss the claim lightly. But we need not take it on their say-so either. The burden of proof is on them to demonstrate *outre passé* flaking cannot result from convergence. And they haven't. As for the other similarities cited, even Captain Cook would recognize them as a result of convergence: it would only be slightly exaggerating to say it is difficult to find a people in prehistory that didn't use red ocher. Its blood-red color has universal (and obvious) symbolism, starting back in Neanderthal times. It's no coincidence that ocher's formal name, hematite, derives from the Greek *haimatites*, or blood-like.[10]

Yet, what ultimately sinks the Paleolithic French connection are not the few similarities, but the large number of Solutrean items *missing* from the Clovis inventory. As Lawrence Straus, Ted Goebel, and I argued in a published rebuttal (Straus is an expert in Solutrean archaeology, Goebel in the archaeology of Siberia), consider what we ought to see if a group of Solutreans somehow had been able to successfully negotiate the cold, harsh, sea-ice choked Pleistocene North Atlantic and beach in North America (and just what were non-seafaring Solutreans doing in boats in the ocean anyway?). Those splashing ashore would have transplanted their unmistakably Solutrean lifeways and toolmaking practices here, traces of which would have been visible in the archaeological record for centuries and, possibly, millennia afterward.

Even so, we have looked in vain in Clovis (or pre-Clovis) assemblages for typically Solutrean cave or portable art (such as engraved stone, bone, or ivory figurines); beveled antler points; spearthrowers (atlatls); bone needles; perforated animal teeth and shells; backed bladelets and true burins; their many hafting designs (bipointed, shouldered, tanged, stemmed); and most obviously, the distinctive and diagnostic Solutrean point (Figure 34). None have ever been found in America North or South, either in Clovis or pre-Clovis assemblages. If Solutrean boat people made landfall in America, they must have suffered instant and almost total cultural amnesia—and, for that matter, genetic, dental, linguistic, and skeletal amnesia as well. Perhaps it was the stress of the voyage?

More likely, Solutrean groups never made landfall in America, in which case the Norse retain their title as the first Europeans to reach America. And *they* left behind unequivocal proof of what was a very brief visit around AD 1000: the settlement of L'Anse aux Meadows, Newfoundland.[11] It's also telling that Viking

FIGURE 34.
Shouldered Solutrean point from La Reira
Cave, Asturias, Spain (left), and a Solutrean
laurel leaf point from the site of Laugerie
Haute, Dordogne region of France (right).
(Photograph copyright by and courtesy of
Lawrence G. Straus.)

Erik the Red lost nearly half his fleet of twenty-five ships just sailing from Iceland to Greenland. That's a good indication of the dangers involved in crossing the turbulent and ice-filled North Atlantic, and the Vikings didn't have to contend with the Pleistocene North Atlantic, when sea ice was more abundant and treacherous, water temperatures much colder, and thermohaline circulation had stalled.[12]

Problems aplenty. But Bradley and Stanford are sticking to their guns. Bradley insists, "If the French Solutrean were found in northeastern Asia, there would be nobody asking . . . whether Clovis was related."[13] Actually, we would be: changing location does not mean abandoning principle. The same rules and expectations apply, whether land or ocean intervenes. Want proof? When Maureen King and Sergei Slobodin announced a fluted biface had been found in Siberia and raised the possibility it was an Asian precursor to North American fluting,[14] questions were immediately asked—by them and others—whether its age was secure, whether it truly was fluted, if this could be ancestral Clovis, and so on. Ultimately, the piece was not accepted as a precursor, but instead as a bit of fortuitous flaking that mimicked fluting.

Nowadays, few if any archaeologists—or, for that matter, geneticists, linguists, or physical anthropologists—take seriously the idea of a Solutrean colonization of America. However, it has achieved Internet immortality among believers in the Lost Continent of Atlantis.

Recall from Chapter 1 that *Homo sapiens* first reached central Siberia about 35,000 BP, but only after about 27,000 BP, and over the next 15,000 years, expanded north and east. But that expansion stalled, and possibly went into full retreat during the LGM, as testified by a scarcity of sites in northeast Asia. Humans may not have arrived in significant numbers in western Beringia until well past 14,000 BP.

The artifact assemblages in Siberian Upper Paleolithic sites include, as Ted Goebel shows, the same broad classes of tools and technologies one sees in early American assemblages, whether Clovis or (granting what little we know) pre-Clovis. In those Siberian assemblages, one can spot bifaces and points, blades (cutting tools), end and side scrapers (handy for hide-preparation work), gravers (tools for incising bone and wood), a variety of bone and ivory implements, tools made on biface flakes and blades, and even the occasional *outre passé* flake. It's not an identical toolkit by any means, but it shows a degree of similarity we expect given the time and space separating the last Siberians from the first Americans.[15]

Which is not to say we haven't sought (and still seek) tighter links. From the moment E. B. Howard found fluted points embedded in mammoth skeletons at Clovis in the 1930s, archaeologists have been trying to track this distinctive projectile point form back to Asia. For decades little was found. Hopes were raised when Maureen King and Sergei Slobodin spotted a stone biface fluted on one face in a museum drawer of artifacts from the Uptar site near Magadan, Siberia (Figure 35).[16] But the Uptar excavations had taken place years earlier, and the potential significance of the biface had gone unrecognized. Its age and context could not be pinned down precisely. Was this ancestral Clovis in Siberia, or a biface accidentally flaked in a way reminiscent of fluting? In the absence of any other fluted bifaces in Siberia, the betting money was put on accidental fluting.

And then there is the Yana RHS site, found on an ancient terrace of the Yana River at 70°N, some 100 kilometers from where it empties into the Laptev Sea. The site has yielded stone bifaces reminiscent of Clovis, as well as bone foreshafts made of woolly rhinoceros horn and mammoth ivory that look startling like specimens found in Clovis

FIGURE 35.
The "fluted" biface from the Uptar site, Siberia. Its flute is more likely a fortuitous technological coincidence than evidence of ancestral fluting in Siberia. (Photograph courtesy of Maureen King.)

assemblages (see Plate 7).[7] Yet, Yana is 27,000 years old, making it 16,000 years older and tens of thousands of kilometers from the nearest Clovis site, with a full-blown episode of Pleistocene glaciation in between. The demands of a Solutrean-Clovis connection equally apply here: great gaps in time and space must be filled before solid links can be forged.

On the American side of Beringia, a possible Clovis precursor is the Nenana Complex, found in the Nenana and Tanana river valleys of interior Alaska. Its artifacts include distinctive Chindadn points (Figure 36), which are small, straight based, and triangular or tear-dropped in shape, as well as large bifaces, scrapers, gravers, and large planes—the latter evidently used for woodworking. The initial radiocarbon ages for the Nenana sites of Moose Creek and Walker Road clustered around 11,800 BP, putting them in Alaska at the right time to be a Clovis precursor. On the supposition its toolkit resembled Clovis in types and technology, it was hypothesized to "represent the same population."[18]

The timing is right, though recent re-dating of several of the Nenana Complex sites showed them to be 500 years younger than previously thought, clustering around 11,300 BP. That makes them roughly contemporaneous with Clovis. Yet, Nenana assemblages lack fluted points. That absence has been downplayed since the small, non-fluted Chindadn points are reminiscent of miniature points from Clovis sites. But that similarity is more apparent than real. Small points only rarely occur in Clovis assemblages, and they are usually fluted—quite unlike the thin, fine, marginally retouched Chindadn points. And (as in the Solutrean case) Nenana sites also contain tool classes not found in Clovis.

FIGURE 36.
Chindadn points from the Walker Road site, Alaska, diagnostic of the Nenana archaeological complex. (Photograph courtesy of Richard Reanier.)

But taking a step back from the tool forms, it is apparent that Nenana and Clovis technologies are broadly similar: both contain elements of biface and blade technologies, unifacial tools made on biface thinning flakes, and bone and ivory rods fashioned in a similar manner. Again, the fit is hardly perfect, but the discrepancies could well reflect different adaptations or disparities in access to useable stone (which likely explains the lack of large, elegant bifaces and blade cores in Nenana sites). In any case, there are more similarities and fewer differences between Nenana and Clovis than between Clovis and Solutrean. Best of all, there are no Ice Age oceans to cross or yawning gaps in time to close.

Archaeological affinities to Beringia and northeast Asia match where Turner's dental evidence places the ancestral Americans. Recall he identified them as descendants of Sinodonts, and not the Sundadonts of Southeast Asia, whom he expressly barred from the New World (Chapter 5). But recall, too, that some studies of early American craniofacial patterns point to affinities with populations from Southeast Asia (the Ainu, Polynesian groups, and others) and western Eurasia. It's dangerous (and frivolous) to jump to the conclusion the first Americans' heads came from one place and their teeth from another, nor do evolutionary principles allow that leap, especially given the analytical baggage these studies tote: few, fragmentary, and unrepresentative samples of crania, and questionable groupings of teeth. When there are representative samples, and a better understanding of the evolutionary processes in play among small and widely scattered populations, that will be the time for jumping to conclusions—if necessary.

The genetic evidence (and that includes autosomal DNA) also points unerringly to Asia as the homeland for the first Americans. American Indian mtDNA shows affinities to Asians who occupy a broad swath from the Altai Mountains to Japan and Korea. Like the situation in the Americas, haplogroup frequencies there vary by region: almost all Siberian and East Asian populations, for example, lack haplogroups B and X. The two major NRY haplogroups have an even wider distribution outside America, haplogroup Q occurring at varying frequencies throughout the world. Even so, Zegura and colleagues found haplogroups C and Q intersecting in the Altai region, where (as noted) there are also populations harboring various of the mtDNA haplogroups. What's not seen in the Altai region, however, is a distinctive autosomal marker—D9S1120 in the DNA nomenclature—that occurs throughout the Americas, but only in a couple of Asian populations (Koryaks and Chukchi), indicating it was part of the ancestral founding population, but one that is evidently no longer represented across Asia—if it ever was (its presence there today might be from a more recent back-movement of people from America to Asia).[19]

Of course, and as earlier emphasized, Asian populations have hardly stood still since the Pleistocene. The rich diversity of genetic lineages in the Altai region could result from post-Pleistocene gene flow of multiple Eurasian populations forming, as Connie Mulligan and colleagues describe it, "a genetic melting pot [that] could easily contain all the elements common in Native Americans and falsely give the impression of being *the* ancestor."[20] Again, merely because the closest living mtDNA or NRY relatives of American Indian groups live in those parts of Asia today does not mean they were living there in Pleistocene times. We await proof on this point from *a*DNA.

DEPARTURES AND ARRIVALS

On the matter of timing there is agreement, but again mostly in general terms: all lines of evidence point to a late Pleistocene departure for the New World. Turner's dento-chronology puts the split between Amerind and Northeast Asian populations at roughly 14,000 years ago, in time for a late Pleistocene Clovis arrival. Not a bad number, but dentochronology gives discrepant results in other cases, and cannot be wholly trusted in this one.

Greenberg, though well aware of the limits of glottochronology, but with an abiding faith his language families correlated precisely with Turner's dental groups, likewise put the arrival of ancestral Amerinds in Clovis times, though expressed a willingness to extend the date if needed. Others saw a need. Based on presumed rates of linguistic change and the diversity of languages, Johanna Nichols argued a Clovis antiquity was not even in the right ballpark, and instead proposed there had been a single entry 50,000 years ago, or multiple entries beginning at least 30,000 years ago. When asked why archeologists hadn't found such old sites, Nichols replied, "As a linguist, that's not my problem."[21]

Estimates based on molecular clocks vary, and in many cases have large error bars attached to them (recall Tables 8 and 9), depending on which source is consulted, which method was used, and a variety of other contingencies. This is especially true of mtDNA; age estimates are tighter over on the Y chromosome (ranging from 14,000 to 18,000 years), but it's not irrelevant to add there have been far fewer studies calculating ages from Y chromosome DNA. In any case, none of these estimates directly dates the migration, only the emergence of particular molecular lineages, which may or may not coincide with divergence.

There is abundant evidence of Clovis in North America soon after 11,500 BP. A single site, Monte Verde, gives secure evidence of a still-earlier human presence. Many more are reportedly earlier, but are on less secure grounds. Given the location of Monte Verde, and its distance from the presumed Beringian entryway, we assume people arrived in the New World much earlier than the 12,500 BP age at MVII implies. How much earlier we cannot say, although from what we know of the timing of the opening of the coastal route, it was possibly a couple thousand years earlier. If they came down an interior route before the ice-free corridor closed, then it might have been earlier still. Importantly, there's no reason why both routes could not have been used by early colonizers coming at different times—if there had been more than one early migratory pulse (more on that shortly).

Ironically, the archaeological record does provide one well-defined, easily followed trail of a Pleistocene population migrating south from Alaska. The Paleoarctic Complex is marked by stone microblades (Figure 37): these are 2–3 centimeters long and barely 4–8 millimeters wide; pressure flaked off small, wedge-shaped microcores; and presumably slotted into wooden, antler, or bone handles to form composite tools. Microblade/microcore assemblages arrived in southern Siberia about 22,000 years ago, and from there were carried rapidly north and east. They were in far northeastern Asia and western Beringia by 10,800 BP, and within a couple of centuries were being used in interior Alaska and the western Yukon Territory. From there the trail moves south down the coast and interior boreal forests. Microblade/microcore assemblages arrived in Juneau, Alaska, around 10,000 BP, and then at successively younger sites in central British Columbia (about 9700 BP), Prince of Wales Island (about 9200 BP), and the Queen Charlotte Islands (about 7000 BP), finally reaching the vicinity of the Strait of Georgia and the Columbia River in Washington around 5000 BP.[22]

This time-transgressive expansion of the Paleoarctic looks exactly like what an archaeological migration ought to look like. Unfortunately, it's not the migration we're looking for. The bearers of the microblade/microcore technology could not have been

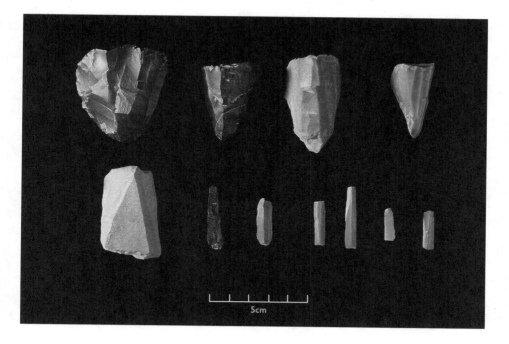

FIGURE 37.

Microblades from the Panguingue Creek site, Alaska. These would have been set into bone or antler handles, to be used as composite tools. (Photograph courtesy of Richard Reanier.)

the first Americans, for this technology does not occur in either pre-Clovis or Clovis sites, and arrived in North America too late to have been carried by the first Americans. Indeed, the southward advance of the bearers of this technology may have been checked when they reached areas already occupied.[23]

What of the tracks of the first Americans? Those have yet to be found, and may never be, given the processes that conspire against us, and the fact that we seek what may have been a fleeting movement on now-vanished coastline. For if the first Americans came down the Pacific coast, hugging the seashore, rising sea levels may have since drowned their footsteps (or boat landings); even where isostatic rebound and tectonic activity have uplifted the Pleistocene coastline, dense forest and under-brush hinder archaeological survey (Chapter 4). Nonetheless, there is plenty of reason to go look, for it has already been shown that humans occupied Prince of Wales and the Queen Charlotte islands along the British Columbia coast as early as 10,300 BP, and more importantly, it has been shown these places could have been occupied a great deal earlier.[24]

If colonists came via an interior route in pre-LGM times, all traces would have been obscured by later glacial action. Had they moved through the ice-free corridor soon after it opened, they likely moved quickly: hunter-gatherers tend not to dawdle in harsh and marginal environments, as the corridor was in the immediate wake of deglacia-tion.[25] At the moment, the earliest sites in the corridor region—Charlie Lake Cave in British Columbia, and the Vermillion Lakes site in Alberta—are dated at 10,500 and 10,800 years old, respectively. Although Charlie Lake Cave has yielded a broken and beat-up fluted point, it's not quite a Clovis point.[26] In fact, neither site is old enough to mark the southbound trail of Clovis people, let alone ancestors of Monte Verdeans.

In the absence of actual archaeological evidence of the path of the first Americans, evidence from languages and genes has been pushed into the breach. It's not a very satisfactory solution.

The argument from languages, ironically, is made by archaeologists rather than linguists, and is based on the simple observation there are more Native American languages along the Northwest and California coasts than in any other area of North America. That's said to imply a great time depth for human occupation, and thereby "strongly supports" a coastal entry.

To be sure, the greatest numbers of languages (nearly 40% of North America's total) were spoken along the West Coast.[27] Yet, as linguists well appreciate, the abundance and distribution of American languages is hardly a function of time alone (if at all), and owes more to geography, climate, population density, economy, and other factors. Thus, areas rich in resources (coasts, lower-latitude areas) allow large populations to con-gregate in small areas, which can produce great language diversity (the best example: Papua New Guinea harbors 862 languages, fully 13.2% of the world's total, on just 0.4% of the world's land area). In the cold and dry North American Arctic, with far fewer people inhabiting much larger regions, there are fewer languages spoken per unit area. Likewise, regions overrun by expanding empires have lower linguistic diversity because empires impose languages—think the Roman Empire and Latin—thereby reducing the number of indigenous ones.[28]

Consequently, the modern diversity or distribution of languages may say little about the prehistory of colonization, since these patterns may have little to do with when people arrived in a region, or where or how they moved over time.

And something more: Native Americans were hard hit by infectious epidemic diseases (smallpox, the most deadly among them) introduced by Europeans in the decades and centuries after AD 1492. Regions that at the time of European arrival were home to dense populations, such as the Mississippi and Missouri river valleys, were especially devastated since their populations were thick on the ground. The Spaniard Hernando de Soto wandered through what is now the southeastern United States between 1539 and 1542, and described a region teeming with people. When the Frenchman René-Robert de la Salle returned a century later, he was astonished to find virtually no one home. The very few languages recorded in that region serve as a grim illustration of population devastation. In contrast, a large number of languages are known from the Pacific Northwest and California because intensive linguistic fieldwork took place here before diseases took their full measure (particularly among remote inland groups).

All this makes it doubtful whether the number and distribution of languages bears any relationship to the length of time a region was inhabited, let alone to fleeting migration routes, particularly those taken tens of thousands of years earlier across a radically different, partly ice-shrouded landscape by people who may have been historically and linguistically unrelated (save in the most general sense that all were ultimately Asian in origin). It is probably no more realistic to infer Pleistocene migration routes to North America by the number and distribution of modern language groups, than it would be to infer de Soto's route in the sixteenth century by looking at the number and distribution of Spanish speakers in the Southeast today. And at least we know de Soto spoke Spanish.

Be wary, too, of the claim that patterns of mtDNA haplogroups along the west coast of North America "appear to support the hypothesis of a population expansion out of Beringia associated with gene flow along the west coast."[29] We cannot assume that migrating groups, like Hansel and Gretel, left behind a trail of genetic breadcrumbs—that is, people permanently settled on the trail of their Pleistocene ancestors. The presence of a haplogroup in a particular region does not mean its bearers or their ancestors have always been there, just as its absence does not mean their ancestors never passed through. Haplogroup B is proof of that: it is scarce among Bering Strait populations, yet present across the rest of the Americas. Unless its bearers sailed directly across the Pacific in Pleistocene times (and that's *very* doubtful), individuals possessing haplogroup B must have traversed Beringia without settling down.

Having failed to catch the first Americans in the act of colonizing, archaeologists and others have attempted to simulate the process by which the Americas were colonized, to gain a sense of where and how the first Americans may have entered the continent and dispersed once here. It's colonization by computer, and is done by building an equation with variables such as the number of individuals in the colonizing group, their rate of population increase, the distances they moved annually, what triggered the movement (usually, population growth filling up the home territory), what kinds of foods and other resources were available in different environments, what might guide movement (a path of least resistance, major drainages, or even fanning out across the landscape), how fast colonizers could or would move along different routes, and so on. Then, with the aid of the computer, one starts the algorithm iterating: a population arrives at time x, for example, and within y years has increased to the point of straining the local resources, and so moves z kilometers in the direction of the closest available resource-rich territory. That process is then repeated in the virtual world many times over, and after n years, the descendants of the initial group have filled the New World. It's not SimCity™ exactly, but close to it (SimColonization?).

David Anderson and Christopher Gillam simulate the process of expansion into the New World under the premise that humans, like water, flow downhill.[30] Or, more properly, that people will take the easier rather than the more strenuous route when moving across a landscape. From the point of view of colonizers arriving in western Alaska, Anderson and Gillam ask what would have been the easiest route of entry, bearing in mind they did not know they were supposed to be headed to Monte Verde, but were simply moving ever outward along a path of least resistance? To find the answer, they compiled high-resolution digital elevation maps of the New World, and analyzed them using Geographic Information Systems, which enabled them to calculate the relative ease of different routes through the continent. In doing so, they found that the least-cost pathway from Alaska to *all* land south of the ice was via the interior route—*if* the colonizers were walking. A coastal entry was never a least-cost solution, owing to its highly dissected terrain. Of course, if colonizers had boats, then the coast becomes a viable option (why walk when you can paddle?).

Todd Surovell likewise concluded a coastal colonization was implausible, based on a model driven by *return rates* (how much food can be acquired from a habitat in a given amount of time?).[31] Colonizers increase in number (at the rate of 3% per year, he assumes) and move when the population reaches the point where return rates diminish, and the group as a whole would be better off if some moved on to the next available unoccupied habitat to make their living. By the terms of Surovell's model, people can move down the coast, but if they did so all the way to Monte Verde, then they also should have "backfilled" interior North America with substantial pre-Clovis populations. Since in his view no such population exists, he reasons they must not have come via the coast (of course, there's no evidence such a population came via an interior route either, but leave that aside).

Weighing in on the genetics side, Alan Fix begins with the observation that mtDNA haplogroups A–D and X are widespread on the North American continent, but in varying frequencies. He then asks whether a single migration down the coast could account for that continent-wide distribution (granting changes in gene frequencies over the millennia since that pattern was established). By his model, and assuming the colonizers had boats, the answer is yes. A starting population of 25 females harboring all five mtDNA haplogroups and increasing at 0.7% per year, moving south each time the number of females reached about 175, would spread down the coast at the rate of 4 kilometers/year. Within 3,000 years, they could travel from Anchorage to Panama, jump the isthmus, and then make their way up the Gulf and Atlantic coasts, and deep into the Mississippi Valley. The end result would be a North American population with haplogroup frequencies that vary much as they do at present.[32]

All these are elegant models, but do they answer the question of whether the first Americans came via the coast or not? Not really. The merit of any simulation rests on whether the model incorporates relevant variables, and realistic values for those variables. Unfortunately, there's little hard data for many of the variables—that's why the process is being simulated—so the models instead assume a value or range of values to approximate an on-the-ground reality. Surovell pegs population growth at 3% per year; Fix, a population geneticist, dismisses that as too high and uses 0.7%. Both their results would change significantly were they to trade numbers. But then others have argued colonization was not driven by population growth at all. For that matter, because of computational complications, there are limits to how many variables can be incorporated in a model, and it would be conceit on our part to believe we have included (or even understood) all the variables relevant to the colonization process.

But most important of all, a simulation model, however elegant, should never be confused with actual evidence.[33] Relying as they must on a series of assumptions (even plausible ones), a model cannot prove a route taken: it can only show whether a hypothesis seems reasonable given its ground rules. Can it prove a hypothesis wrong? No, though it can show it unlikely or implausible. As always, the proof must be found in the ground. A few early, well-dated archaeological sites along either route would do the trick.

COLONIZATION'S TEMPO AND MODE

Joseph Greenberg was bemused by fellow linguists who insisted it was impossible to group American Indian languages into fewer than 150 or 200 families. If there were that many, each representing a separate migration, he joked it would have required "a traffic controller at the Bering Strait"[34] to keep them from bumping into one another en route to the Americas. Better, he thought, to reduce the number to a more manageable three: no stoplights needed. Turner's teeth go along with that, but not the craniofacial evidence, which at the moment sees just two migrations, though not always the *same* two migrations. Geneticists variously estimate there were anywhere from one to four separate migrations to the Americas.

In the absence of an answer about the number of migrations, perhaps it's time to rethink the question. Rather than try to gauge how many migratory waves are necessary to account for known patterns of diversity, whether genetic, linguistic, or anatomical, let's instead look from a Pleistocene perspective.

We know humans reached northeast Asia before the LGM, and were poised to depart far eastern Siberia by about 14,000 years ago—or possibly earlier (given what we don't know of the archaeology of Siberia). We also know (Chapter 2) that throughout the late Pleistocene, there were no significant barriers to cross-Beringian traffic: no glaciers, no looming mountains. And although it was cold, dry, and virtually treeless from before the LGM until at least 13,500 BP, that would not have stopped hunter-gatherers who'd spent their lives in the frozen north. Like their brethren on the treeless central Russian plain, they could have used animal dung, fat, and bones to fuel their fires. Would-be colonizers were free to move across the land bridge *and back* over tens of thousands of years.[35]

Getting south from Beringia may have posed a more formidable challenge, at least given our current understanding of the timing of availability of coastal and interior corridor routes: the interior route apparently was closed between about 28,000 and 12,000 BP. Long reaches of the outer coast were evidently ice free between 16,000 and 14,500 BP, and permanently open to the traveling Pleistocene public only after 13,400 BP. We are entitled to hedge a bit on the dates, given how they have changed over the last several decades.

Regardless, one point is clear: during the Pleistocene, there were several wide-open windows of opportunity and routes of travel by which to migrate from Asia to America (Figure 38). So instead of conceiving of early migration as a single pulse by a small group made but once, is it possible to think of it as Hrdlička did, as a "dribbling over from northeastern Asia, extending probably over a long stretch of time"?[36] Perhaps. But that depends on when people reached Siberia and western Beringia, on the size of the parent population(s), on the environmental, climatic, and social catalysts that might have spurred migration, and so on. These matters are currently poorly known.

Linguists have not precluded the possibility of multiple migrations, for most are unwilling to assume—as Greenberg does—that the vast majority of American languages can be telescoped back into a single ancestral dispersal. Multiple migrations from unrelated source populations would better account for the diversity most linguists see in the languages of the Americas.

On the other hand, geneticists at the moment are less inclined to accept the idea of multiple migratory pulses, and for seemingly good reason: given the finite number

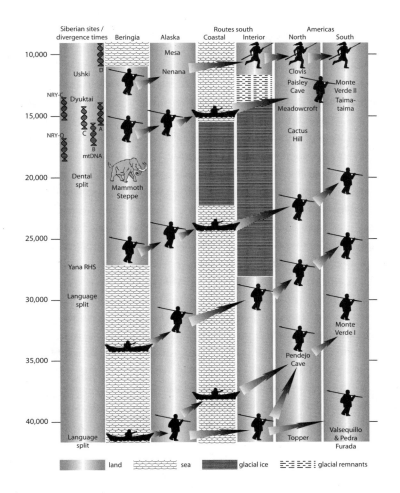

FIGURE 38.

Time chart for the peopling of the Americas, based on chronologies of dental, linguistic, and genetic lineage divergence (from text and Tables 7–9); the age of archaeological sites in Siberia; the timing of Beringia and of the coastal and interior routes south from Alaska; and the ages of archaeological sites and complexes in Alaska, North America south of the ice sheets, and South America, including a number of unconfirmed pre-Clovis claims. Ages are in radiocarbon years BP. The arrows and figures are not intended to convey actual migrations, but instead to show possible times and routes of travel, and how the timing of the migration could have constrained which routes were available (arrows that span either the coastal or the interior route indicate that route was not taken at that time). Running figures convey faster migrations—as supposed for Clovis. The scenarios currently favored are the top two sets of arrows and figures: an earlier entry down the recently deglaciated coast in time to reach Monte Verde by 12,500 BP, and/or a later entry down the newly opened and passable ice-free corridor at the outset of Clovis time. Obviously, as chronologies are refined and older sites are discovered and accepted, the favored scenario(s) will change. (Adapted from Meltzer 1993b.)

but widespread distribution of founding haplogroups and haplotypes in the Americas, it seems unlikely that separate migrations from Asia would introduce the same set (Chapter 5). Nor should these haplogroups, as Jason Eshleman and colleagues indicate, show the same level of sequence diversity—as they apparently do.[37] Strong rebuttals both, but perhaps not entirely fatal to the idea of multiple migratory dribbles.

Certainly, the recent findings from aDNA confirm suspicions that we have yet to record the full diversity of haplogroup and haplotype sequences in America past and present. Consider the Siberian source population: it is assumed that because modern Asian populations possess a large number of haplogroups in addition to A–D, then so, too, did ancient populations. Of course, there is no aDNA evidence on that score. Moreover, Eshleman and colleagues point out that "a random selection of even a small group of emigrants from Eastern Siberia today would have a high probability of including members of haplogroups A, C and D."[38] Those common haplogroups would likely show up in a random Pleistocene selection as well, perhaps even if these occurred repeatedly over tens, hundreds, or possibly even thousands of years, as successive groups departed for America.

So, too, there is the interesting but evolutionarily complicating possibility of earlier migrants returning to Siberia from Alaska: geneticists are now seeing evidence of just such back-migrations in the occasional occurrence of American Indian markers in Asian populations.[39] Those back-mixes appear to have occurred recently. Had any occurred when Beringia was open for pedestrian traffic, groups may have mixed with distant and possibly unrelated populations that by then had arrived in Siberia, becoming yet another source populations for migration(s) to America.

Geneticist Theodore Schurr observes, and rightly so, that "if multiple expansions of human groups occurred within a limited period of time, the demographic signals of these events may not be detectable." Twelve thousand years or more after the fact, over which populations mixed and moved, and experienced gene flow and genetic drift, the lines originally separating a stream of small and related (or even unrelated) groups are easily blurred. It might prove impossible using modern DNA to distinguish a single migration with multiple, contributing sources from repeated splitting from the same ancestral population, or from multiple, closely spaced dispersals from *multiple* source populations in Asia (Chapter 5). Studies of aDNA might help bring those original lines back into focus.[40]

Kari Schroeder and colleagues likewise admit "the number of migrations cannot be inferred from genetic data," but then turn the question on its head to ask, instead, are all Native Americans descended from the same founding population? There are certainly strong hints that was the case: there's D9S1120, that ubiquitous and almost uniquely American autosomal marker; there are within haplogroup C several founding haplotypes (C1b, C1c and C1d) so widespread as to indicate common ancestry; and there are sure signs in mtDNA, NRY, and autosomal DNA of a population bottleneck (Chapter 5). All of these point to a shared genetic prehistory and descent from a common founding population, but as Schroeder and colleagues add, by itself that does not reveal the number of source populations contributing to the founding population—the streams flowing into a Beringian funnel—nor exclude the possibility of small genetic contributions from other populations, which about 400 generations later hardly appear visible.[41]

The archaeological record provides some insight into the number of migrations, though it is a fairly blunt instrument. There are, for example, no obvious historical or technological links between the pre-Clovis-age complexes of South and North America: they form no coherent patterns, share no specific technologies, and are marked by no common artifacts. On the other hand, perhaps there is no reason to expect similarities among artifacts so widely separated in time and place, especially since it is not yet apparent whether all these sites are credible.

As for Monte Verde (which is credible), its toolkit seems very different from Clovis in North America, and from Clovis-age South American archaeological complexes (again, there is no Clovis, strictly speaking, in South America). Perhaps Monte Verde and Clovis represent two distinct archaeological traditions and separate migratory pulses, as opposed to an outgrowth of the same historically related colonization. At the moment, the call could go either way.

Filling in the gaps in space and time between Monte Verde and Clovis—more pre-Clovis sites in North America will surely help—may make the differences between the two disappear (or, perhaps, grow larger). Then it will become possible to determine whether they represent the same or separate migrations. For the moment, the best we can say is that there are at least two Pleistocene-age migrations into the Americas: Paleoindian and (later) Paleoarctic (this does not include later Holocene migrations by, say, ancestral Eskimo groups). It is possible there were even more migrations, major or minor, including some that may not have been successful. But to show that, we will need to learn a great deal more about the earliest archaeology of the Americas.

Given the timing of the opening of the ice-free corridor, and the sudden and widespread appearance of Clovis soon thereafter, the early peoples they represent could have come down the newly opened corridor, then radiated out across North America and even into South America in a matter of centuries (in actual calendar years, the radiation may have taken longer since radiocarbon calibration during this period is still imprecise). That's far faster than the current American record for prehistoric migration in an empty niche by hunter-gatherers, now held by the Thule Eskimo, who, starting around AD 1000 (or possibly later[42]), flashed from Alaska across the northern maritime region to Greenland, a distance of more than 5,000 kilometers in a few centuries (there's more on the Thule below and in Chapter 7 since they represent in many ways the closest and most interesting comparative case).

But then, Clovis might not represent a separate migration. As previously noted, there is no well-marked trail of Clovis and Clovis-like fluted projectile points leading from

Alaska down to the Lower Forty-eight, which one might hope to see from the movement of a people, even one moving quickly. And if Clovis does represent a separate and later migration of people, the fact that they moved through North America so quickly suggests they did not encounter many (or any) other people along the way. If so, what had become of earlier, pre-Clovis colonizers? It's a fair question, but one at the moment without a fair answer.

But perhaps the sudden appearance of Clovis and the opening of the ice-free corridor are mere coincidence, and Clovis represents instead a new technology that developed in America and spread widely and rapidly among a preexisting, pre-Clovis population. Given that Clovis sites from west to east are nearly contemporaneous, the spread must have been lightning fast, on a scale that would not be seen again until the invention and spread of the hula hoop, and of a technique (fluting) that—like the hula hoop—"was not all that revolutionary in the first place and that died out relatively soon thereafter"[43] (fluting, however, enjoyed a much longer popular run than the hula hoop).

At the moment, the evidence leans against the idea of Clovis as the diffusion of a preexisting technology. Diffusion demands a faithfulness in copying the Clovis technology by many widely scattered groups, which seems unlikely across so large a span of space and time. Further, we ought to see not just a substantial pre-Clovis presence in North America, but also evidence of Clovis technologies being grafted onto (and replacing) "fluteless," pre-Clovis assemblages. We see neither. Diffusion seems less likely than an on-the-ground, rapid movement of people.[44]

On the matter of speed, very recent mtDNA evidence points to a long pause in Beringia, followed by a rapid expansion across the Americas, notably in the three founding haplotypes (C1b–C1d) within haplogroup C. That all three apparently arose in Asia, show similar molecular ages, and yet are widely and uniformly distributed across this hemisphere—as opposed to exhibiting a "nested" structure that would imply they had derived from one another in a serial fashion—points to a "swift pioneering process, not a gradual diffusion." In haplogroup C, it would appear, the lineages took a long time to evolve in Beringia (the 10,000–20,000-year standstill mentioned in Chapter 5), but a very short time to expand.[45] In contrast, autosomal DNA displays a north-south gradient of genetic diversity, but whether that indicates a more gradual and progressive splitting as the ancestral group moved deeper into the Americas—or even says anything about the speed of expansion—is unclear.[46]

CAN WE ALL GET ALONG?

There is only the broadest agreement among the different disciplines on the central questions related to the peopling of the Americas. But perhaps that's to be expected, given their very different approaches. Linguists and geneticists view the process through the present, relying on known languages and DNA of living Native Americans (aDNA has so far played a limited role in this investigation). From this they infer population histories, assuming that modern American Indians are descendants of the first Americans (we have no evidence, of course, to indicate otherwise).

Archaeologists and physical anthropologists, working with sites and skeletal remains, approach from the opposite direction. But we see only what the Pleistocene left behind, often little more than isolated snapshots of sites and skeletal remains, which do not occur in sufficient numbers to allow many conclusions about the populations from which those samples were drawn.

Accordingly, there are significant differences in data and method, and advantages and disadvantages to each. That's obvious.

What is often less obvious, especially from the outside looking in, is whether those discipline-specific results stand on their own merit. Science in the twenty-first century is a highly insular affair, and it takes time for knowledge from one field to fully penetrate another. Even with all the linguists' ear-piercing criticism of Greenberg's method of mass comparison, it was more than a decade before geneticists examined whether mtDNA and NRY haplogroups actually coincided with Amerind, Na-Dene, and Eskimo-Aleut, as opposed to merely assuming they did. And, of course, they didn't.

What this means is that although each discipline has a place at the table, no one gets a free pass: the results of research in one area are not inherently superior to another (notwithstanding Jantz and Owsley's claim that cranial morphology will be "more informative" than genetics), nor can we simply pick results we like. Equally important, we have to clarify just what different disciplines can and cannot reveal. Take the matter of chronology. Geneticist Antonio Torroni can *say* that "genetics *confirms* the first Americans were pre-Clovis,"[47] but can we really *know* that? A genetic haplogroup cannot be directly dated; only archaeological remains can. And at the moment, archaeologists are in the best position to *know* when the first Americans have arrived, but we're not yet able to *say*. Yet, by joining forces, we can perhaps bracket an answer.

Consider what archaeological ages represent. At the beginning of American prehistory, there were so few people spread across such a vast area they could not have left much of an archaeological imprint. The odds of our finding their sites are correspondingly slim, and people likely got here much earlier than we see them archaeologically (Chapter 4). Unless we are lucky enough to find the very first site in the Americas and to know we have found it, the oldest site we do find will at best provide a minimum age for the peopling of the Americas.

On the other hand, molecular clocks provide a maximum age estimate for the peopling of the Americas. This is so because—leaving aside all the technical concerns about the ticking of these clocks (Chapter 5)—if a haplogroup occurs in both Asia and America, it must have arisen within the ancestral Asian population prior to the departure of ancestral Native Americans. That being the case, and using the oldest ages currently provided by genetics, the first Americans must have arrived sometime after, say, 40,000 years ago, but before 12,500 years ago (Monte Verde MVII). A broad window, to be sure, but we may soon close it considerably if very recent molecular estimates converging on a time range for both mtDNA and NRY of 18,000–14,000 years ago hold up.[48]

CONTINUITY AND DISCONTINUITY

Whether they do depends in large part on whether we have an accurate count of the founding lineages on which the molecular clock runs, and that in turn depends on whether there has been lineage loss over time. There is strong reason to suspect there has been. Ancestral Americans were particularly susceptible to mtDNA and NRY lineage loss at two points in time: the first was immediately after their Pleistocene arrival. Small, isolated, rapidly moving bands, with attenuated ties to distant homelands or kin, and without nearby populations with which to seek mates, could easily disappear in a few tens or hundreds of years for want of sufficient numbers, or an ability to cope with events in their environment. Likewise, relatively rare haplogroups can easily disappear (haplogroup X, for example, very nearly did). There is evidence from mtDNA among North and South American Indians for at least a 'moderate-intensity' bottleneck and the potential loss of some genetic diversity.[49] But that's not surprising: almost certainly over the grand sweep of world prehistory, groups vanished without descendants. Recall that the Norse reached Newfoundland 1,000 years ago, only to soon beat a hasty retreat in the face of extreme circumstances, not least hostile Native "Skraelings" (likely, ancestral Beothuk Indians). One would assuredly look in vain for genetic evidence of the Norse presence in Newfoundland. The first Americans did not have to contend with hostile Natives (no one else was home), but as newcomers, they faced significant adaptive challenges (Chapter 7).

The second time American Indian mtDNA and NRY lineages were likely to have vanished was in the centuries after European contact, which caused a massive, centuries-long population decline as a consequence of introduced infectious diseases (more on this in Chapter 10). Although these terrible scourges devastated Native American numbers, the major mtDNA and NRY haplogroups (A–D, and C and Q) survived, though potentially in very different frequencies than prior to 1492.[50] It is apparent from aDNA that other minor DNA haplogroups and haplotypes did not. Languages were assuredly lost, too, as the blank areas on the map of American languages attests (Figure 30).

Before proceeding any further, let me be clear on a couple of points: first, lineage loss detected in aDNA is *not* evidence American Indians are not descendants of the first Americans. Nor, in fact, does it preclude genetic continuity in autosomal DNA (again, one has many more autosomal ancestors than mtDNA or NRY ancestors). What it *does* mean is that these single locus markers—the five mtDNA and two Y-chromosome founding haplogroups of modern American Indians—may not be fully representative of Pleistocene or even pre–European contact mtDNA and NRY.

Second, it is only in Pleistocene archaeological sites or skeletal remains that evidence can emerge of an early but ultimately unsuccessful migration to the Americas in the form of artifacts or skeletal forms different from those that followed. Although there are assertions early American crania are morphologically distinct from those of modern Native Americans, we cannot put stock in these claims, plagued as they are by severe sampling and analytical problems (even the dental evidence, problematic though it is, testifies to continuity between early and late Americans[51]).

Which brings up once more a matter of terminology. It used to be that the first Americans were referred to as *Paleoindians* (from the Greek *palaios* [παλαιός or old + Indian). In the last decade in some circles, there has been a none-too-subtle change in language: Paleoindian is now insistently referred to as *Paleoamerican*. The terms may be synonymous in a very general sense. If the ancestors of American Indians are the first Americans, Paleoindians by definition are Paleoamericans. But the explicit effort to substitute one for the other is not just a matter of synonymy. The change corresponded with the Kennewick discovery and ensuing legal battle, and was made most visibly in the publications of the *Center for the Study of the First Americans*, whose director (the late Robson Bonnichsen), led the lawsuit against the federal government to gain access to the Kennewick remains. As David Hurst Thomas writes, naming something is not always a neutral act, and it's not in this instance.[52] Calling the first peoples Paleoamerican rather than Paleoindian is not just a statement about the genuine uncertainty of identifying ancient remains with modern tribes; it subtly implies the first people to the New World were not ancestors of the American Indians.

Although we may not yet know the population history of the peoples of the New World, and there remains debate about the precise number and timing of migrations, at the moment all indications are that today's American Indians are descended from the first Americans. *Paleoindians* they were and should be until reliable scientific evidence indicates otherwise. Using the term Paleoindian helps us not only appropriately recognize that ancestry, but it also serves as a reminder that the past is different, and that we are speaking of peoples who—though related to American Indians—lived at a very different time and in a very different kind of place. There is reason aplenty to expect their lives to have been lived in very different ways (Chapter 7).

SEEKING CONVERGENCE

All of which raises a final, much larger question: should we necessarily expect convergence between archaeological and non-archaeological approaches? Perhaps, but again only if we see something akin to what Goldilocks saw: that each group had its own, and only its own, artifacts, anatomy, language, genes, and teeth, which co-evolved over time. Sometimes we do.

Take the case of the Eskimo: their languages, as even warring linguists agree, form a continuum of dialects that stretches from the Seward Peninsula in Alaska across arctic Canada to the coast of Greenland. Mutual intelligibility between contiguous dialects is high, though speakers at the extremes of the range might have difficulty understanding each other.[53] Similarly, biological and genetic evidence (including mtDNA, Y chromosome, and autosomal DNA) indicates Eskimo peoples, regardless of geographic location, are quite similar genetically. That similarity points to a recent and rapid expansion of their ancestral population.[54]

How recently and rapidly can be tracked on the ground. The ancestors of Eskimo peoples—known archaeologically as the Thule—are first spotted on islands in the Bering Sea and around Seward Peninsula about AD 600. They were skilled bowhead whale hunters (among other things). As the climate warmed around AD 1000 and reduced summer pack ice across the high Arctic, they set off in pursuit of western bowhead whales migrating east into feeding grounds of the Beaufort Sea. Over the next several warmer-than-average centuries (the Medieval Warm Period), as whale populations expanded still farther east (ultimately linking up with North Atlantic pods), the Thule went right along with them, eventually reaching northwestern Greenland . Their language, genes, and material culture converge on a common population prehistory, and explain the biological, linguistic, and cultural similarities of Eskimo peoples between the Bering Strait and Greenland.[55]

About the same time the Thule were expanding across far North America, ancestral Polynesians were sailing double-hulled canoes eastward across vast stretches of the remote Pacific, and settling distant and isolated islands like New Zealand, Hawaii, and Rapa Nui. One thousand years later, as Patrick Kirch and others have shown, the linked linguistic, genetic, and material culture imprint of that expansion from a common ancestor is still visible.[56]

Among the colonizers of the high Arctic and the remote Pacific islands, artifacts, language, and genes co-evolved as descendant populations dispersed over time and space. That's why Captain Cook so readily spotted the similarities between Alaskan and Greenland Eskimo, and among the residents of Tahiti, New Zealand, and Rapa Nui. That these two cases meet the Goldilocks standard is not mere happenstance: the high Arctic and the Pacific were both peopled in just the last few thousand years, with populations expanding into remote and largely uninhabited regions (entirely uninhabited, in the case of the far Pacific), leaving the archaeological tracks of their dispersals still relatively fresh and untrampled on the landscape. And they remained largely isolated by virtue of their remoteness, which also explains why it wasn't until the late eighteenth century that Europeans made lasting contact, and why the echoes of their dispersals are still within reach of historical linguistics, genetics, and even of memory and folklore.[57]

It may be there was comparable convergence early in the peopling of the Americas, when there were still few of them, and they were maintaining contact even as they radiated out across their new landscape. Yet, as time went by, and populations grew, split, reached the far corners of both continents, and along the way developed new dialects and languages, artifacts, and their own distinctive genetic lineages, the situation inevitably became more complex. Matters would hardly get simpler over the next 400 or so generations as populations encountered one another and exchanged artifacts, words, and genes, and as culture, language, and genetic lineages evolved (or were lost).

Once isolation breaks down and interaction re-commences, there is no longer a reliable correlation between artifacts, language, teeth, and genes. People can quickly learn a new language or adopt new artifacts; they cannot quickly "learn" new genes or teeth. And that is why Hunley and Long found that striking mismatch between language families and haplogroups on both the mtDNA and NRY sides of the aisle. There's no reason to assume the genetic ancestry of a group will yield the same result as a reconstruction based on its linguistic history.

And what of linking modern languages or genes with archaeological or skeletal remains from the Pleistocene? Strictly speaking, the archaeological record—artifacts, dental and skeletal remains, sites—only speaks to the appearance of a colonizing group. At the risk of stating the obvious, artifacts, bones, and teeth speak no language and have no genes (unless we recover aDNA). Archaeologists have trouble enough deciding which stone tool assemblages belong together; we are even harder pressed to link archaeologically detectable patterns with those identified among modern languages or genes.

Here, again, is the great potential of aDNA. If we can recover aDNA from Pleistocene dental or skeletal remains, and see those same haplogroups in modern populations, we have the potential to firmly link the most ancient and most modern Native Americans, determine the number of migrations, from whence and where they came, or even the route(s) traveled—assuming aDNA is recovered from remains found along the way. As more ancient and autosomal DNA work is conducted—and it's important to emphasize that it's going to take many large-scale genetic analyses to produce a consensus picture of the peopling of the Americas based on genetic data—the picture of genetic diversity will likely become more rather than less complicated, but at least our molecular clocks will run on time.[58]

Until then, questions about how far into the past we can follow the trail of modern languages or biological populations, or what we can we say about the biology or language spoken by Clovis or pre-Clovis peoples, or more broadly, whether American Indians are descended from groups we see archaeologically as Clovis or pre-Clovis, will remain loaded questions and strike at the heart of the highly contentious legal, political, and emotional issues related to the identity and "ownership" of ancient skeletal material. But that's an issue for later (Chapter 10). Now that the first Americans have arrived, it's time to investigate their exploration of what was a truly New World.

7

WHAT DO YOU DO WHEN NO ONE'S BEEN THERE BEFORE?

The helicopter pilot pointed to the Arctic tundra below, where a pair of grizzlies were ambling in the distance. His voice came over my headset: "Did you want to go down for a closer look?"

"Sure."

We dove for the deck and were soon skimming along fifteen feet above the surface, headed for the grizzlies. The wind was against us, but the bears soon enough heard the thumping rotors and raised up on their hind legs to look around. The instant they spotted us, the female started galloping away. The big male took off running, too—straight for us.

Right. I'd seen enough. At the last minute, the pilot shoved the collective, and up we soared in a stomach-churning climb. Though we passed over the grizzly with room to spare, I looked down in amazement as he leaped and swatted at the helicopter as we flew past. The snarling maw of an angry grizzly bear is terrifyingly large when it fills the Plexiglas window right below your feet.

Up to that moment, as a rookie to archaeological fieldwork in remote Alaska, I'd not taken very seriously the fact that when surveying for sites, we always went out in pairs, or that one person always carried a twelve-gauge shotgun primed with lead slugs in case of a grizzly encounter. And I'd laughed at the ghoulish jokes: you don't have to outrun the bear—just your hiking partner. And the gun wasn't to shoot the grizzly—it was to kneecap your buddy to guarantee you could outrun *him*.

But seeing that this grizzly bear—weighing more than 1,000 pounds, moving at top speed, and capable of giving us a good rattling had he gotten his paw on the helicopter's skids—thought nothing of charging a roaring, jet-engine helicopter made it all too clear that when on foot, humans are not necessarily at the top of the food chain on this landscape.

Which got me to wondering: what must it have been like more than 12,000 years ago for those Ice Age peoples who first colonized the Americas? Although adept and well adapted to hunter-gatherer life, they were armed with little more than wooden or bone spears tipped with stone projectile points. They were assuredly predators; occasionally, they could be prey. More challenging still, they were colonizing a landscape far different and in many ways far more complex, diverse, and dynamic than at present. Although they were seasoned veterans of life in Siberia, as these wide-ranging peoples made their way across to Alaska and turned south into what was truly a New World, the land and its resources became increasingly unfamiliar. More than that, they were facing those adaptive challenges alone. It must have been a strange sensation to look around and realize it had been months, years, decades, or even longer since they had last encountered people outside their band, or seen telltale human signs like smoke on the distant horizon, or evidence of a freshly killed and butchered animal.

Those of us who spend time in wilderness today are afforded only the tiniest glimpse of what life might have been like for humans on a continent empty, unfamiliar, diverse, and rapidly changing. Yet, no matter how remote an area we might find ourselves, we are never truly alone nor, for that matter, in unknown terrain. Virtually every inch of our planet has been charted if not trod upon. We can carry survival gear and food, guides to plants and animals, maps to water and trails, and instruments that can tell us precisely where we are, or enable us to talk to others in distant cities or flying 30,000 feet overhead. In most any emergency, we can have vital supplies brought in, or people carried out. We rarely have the sense of being completely out of reach, or having to survive on our wits and by our hands—except, of course, when things go badly wrong.

The Pleistocene colonizers of the Americas had no such advantages.

PEOPLE WITHOUT NEIGHBORS

We're about to venture farther into unknown territory and explore questions about why people may have left Asia for America, what kinds of adaptive challenges they faced, and how they may have met those challenges and colonized a continent, possibly at record speeds. We won't have much help: although there is a rich body of anthropological knowledge about modern hunter-gatherer adaptations and processes of migration and colonization, much of it is irrelevant.

The reason? Our world filled up long ago. Nowadays, migration mostly involves moving relatively short distances over brief periods of time into already-occupied landscapes—hardly the scale or circumstances of the colonization of America, which played out much longer and over a much larger area. Moreover, all anthropologically known hunter-gatherers have neighbors, near or distant, on whom they can rely for information, resources, and potential mates (the latter are especially important, since one can always gain information and resources on one's own), or against whom they may compete. Either way, having neighbors fundamentally changes the decision-making calculus of landscape use. Finally, our anthropologically known hunter-gatherers possess deep knowledge of their landscape: not complete knowledge (no group ever has that), but knowledge sufficient that it has reduced much of the uncertainty in their lives. The process of hunter-gatherers learning a wholly new landscape has rarely, if ever, been recorded.[1] By the time anthropologists arrived on a scene, most indigenous peoples were already long resident on their landscape, had mapped its major features, knew what resources were available and where to find them, and were efficient foragers.

All that said, we can still use anthropological knowledge of the processes of human migration and adaptation observable in the present as a key to explaining the past: it's the uniformitarian principle, borrowed from geology and applied to archaeology. That doesn't mean the present and past are identical, but rather that ideas and explanations about the past should begin with (and must not violate) what we know of, say, general principles of human adaptation, or the limits of mobility or demography. The discussion below aims to do that. And, of course, we can supplement the anthropological record with what has been learned archaeologically of other cases in the near and distant past of hunter-gatherers colonizing new lands.

We will likely never find the very earliest sites in the Americas (Chapter 4), nor see the traces of the very first people who, say, encountered an angry rattlesnake (no Siberian *ever* experienced one of those), ate a poisonous New World plant, or looked around one day and realized they were alone or lost. Even so, we can broadly surmise how they may have responded if and when they did, and put such notions to the test against the archaeological record of the first Americans—or at least that of Clovis (there's not enough of a pre-Clovis record to allow such a test). Clovis was the first large-scale occupation of North America; possibly represents a separate migration; had access to the continent that appears to have been essentially unrestricted; and even if not the first to arrive, was the first to make substantial inroads (Chapter 6). As such, understanding Clovis can yield insight into the adaptations of early colonizers on what was at the very least a new and underutilized landscape.

WHY MOVE?

Let's start with motivation. What would prompt a band of hunter-gatherers to leave Siberia for the New World, or once here, expand the length and breadth of the continent? Studies of modern peoples who migrate find they routinely know something about what the destination has to offer, and decide to move or not accordingly. Indeed, it is an anthropological truism that people are unlikely to migrate into areas about which they have no prior knowledge. Yet, at a certain point in the peopling of the Americas, there was little choice in the matter. *Someone* had to be the first to set foot on this continent.

History records many reasons why groups venture from their homeland, and these can be divided into two broad categories: negative factors in the home region that *push* a people out, and positive attractions in a distant area that *pull* them in. The push side of the ledger includes a decline in resources, environmental or climatic stress (such as drought), overpopulation, exile, strife, or warfare. Groups might be pulled from their homeland by the lure of preferred resources (such as in the example of the Thule whale hunters); by the seeking of new economic or trade opportunities or adaptive "insurance" (ascertaining where to go when current circumstances deteriorate); or by motives as intangible as curiosity, a sense of adventure or wanderlust, the joy of discovery, or the journey to gain prestige or mates.[2]

Of the peopling of the Americas, we cannot say which push and/or pull forces were motives, but perhaps we can identify which were not. First and foremost, it seems highly unlikely that siphoning off overpopulation in their Beringian homeland—one of the most common forces pushing historic population out-migration (think of the post-sixteenth-century desire to escape crowded Europe and make a new life in America)—was a factor in the colonization of the Americas. There is no archaeological evidence of population crowding in Siberia or Alaska: quite the contrary, in fact. And the first Americans traveled much farther and much faster than they had to if they were merely looking for uninhabited land to settle.[3]

Likewise, we have no reason to believe exile, strife, or warfare played a role. Conflict that creates refugees is a much-later phenomenon in prehistory, initially occurring when village-based farmers arc past the tipping point when populations have expanded to the degree they must compete against others for food or the arable land to grow it.

Still, there are push factors that might be relevant to the peopling of the Americas. Northeast Asia and Siberia in the late Pleistocene were not easy places to live, and harsh glacial climates may have driven human foragers out of the region. They appear to have fled during the LGM, but toward the south, and not evidently toward the east and the Americas. Other climate changes at later times may have played a role. But caution is in order here: throughout prehistory, major changes in climate have occurred without flinging colonizers outward, and many colonization events had no environmental trigger.[4] We must be mindful of the fact that people can be agents of their own actions.

Another potentially important impetus was economics, though this is not to be thought of in modern market terms, such as setting up shop in a nearby town to make and sell widgets because the home widget market is saturated. Rather, it is akin to cases where, for example, only the first-born child inherits family land. Later-born children who receive nothing must seek new territory where they can found their own house and lineage, and assure their offspring access to quality resources. Polynesian oral traditions of ancient exploration of the Pacific abound with stories of younger brothers who left to colonize new islands.[5]

In terms of pull variables, groups headed toward the New World might have hoped—but could not have known—they would find prestige or prized resources or trading partners here. Obviously, for the first Americans, there were no trading partners to find. But perhaps the unknowing hope there were such partners might have been motivation enough. People setting off for new lands did not have to be right about what they might find in order for it to be a reason to go looking for it.

For that matter, one should never downplay curiosity, wanderlust, the joy of discovery, and the like, even though such motivations are impossible to see in the archaeological record. Over the tens and hundreds of thousands of years modern humans spread around the globe, surely some were motivated by little more than wanting to see what was over the next hill.[6]

Curiosity, in fact, has an adaptive underpinning. Hunter-gatherers live off the land, but conditions change over the course of a year, several years, and longer. Plant and animal density and distributions fluctuate, particularly as a consequence of human predation or exploitation. Knowing this, hunter-gatherers routinely scout out places they can go when local conditions begin to sour (such as when the firewood supply is exhausted, the animals are hunted out, or the plants are no longer in season).

MOTIVE TO METHOD . . .

So how do groups move across empty and unfamiliar landscapes? Clovis poses a challenging case, since its spread occurred quickly across a wide geographic area, possibly in a matter of centuries and certainly no more than a millennium—or so it appears, judging from its narrow range of radiocarbon ages and the continent-wide similarity of Clovis and related toolkits.

As a very rough measure of Clovis movement, dividing their maximum geographic range (in kilometers) against the duration of the Clovis period puts their expansion at a net 10–20 kilometers per *year*.[7] That's not much by our measures: many of us have longer daily commutes. But it's extraordinarily fast by hunter-gatherer standards, and furthers the suspicion Clovis people encountered few (if any) others along the way. Yet, the question of how fast they moved is not nearly as interesting as how or why they moved so far, so fast, in a new (or at least people-free) landscape.

There are examples of archaeological and ethnographic groups moving that quickly across landscapes and, for that matter, seascapes, but few of pedestrian groups moving that fast across an empty landscape, let alone an empty continent. The closest comparisons are to the colonization of the high Arctic, and that's no surprise: that's an environment that lends itself to lightning traverses. Consider the cases of the Paleo-Eskimo and the Thule.

The Paleo-Eskimo (badly named: there is nothing to indicate they are ancestral Eskimo or spoke an Eskimo language; today's Eskimo refer to them as the *Tuniit*) moved across the high Arctic from Alaska to Greenland between 5000 and 4500 BP. Their origins are uncertain, but they carried distinctive stone tool assemblages, found in sites that resemble those of the Siberian Neolithic, where they seemingly originated. These were groups that exploited seasonally freezing Arctic coasts and especially the adjacent interior highlands. They traveled quickly and left little behind: their sites are few, widely spaced, and sporadically occupied by what must have been small populations. As with Clovis, vast stretches of ground were crossed but not occupied.[8]

How the Paleo-Eskimo fueled themselves is puzzling, given their lack of time to accumulate knowledge of complex environments and resources. Robert McGhee offers the tantalizing hypothesis they could do so by targeting musk ox, which had evolved the unfortunate defensive strategy of responding to predators by bunching together, with rear ends inward and massive heads and sharp, hooked horns aimed outward. That strategy may have been effective against wolves, their longtime natural enemy, but proved ill adapted when a new predator came on the scene. To humans armed with a new and deadly weapon—the bow and arrow—a circle of musk ox was a hunter's dream: large, stationary targets that could be shot at relatively close range, and which provided ample food.[9]

The descendants of the Paleo-Eskimo span the northern portion of the continent, then ultimately disappear from archaeological sight, their place on the landscape later taken by the Thule, who *were* ancestral Eskimo. The Thule also radiated quickly across the high Arctic, but later in time (Chapter 6). And like the Paleo-Eskimo, the Thule were apparently following a prey species: bowhead whales, a relatively slow-swimming and placid species that could yield upward of 50 tons of meat and blubber. The Thule had hunted bowheads for millennia, mostly around the Bering and Beaufort seas. There whales and people may have stayed, save for the Medieval Warm Period, which enabled the whales to expand east through now-unfrozen seas. The Thule hunters went along for the ride, metaphorically speaking, along the way leaving artifacts and sites that are remarkably similar from Alaska to Greenland.[10]

There are obvious differences between the Paleo-Eskimo and Thule cases, and that of the colonization of the Americas. The later peoples were radiating outward in a familiar, relatively homogeneous, and (in the case of the Thule) newly expanding Arctic niche. Clovis groups were pioneering a vastly larger, trackless, and increasingly exotic and highly diverse landscape, one that ranged from open northern grassland to closed southern forest. No matter how varied the Late Holocene Arctic might seem (and no doubt it seemed varied to those who lived there), it was nowhere as varied, or as environmentally and climatically unsettled, as late Pleistocene North America (Chapter 2).

The whale-hunting Thule also had at their disposal watercraft for long-distance travel, making it easy to move far and fast. Paleoindians surely had rafts as well. They did, after all, cross the Mississippi River at a time when it was swollen with glacial melt-water, and they reached the Channel Islands of California. But boats would hardly have helped them rapidly colonize the Great Basin, the high plains, or any of the other areas of the continent where they camped far from any lakes or navigable rivers. The move from Alaska to Greenland across the northern maritime region was hardly as compli-cated as the one from Alaska to Panama.

But there is one apparent similarity in the three cases: the size of their prey.

. . . AND METHOD TO MAMMOTHS

Robert Kelly and Lawrence Todd argued in a very influential paper that Clovis groups moved quickly into and across the Americas by following the big-game animals of late Pleistocene North America. It's an argument since elaborated by others, and it has obvi-ous appeal. Preying on large animals, and especially targeting the Pleistocene mega-fauna, would have made it possible—even obligatory, they say—for the first Americans to move rapidly across long distances and through the continent, and it would have enabled them to easily hurdle any ecological boundaries. The journey from open plains into closed forest would not have required a change in adaptation. Just take aim at mast-odons, now that their distant elephant cousins (the mammoths) had been left behind.[11] This idea, taken to its extreme, roots Paul Martin's belief Clovis groups hunted the several dozen genera of Pleistocene megafauna to extinction.

Could the colonization of the Americas have been that straightforward? Unfor-tunately for them (and us), probably not. Not all animals, let all alone all elephants (or mammoth and mastodon, in our North American Pleistocene case) are alike in their behavior. Wildlife biologists and skilled hunters alike know that finding animal prey requires knowledge of how they behave on the landscape, and that varies by the animals' age and sex; the size and composition of the herd (and where they fit in the herd hierarchy); whether animals are breeding, pregnant, or lactating; the season; the availability of forage and water; the topography; exposure to predation; and so on.[12] As archaeologist (and lifelong hunter) George Frison observes: "To suc-cessfully match wits with wild animals with the intent to kill them requires a thor-ough knowledge of the hunting territory and the behavioral patterns of the species residing within it."[13]

A hunter might be able to bring the same weapons to the hunt. A mammoth or mastodon could be felled just as easily with the same type of spear point. But how different species responded to stalking or attack surely varied. While accompanying G/wi hunters of the Kalahari Desert, George Silberbauer observed that "the bow and arrow are versatile weapons, but switching from one prey species to another involves more than aiming at a different animal. Each species calls for a different approach, target selection, and because of differences in susceptibility to arrow poison, a differ-ent method of pursuit after the arrow has struck home."[14] Hunting a lone mastodon browsing in the forests of eastern North America would have posed a very different challenge from hunting mammoth (a herd animal) grazing on the open plains, or for that matter, from hunting bison, deer, moose, or any other large animal, each with its own idiosyncratic behaviors—which probably explains why megafaunal kills sites are so rare (Chapter 8).

But let's not stop the pursuit of big game just yet, for mammoth and mastodon were occasionally taken by Clovis groups, and there's reason to suppose hunting big game

FIGURE 39.
Excavations in the Late Pleistocene levels at the Broken Mammoth site, Alaska. (Photograph courtesy of David R. Yesner.)

was important, even if it did not serve to propel Paleoindians across the continent. The reasoning goes like this: colonizers newly arrived in North America were "pre-adapted to hunting," most often large game.[15] But that requires qualification.

One of the important facts to emerge from David Yesner's excavations at the Broken Mammoth site (Figure 39) on the Tanana River in central Alaska, dated to 11,800 BP, is what other game was available and exploited in the far north. Among the approximately 10,000 bones recovered at the site were those of bison, elk, caribou, moose, mountain sheep, ground squirrel (the most numerous of the small mammals), hare, marmot, river otter, and collared pika. But mammals were not the only thing on the menu: so were birds (fatty birds, too). Most of the bird bones (65%) were from tundra swan, but there were also three species of goose (Canadian, Snow, and White-fronted), five species of ducks, and willow ptarmigan (not only does this tell us something about the menu, but it also indicates the North Pacific avian flyway had by then been reestablished in the wake of deglaciation). And then, too, there were salmon and grayling pulled from the Tanana. Ironically, there were no mammoth eaten at the Broken Mammoth site. Its name came from ivory fragments the inhabitants of the site must have found on the landscape and collected as keepsakes or raw material for making tools, for this was old ivory, predating the site occupation by nearly 4,000 years.[16]

The picture that emerges of high Arctic subsistence, even in Pleistocene times, is one of broad-based diets. Meat was a large part of it, and not by fad or choice but by necessity. In the Arctic, there were not many other subsistence options available. Edible plants were and are rare, and only seasonally available. That said, even obligate Arctic hunters in historic times made use of about 140 species of plants.[17]

When the first Americans arrived south of the ice sheets in unglaciated North America, they would have found an environment teeming with animals, and a far richer supply of carbohydrate-yielding plants, allowing a wider range of food choices. Even better, the animals had never before peered down the shaft of a spear, and initially were surely naïve about the danger humans presented. Faced with a full menu of possible entrées, the first Americans likely did what hunters the anthropological world over do: they ranked their targets based on their expected return rate. High-ranked resources, the ones worth pursuing, are those that provide the greatest payoff for the effort expended.

As a general rule, resource ranking of animals roughly correlates with body size: large game tends to be a high-ranked resource, smaller game lower ranked. There's far more return for the effort of killing a large animal like a bison, say, than a small one like a prairie dog. It takes an awful lot of effort and time, not to mention an awful lot of prairie dogs, to match the food and protein equivalent of a single bison. Yet, like all general rules, as Douglas Bird and James O'Connell explain,[18] this one has exceptions: the very biggest of the big game (whales, say, or mammoth) might not be high-ranked prey under all circumstances because of the cost and risk of pursuing and killing these very large animals. Is your dinner worth your life? Not on a regular basis, though there are payoffs to occasionally going after even the biggest game. After all, showing off one's hunting skills has its benefits, too (Chapter 8).

At the other end of the body-size scale, lowly grasshoppers can be high-ranked resources during seasons when they are so abundant a forager can net in an hour enough insect protein to feed a family.[19] And some resources change in rank as environments change (as, for example, when drought sets in, and seeds become a viable option), or for want of other higher-ranked foods, or with changes in technology (the invention of toggle harpoons promoted the rankings of many marine mammals).

Given a choice between pursuing a high-ranked resource and a low-ranked one, hunter-gatherers will pursue the higher-ranked one. They do so even if the lower-ranked resources are more abundant, simply because the costs of going after lower-ranked resources—capturing, killing, and butchering all those prairie dogs—tilt the cost-benefit equation in favor of the one-stop shopping offered by a bison. The number of lower-ranked resources in the diet is thus not a consequence of their abundance, but instead of the abundance and availability of higher-ranked resources (a good thing, too, else we humans would have spent an inordinate amount of our evolutionary history eating beetles).

The *diet breadth* model predicts that hunter-gatherers do not simply eat whatever they encounter, but instead decide on which food(s) to take based on the *search time* that must be spent looking for food, and the *processing time* needed to pursue, capture, harvest, and process the food for eating. The optimal diet is a compromise: as more (and lower-ranked) prey are added to the menu, search time decreases (more potential targets), but in a see-saw fashion, processing time increases (adding prey requires different tactics to track, kill [or harvest], and prepare, and possibly additional, even specialized gear). The goal, often sought though not always realized, is to find the mix of resources that minimizes both search and processing time.

What of our newly arrived American colonizers? The diet breadth model, as Kelly shows, predicts that hunter-gatherers will decide on which food to take based on a knowledge of the quality of different foods, and on a knowledge of resource densities (and hence search costs) and processing costs (return rates). The hunting of the naïve prey was likely good at first, probably so good that the high-ranked resources in an area would soon decline in the face of hunting pressure, and once the prey realized this new predator was something to be avoided (this is called *resource depression*).[20] Faced with diminishing returns, colonists could choose to move to unoccupied areas where familiar game animals were present, or stay in place and expand their diet by working their way down the food chain.

But when to stay and when to leave? *Patch choice* models hold that foragers will stay in a place as long as the expected return rate is above the average return rate for the environment. Once it appears better return rates can be had elsewhere—when the grass, it's realized, really *is* greener on the other side—and after factoring in the cost in time and energy of moving, the original place will be left behind. But whether moving or staying, the forager needs information: What other resources are available in other patches? What other, possibly less-familiar and lower-ranked animal and plant resources are available? Where are they?

LEARNING LANDSCAPES

These models assume a forager knows something about resource availability across a wide area, and can make informed decisions about what to pursue, how long to stay, and when to leave. Gaining that knowledge requires extended periods observing how the animal behaves over the seasons and in different settings and circumstances: the more time that's invested, the more reliable the chances of success.[21] Colonizers new to a landscape haven't had the luxury of long experience learning their prey, and early on would not have the knowledge to anticipate when and where their prey would be found. Nor can they ask the locals for help: there aren't any. Each new habitat colonizers entered would, for a time, have been unpredictable and perhaps unreliable.

This is not to say—as some have accused archaeologists of saying—that the first Americans were "timid, tentative, and diffident foragers [who were] baffled" by this New World.[22] They were surely quite the opposite. The first Americans made it here and thrived because they were anything but cultural dopes. Granted, they started with a proverbial blank slate, but they were not completely naïve either. They must have possessed well-honed hunting strategies, a general knowledge of animal behavior, a broad familiarity with plants, geological prospecting skills, toolkits adaptable to various contingencies and, in all likelihood, a long history—by virtue of the vast and empty northern landscapes they'd already pioneered—of being opportunistic in their adaptations, and of maintaining a flexible social organization, long-distance kin relations, and a tradition of information sharing whenever possible (all vital when there are few of you on the landscape).

Likewise, though they were entering new habitats as they moved across the continent, as pedestrian hunter-gatherers, they would have done so (literally) step by step: they could see in the distance, for example, periglacial environments give way to northern grasslands, and thus would have had time to scout what was ahead and anticipate some of its adaptive challenges. Few colonizers (and the first Americans were not among them) are suddenly thrust into utterly unfamiliar environments poles apart from where they've been. The English in the late sixteenth and early seventeenth centuries in America came close, and nearly perished as a result (it hardly helped that the Roanoke Island and Jamestown colonies both had the monumental bad luck of arriving and trying to establish themselves during the worst years of an extraordinary drought).[23]

What America's first peoples therefore needed to build on their prior knowledge so as to maximize their chances of success were "maps" of near and increasingly distant lands to know *where* to move; an understanding of the seasons and climate to know *when* to move; and a familiarity with a region's animals and plants as well as the location

TABLE II Components of landscape learning

	Geography	Climate	Resources
What humans bring:	sense of direction ability to store information ability to integrate time, motion, and position while moving generic navigation knowledge (celestial markers, etc.)	generic climatic knowledge (seasonal changes, etc.) ability to store or capture information human generation time	generic knowledge of geology, hydrology, plants, and animals ability to store information ability to relocate resource occurrences
What humans look for:	prominent landmarks geographic features (rivers, mountain chains, etc.) spatial patterning to natural features	scale of climate change relative to generation or residence time (detectability) duration, periodicity, and amplitude of climate changes (predictability) variability of climate (stability)	relative movement of the resource (fixed/not fixed) relative stability and availability of resource (reliability) abundance, behavior, and/or distribution of resources (predictability)
How humans learn across space and through time:	begins at the most general landscape features, then becomes more spatially specific and refined over time ("megapatch" to "patch") begins on landscapes easiest to traverse and navigate, then later on landscapes that impede travel or wayfinding may involve artificial signage/waypoints (caches)	begins with specific features (local weather), then becomes more general and cumulative over time (climate) predictability increases with residence time, depending on periodicity and stability of climate learning may have upper limits, depending on detectability, residence time, and capture capacity (one cannot "see" a glacial period)	early, redundant use of fixed, permanent resources (stone) learning begins at general level (animals of same size class; plants of same family or genera), and becomes more specific knowledge increases with residence time but may also require experimentation may have gender correlates depending on relative labor investment in hunting vs. gathering

NOTE: From Meltzer 2003.

of other necessities like water and stone to know *how* to move. To be a successful colonizer of the Americas meant using the neural hardware and cognitive software with which all humans come equipped, taking cues from the environment, and above all, paying attention and learning (see Table 11).

FINDING THE WAY

Much of our modern world is covered in networks of roads laid out with signposts of distance and direction, even entering eastern Beringia (Figure 40). There's little incentive, as geographer Reginald Golledge observes, to pay close attention to the environment we pass through, and most don't. We pick routes that may not be the shortest or most direct, but are good enough to "get me there anyway."[24] Ineffectiveness and inefficiency seem to characterize our movements across space, yet the consequences may be little more than socially awkward moments, like arriving late at a dinner party.

In times past and on unknown landscapes, the penalty for geographic inattention may have been more severe, as when hunting parties became lost and failed to retrace their steps, or ocean-going colonists were not as prudent as they needed to be. Being lost on land was not always fatal, of course, nor would it carry the same costs as being lost at sea. Still, even in places like the lush tropics, becoming lost can be fatal if one fails to recognize the potential foods in that setting; the Spanish Conquistador Gonzalo Pizarro nearly met his demise in the upper Amazon in 1540–42 for just that reason. In the unforgiving desert or high Arctic, the danger is much greater. Hunter-gatherers there take great measures to avoid getting lost and rarely do, save when ill or overtaken by bad weather.[25]

Hunter-gatherers hedge their survival bets by learning the landscape. Working among the Nunamuit (Eskimo) of northern Alaska, Lewis Binford observed that an individual hunter might annually range over about 5,400 km^2 of territory, while together the members of the band might annually move over as much as 25,000 km^2. The knowledge of what was seen and experienced would be widely shared. Accumulate those movements over a lifetime of hunting and shifting settlement, and an individual Nunamuit might ultimately travel over an area of about 300,000 km^2, and if an individual listens carefully to stories told by others, he or she can know about places without ever having seen them.[26] It's little wonder that at the turn of the twentieth century, an Eskimo living in the high Arctic drew maps for an anthropologist that detailed topography, landmarks, and places over an area of approximately 650,000 km^2, parts of which he knew only by hearsay, but accurately depicted nonetheless. This is geographic knowledge on a vast scale, but given the sparseness of the Arctic environment (not much food per kilometer here), it is on the high side for hunter-gatherers. Even so, G/wi hunters of the Kalahari have detailed knowledge of areas covering 20,000 km^2, and similarly scaled geographic knowledge is possessed by native Australians.[27]

FIGURE 40.

Pointing the way to the New World: a sign in downtown Barrow, Alaska, at the gateway to the Americas. The first peoples of the New World had no such wayfinding help, but nonetheless were likely skilled at finding their way out (and back) across unknown landscapes. (Photograph by David J. Meltzer.)

Within these respective areas, the landscape was known and named with remarkable detail, all without benefit of maps or instruments (a traditional knowledge that is fast disappearing in the high Arctic because of the introduction of GPS technology[28]). In their absence and in the face of utterly unknown terrain, colonizers could use a variety of topographic markers, environmental cues, generic knowledge, and the cognitive "software" we humans have developed over evolutionary time: a sense of direction; the capacity to store information; and the ability to integrate time, motion, and relative position as one moves. Wayfinding, our universal ability to "determine a route, learn it, and retrace or reverse it from memory," enables us to travel across a landscape without getting (too) lost, while constantly maintaining an awareness of a position relative to a starting point, and do so despite the well-documented tendency of humans in featureless environments to veer (by an average of about 18°).[29]

Studies of wayfinding suggest that initial colonizers on an unknown landscape would likely be guided by features that are large, readily visible, or otherwise distinctive: prominent landmarks, or major topographic and geographic features (mountains ranges, rivers, coastlines). This, of course, varies by landscape. Among the tundra-dwelling Eskimo, directions are figured by the position of the sun and stars, and by the prevailing wind direction (marked by snow drifts or sand dunes); their forest brethren orient by topography, rivers, lakes, and other such features.[30]

The less familiar the forager is with an area and the greater its size, the coarser the resolution of the cognitive "map" of the landscape. Over time, however, knowledge of the region will come to include highly specific references to particular places, their characteristics, their use, and their identity (social, geographic, historic, ritual, and so on), for the events of a place give it meaning.

Human behavior being what it is, mapping a new landscape was likely a staged process, initially involving scouts or hunters pursuing game while probing into unfamiliar terrain. The oldest archaeological site in an area may not represent the first settler, but instead the first person to scout around.

Not all landscapes are alike in what Golledge terms their "legibility,"[31] their appeal, or the ease with which they can be learned or traversed. In some instances, wayfinding is easy but travel is not (e.g., across mountainous terrain); in other cases, travel is easy and wayfinding not (vast, flat, featureless plains); in still other cases, both travel and wayfinding can be challenges (heavy forest) (Figure 41). Had colonizers new to a continent followed paths of less resistance, as Anderson and Gillam suggest (Chapter 6), they might have threaded their way across the continent, through areas of relatively low-lying topography, along major river valleys, skirting glacial and pluvial lakes, or moving along the coastal margin.

In addition to the many and often highly visible or prominent natural features humans may have used as landmarks for their emerging cognitive maps of the landscape, colonizers likely also created artificial markers to help find their way. These would be especially important on geographically monotonous landscapes like flat, featureless grasslands. Francisco Coronado, who in 1541 led a small army onto the high plains of western North America in search of a rumored City of Gold (Quivira), nervously described the region as so "bare of landmarks [it was] as if we were surrounded by the sea." And though he traveled with a few crude "sea-compasses," every morning his men still got a fix on the sunrise, after which arrows were shot in the direction they were heading in order to maintain their bearing, and piles of bones or buffalo chips were left behind to mark their trail. Even so, Coronado's entourage meandered, and regularly leaked stragglers and lost hunters.[32]

FIGURE 41.

Landscape illustrations of the relative ease or difficulty of travel and wayfinding. These vary in degree of difficulty depending on the topography, landscape cover, orientation points, and the like, as shown in these few examples. (Photographs by David J. Meltzer and Ethan I. Meltzer.)

COPING WITH CLIMATE

Coronado also had the dubious honor of being the first European to experience the fierce wrath of a plains hailstorm, which hit his company in the spring of 1541 and destroyed tents, scattered horses, dented helmets, and broke most of their water containers (pottery and gourd canteens). It was the last straw: a food shortage, doubts about whether Quivira truly existed, and strong misgivings about the Natives who promised to take him there, prompted Coronado to send most of his army back to New Mexico. With a small force, he continued north and found Quivira, but it proved to be only a small, gold-less village on the plains of central Kansas. And at that moment, he illustrated the consequences for would-be colonists of climatic ignorance: he stayed in Quivira only briefly, then beat a hasty retreat, fearing the oncoming winter would be early and arctic, and trap his force amid Natives growing increasingly hostile.[33]

Coronado was right about the Natives, but he had no idea what winters were like in central Kansas. All he had was a brief glimpse of its summer weather, and he knew nothing of its climate. That distinction is important: *weather* is the sum total of atmospheric variables (temperature, humidity, precipitation, wind, and so on) for a brief period of time. *Climate* is the cumulative picture of weather over a longer period of time, and of how it changes seasonally, annually, and over even longer spans (Chapter 2).

Humans adapt to the weather, but to be successful over the long term, they have to know something about the climate over the course of the seasons.

Colonists expanding into new ranges will not suddenly find themselves in radically different climate, but as they shift latitude or altitude, enter regions dominated by different air masses, or expand into interior areas from the coast, they can encounter conditions that differ in subtle and sometimes not-so-subtle ways. Those varying climates are part of the environmental stage on which adaptation and colonization is played out.

Climate also varies over time, occasionally in relatively predictable cycles but mostly not. Some of that variation occurs on a scale detectable to humans, as for example our recurring El Niño. But no one lived long enough to notice, for example, the onset of a glacial period (though it has been said the Younger Dryas came on fast enough to be noticed by people [Chapter 9]). Colonists brand new to a landscape are mostly in the dark about the local climate, having not been on the ground long enough to experience more than seasonal or annual changes. Knowing what climate will bring requires residence.

All of which means that when climates are highly variable (as they were in late Pleistocene times), more time must elapse before an environment can seem less uncertain. Colonists who arrive during climatic extremes (unusually heavy rainfall or prolonged drought) will lack a sense of whether the conditions they are experiencing are typical or unusual, short- or long-lived, and predictable or not.[34] The Norse experienced that: their tenuous toeholds on Greenland and Iceland, established during the several-centuries-long Medieval Warm Period, were shaken loose when the climate changed for the worse, and their meager farmsteads failed. We assume, however, more mobile colonizers with more flexible survival strategies were better able to roll with the punches that nature threw their way.

WHAT FEEDS AND WHAT CURES, WHAT HURTS AND WHAT KILLS

Having made it this far, the first Americans had already figured out fire, shelter, and clothing. Once here, they needed to find the materials (stone, wood, bone) suitable to make or maintain those items, as well as meet their daily requirements of food and water. Learning where, what, and how to locate those critical resources could be relatively fast and easy, or not.

Outcrops of stone suitable for tool manufacture, for example, don't move, and once found become known and predictable points on an emerging map of a landscape (Figure 42). And those can be relatively easy to find: the famous Alibates agatized dolomite outcrop near Amarillo, Texas, prized for its high-quality stone by hunter-gatherers since Clovis times, is bisected by the Canadian River, which carries cobbles of the stone as much as 600 kilometers downstream of the outcrop. Anyone seeing those cobbles in the river gravels would know it was merely a matter of following the stone trail back to the outcrop source. It's no surprise Clovis-age sites often occur near where major rivers and streams (which we presume were Pleistocene travel corridors) intersect geological outcrops of high-quality stone.[35]

Locating water can be relatively easy as well, even in dry environments: there, one follows game trails or the flights of birds. Finding it in the same spot again is hardly guaranteed, however, at least not in semi-arid and arid lands where water is neither permanent nor reliable, or during drought when springs and lakes disappear from the surface, and stream flow diminishes. Still, where water might be found beneath the surface or will return after heavy rains generally does not move, and that can be predicted, too. Surely the first Americans possessed the knowledge of geology and hydrology to enable them to find water—there's even tantalizing but unconfirmed evidence of well digging in Clovis times (Chapter 8).

Neither water nor stone were resources that required new knowledge to be exploited. Once found, stone could be flaked into useable tools, and water collected in skin pouches and transported. There wasn't much to learn.

FIGURE 42.

Inspecting outcrops of high-quality, knappable quartzite in the Gunnison Basin, Colorado. Outcrops like this, where the stone occurs in large blocks, were excellent places to acquire raw material in sufficient supply and size, and could be readily located by following the trail of stone washed downhill or downstream. In this case, the quartzite is more resistant to erosion than the surrounding rock, and so forms a bench easily spotted from a distance. (Photograph by David J. Meltzer.)

Plants and animals, of course, are far more ephemeral and less predictable on the landscape: game-rich areas may lack edible plants; plants may not be edible all times of the year; and animals (especially larger ones) can wander away from a previously grazed habitat, or change their range altogether. Some resources might also require considerable study to ascertain their use(s).

This was particularly true of plants. In North America alone, there are an estimated 30,000 species and varieties of plants, and those in temperate and near-tropical areas were surely unfamiliar to bygone Siberians. The process of learning about specific plants, whether they had food, medicinal, or other value, or possibly were poisonous, must have entailed *observation* (watch to see if other animals eat them), and possibly considerable *experimentation* (try it . . . and see what happens).

Yet, daunting as the overall task surely was, the process of botanical learning was ultimately accomplished in prehistory with astonishing success: as Daniel Moermann observes, "No one has ever found a plant native to North America with any medicinal value not known to and used by American Indians."[36] In Moermann's view, intense periods of plant research ("research" being an appropriate term here) would likely occur under two conditions: when new diseases appeared, and when there was a substantial environmental transformation. Colonizing the Americas met the first condition and possibly the second, with medicinal plants perhaps being needed when the tropics were entered for the first time.[37]

Which is not to discount the role of plants as food. As already mentioned, the first Americans were pre-adapted to hunting, coming as they were from an Arctic environment where being a vegetarian was impossible in the winter and challenging in the summer. Yet, habitual carnivory can sometimes lead to various ills and even death: individuals with protein-dominated diets experience higher metabolism, requiring at least 9% more calories just to maintain basic physiological functions. Too much protein strains the liver's ability to function, resulting in what's called "rabbit starvation" (after the dangers of a diet dependant on this lean animal).[38] Rabbit starvation can be prevented by adding fatty acids or carbohydrates to the menu. This was accomplished in the far north by preying on animals with high body fat content—like the birds and fish consumed at Broken Mammoth—and supplemented by plant collecting in the brief summers. Once in plant-rich, unglaciated North America, colonists were likely also snacking on carbohydrate-rich plants to round out their diet (Chapter 8).

As knowledge of a plant's properties was built through observation and experimentation, there would have been considerable advantage to maintaining it. Otherwise, each subsequent generation would have to scan the plants anew. That seems unlikely, given the vast number of plants to scan. One supposes, instead, that once broad classes were identified—recognizing it's an oak, say, and not a pine—and their general properties learned—oaks produce acorns that can be palatable when leached of tannic acid—this knowledge was broadly shared. Then it was merely a matter of correctly spotting different species of oak in new settings. So it was, for example, that in 1492 Columbus's crew saw few trees on Hispaniola known to them, but they did recognize some pine and palm trees; however, they readily noticed "the palms were extraordinarily high, very hard, slender, and straight [and] Even the fruits they produce in abundance were unknown."[39]

Having knowledge at a more general level would be easier to retain (there are only about 290 families of plants in North America, as opposed to its 30,000 species and varieties) and transmit to subsequent generations, especially if the plant in question was a long-lived perennial. Knowledge at this level would give colonists an entrée into new habitats in which the species might be novel, but where they would at least recognize the plant as familiar. Just how effective this process can be is evident in the rich record of plants and plant parts used for food, medicine, fuel, and raw material by the inhabitants of Monte Verde's MVII occupation (Chapter 4).

Learning about the New World's unfamiliar animals was perhaps less of a challenge (there were far fewer of them), but there was still the task of understanding each species' habits and habitats, and seasonal and longer-term patterns in abundance and distribution. Certainly, clues about animal movements can abound on a landscape, and surely these colonizers were adept trackers—otherwise, they would have perished somewhere in Beringia!

They likely could also count on four-footed help. DNA evidence indicates dogs were domesticated from wolves some 15,000 years ago. Although the clues are faint, it appears dogs accompanied the first peoples in their trek to the New World, since dog remains have been found in several Paleoindian sites. Dogs have held various jobs for humans in the past, besides being our best friends: they helped in the hunt to bring animals to bay, served as beasts of burden, were food sources (regularly in some places, only as an emergency backup in others), and indeed, may have been vital in the exploration of new lands. Dogs figure prominently in many American Indian origin stories, acting as helpmates in ritual passages.[40]

As hunters and their dogs shifted into novel settings, there were animals never before seen or encountered (Figure 43), making humans just as naïve as their prey, though not as disadvantaged for being so (although Lewis and Clark's first encounters with grizzlies provide glimpses of just how disadvantaged humans can be, even when toting firearms).[41]

A generalized knowledge of animal behavior could be extrapolated from one area to another, and hunting skills and weaponry were (to varying degrees) transferable from one target to another. Kelly and Todd rightly argued it is "easier to locate, procure, and process the faunal as opposed to floral resources of a region."[42] That said, it is also true the location of animals is fleeting and unpredictable (while groves of plants, once found, can be readily relocated). Exploiting a new, high-ranked prey species required (as earlier noted) studying its behavior, learning where and when it might be found (its preferred grazing or browsing areas), and how best to approach it.

Animal prey can be highly mobile, moving seasonally as forage conditions change, especially in response to the arrival of hunters on the scene. Prey quickly get over any initial naïveté they might have once the spears start flying. They either change their behavior or move out of reach. Finding fewer targets, the hunters themselves need to range more widely in search of game.

It's a volatile market for hunter-gatherers, and one they monitor by moving: insurance for hunter-gatherers is knowing where to go next, when return rates diminish where they are. Colonists entering a new region, Bird and O'Connell suggest, "should occupy the highest-ranked patch first," all other things being equal (and assuming they can find it).[43] Whether, when, and where to move is decided after solving a complicated equation incorporating knowledge about climate, plants, and animals. Long-time residents of a landscape have relatively complete information about resources on which to base decisions about how long to stay and when to leave.[44] The G/wi, for example, carefully consider

such factors as the severity of the coming winter and the first months of summer, the prospects for rain and the anticipated date of the first rains. A floristic forecast is made for each foreseeable set of circumstances: Should there be a severe winter with sharp frosts but a not-too-severe early summer then such-and-such locality would be best. . . . The band considers not only its next move, but the whole series of migrations in the foreseeable future. The aim is not to plot the coming season's itinerary in detail but to work out a series of moves that will permit the band the widest choice of subsequent sites.[45]

FIGURE 43.
The first Americans occasionally encountered animals they'd never seen before, such as this rattlesnake waiting beneath a bush. Fortunately, a rattlesnake often provides a warning of its intentions. (Photograph by David J. Meltzer.)

But if that landscape is unknown, or populated with exotic plants and animals, knowing when and where the chess pieces need to move is even more challenging.

GENDER NEUTRAL?

It has been argued that aspects of landscape learning are *not* gender neutral among hunter-gatherers. There is in our shared past an evolutionarily derived division of labor, in which males hunted and females foraged. Although the pattern is by no means universal (more of a guideline, say, than a rule), the trend is rooted in the fact that hunting—especially of large game—can be very risky. Women tend to be risk averse, especially when carrying or tending children (though there may be other reasons why, in general, women tend not to hunt). Moreover, as Bird and O'Connell observe, women target resources that daily provide food for themselves and their offspring, while men favor larger prey that can provision many in the band (but not necessarily their own family), and draw attention and higher status to themselves. There are, of course, situations in which women aid or otherwise participate in the hunt, which, as Elizabeth Chilton points out, readily shift with adaptive circumstances.[46]

Regardless, by virtue of this difference, there are or might be gender correlates or patterns in wayfinding skills, foraging knowledge, and the scale of foraging. This is so because finding and killing highly mobile, large animals entails different spatial challenges than does foraging for edible plants. Tracking animals requires the ability to orient oneself in relation to distant points on a landscape (either visible or not), and being able to constantly and accurately update position and location while on the hunt. In contrast, gathering involves locating food plants within complex vegetation communities, tracking their seasonal changes, and maintaining memory for objects and their locations.[47]

From this, we might speculate that landscape learning of routes and resources was similarly gender linked. Much of the "local" knowledge of plants and small animals, and their patterns and distributions, was likely learned, acquired, and accumulated by women (and young children). Broader-scale landscape knowledge, and this includes both routes and resources, would result from men's more extensive, long-distance logistical hunting forays across the area.

Our evolutionary history fails to explain, however, why men won't ask for directions (and *that's* a rule, not a guideline).

TRUST ANYONE OVER THIRTY

Richard Nelson, working among the Eskimo in the 1960s, was struck by how the younger generation believed everything it was told by its elders, quite unlike many of America's youth in the 1960s. He instantly understood why: an Eskimo hunter devoted "a lifetime to learning more and more about the habits of the animals and about the mobile sea ice on which he hunts," and about the landscape he traverses. For Eskimo youth, every bit of information they gleaned from their elders' experience would enhance their odds of success, and might even save their lives.[48]

Yet, one individual or group can only cover so much ground. By talking to others and tapping their knowledge about resources available in other areas, a person can learn about places they've never been, animals they've never hunted, and plants they've never tasted, not to mention draw maps of areas far larger than they had ever seen. Not surprisingly, as Silberbauer observed, when calculating moves on the landscape, all adults in the foraging group participated in the decision-making process.[49]

Colonizers new to a landscape haven't got elders, relatives, or neighbors who can pass along what they have learned over their lifetimes (though it perhaps goes without saying, they haven't got written records either). But they have descendants. As groups moved into the new landscape, knowledge was gained by observation and experience, and "recorded" in folklore, storytelling, and ritual. But how quickly would colonizers learn the routes, climates, and resources of a landscape, particularly a landscape the size of, say, North America? We have no answers. Most studies agree detailed landscape learning is a prolonged process, with estimates ranging from 200–1,000 years. Even the early English colonists in America, with all the relative technological advantages they possessed, took more than a century before they had accurate maps of their region.

Learning is cumulative and ultimately can be sustained over multiple generations, even among people who rely entirely on memory and oral transmission of knowledge. Naturally, there is some information decay, and lessons learned at one time can fail to be passed on to others. That can be seen even as recently the massive failure of Great Plains farmers during the 1930s Dust Bowl (Figure 44). They had arrived in great numbers during a relatively wet period and considered those conditions normal, unaware that a previous wave of settlers had been driven off by severe drought in the early 1890s. They had no knowledge of that previous experience, nor appreciation of the inevitability of drought and the risks of farming in that semi-arid climate. And they could have read about it.

Knowledge grows exponentially, since it can be transmitted by the telling (genetic adaptations take longer to pass along) and thus shared widely within and between generations, thereby rapidly spreading information about areas, events, and conditions otherwise experienced by only a few. To be sure, lessons learned in one area may be irrelevant in different settings, but one never knows.

Landscape learning likely begins at the broadest geographic scale, in what John Beaton calls *megapatches*, which are ecological features on the order of coasts, plains, mountains, deserts, or forests.[50] The first occupants of the Great Plains (a possible megapatch), would have seen large-game animals on a vast grassland. It likely all looked the same to them, for they would not have known and would have to learn there was surprising variability in the grasses on which those animals grazed, and thus differences in

FIGURE 44.

Arthur Rothstein's iconic Dust Bowl photograph of a farmer and sons walking in the face of a dust
storm in western Oklahoma in 1936. Drought and other changes in climate can have a significant
impact on human groups, and knowing that such changes occur, even if they are not predictable,
is an important element of landscape learning—which was neglected or forgotten in this case. This
photograph is apt in another way: the dust storm captured by Rothstein that spring helped expose the
nearby Nall Paleoindian site. (Photograph by Arthur Rothstein, Library of Congress LC-USF34-4052-E.)

their abundance and distribution across the region. Over time, those came into better
focus as colonists saw more of the details within the plains.

In effect, landscape learning of routes and resources probably took place from the
general to the specific in terms of space, but from the *specific to the general in terms of time,*
particularly in the case of climate (which is initially observed as the daily weather).

Landscape learning has costs in terms of time and energy. But those costs are offset
by the benefits of acquiring information that reduces uncertainty and increases one's
odds of surviving and thriving.[51] People can still get lost, can still starve, and can still run
out of food or water. But they are less likely to do so the more they know.

It's fair to surmise that the greatest effort in information acquisition would occur in
patchy and unpredictable environments, the very situation in which the first Americans
initially found themselves. Accordingly, early on in the colonization process, there was
a great advantage to learning the local geography, climate, and resources of these unfamiliar landscapes. Under the circumstances, there was good reason to see what was over
the next hill . . . and the hill after that.

Much of the attention of colonization models, such as the one offered by Kelly and Todd, have focused on how it was possible for colonizers to move so rapidly across the New World. Their answer was hunting large game. Yet, as Christopher Ellis puts it, there's more to being a successful colonizer than "what to eat"—just as important is "who to meet."[52]

We don't know how many people were in the group that first came to America. It could have been dozens, it may have been hundreds (as Jody Hey estimates), and it possibly was in the low thousands—but only the low thousands. This was a colonization, not a swarming invasion. On a continent the size of North America, such relatively low numbers would have translated into an extraordinarily low population density, lower than at any time in American prehistory.

To make that colonization successful over the long term, the group had to reproduce, which meant that as its children came of age, they needed to find mates to have children of their own. The task would have been more or less challenging depending on the group's size, birth/death rates, age and sex composition, marriage patterns, and how rapidly it was moving from its geographic homeland and/or other groups. That, in turn, raises several questions: is there a minimum group size that had to be maintained in order to stave off extinction? How does an individual who is a member of a colonizing band that is steadily moving away from other people find a spouse? And do these demographic demands suggest how colonizers might have spread across the landscape?

As anthropologist John Moore has shown, there is no magic number that guarantees survival. A smaller band (of, say, twenty-five) might seem more vulnerable to extinction, but it can survive as long as birth rates are high, death rates are low, and there's a relatively even mix of males and females of reproductive age (assuming, that is, they can find mates). Likewise, a larger group (in the hundreds, say) can plummet to extinction if its birth rates are low, death rates are high (especially if there is high infant mortality), and there's been a long run of bad luck in the sex ratios of the children—perhaps a string of too many baby boys and not enough baby girls.[53]

But even if the mix of ages and sex ratios is just right (there are enough marriageable males and females), individuals in smaller bands that have pushed far ahead into unoccupied lands will either have to track great distances to find a mate in another group, or else marry within their own band which, depending on how small the band is, might mean incest or polygamy. Neither of those options is beneficial if sustained over the long term: incest can have bleak genetic consequences, while polygamy slows overall population growth (women in polygamous marriages tend to have fewer children than they otherwise might). It's best, then, to participate in the larger effective population (the *metapopulation* of geneticists). Maintaining links to distant bands can be accomplished in several ways, not the least of which are long-distance mating and/or the occasional aggregation (rendezvous) of smaller groups.

Demography and landscape learning are tightly linked for several reasons, not least that the decision to stay in place or move is based on the suitability of a new patch relative to the current one. Those foragers who have more knowledge can better calculate those costs and increase their survival odds. The information thus gained helps increase population growth and reduce mortality.

CONTINENTAL COLONIZERS

How, then, might a colonizing band move into an unoccupied land, maintain their demographic viability, and thereby minimize their risk of extinction? John Beaton envisions two strategies that mark the opposite ends of a continuum of possibilities: Estate Settlers and Transient Explorers. We'll call them the Cautious and the Bold.[54]

Cautious is the band that moves slowly and over relatively shorter distances. They settle into a valley, and explore it and the surrounding region; over time, their numbers build. As this happens, some of the younger members see the area becoming crowded, minor spats become more frequent, and once-plentiful game declines. On their increasingly more-distant hunting forays, hunters spot rich and unoccupied lands down the valley and beyond the edge of the home "estate." A decision is made, and soon a part of the group splinters off, but they don't venture too far. Instead, they stay close enough to maintain contact with kin, keep to an environment with which they are familiar and in which they have become successful, and a place from which—if things go badly—they can always retreat and find relatives who will help. It's a low-risk strategy. Over time, the process of population growth followed by budding repeats itself, and the descendants of the estate ultimately expand to cover the continent. This process does not happen at a terribly fast pace. Beaton envisions groups that "moved only slowly across the latitudes and longitudes, each daughter colony being spawned by the overflow of a saturated estate." Since colonization is ultimately fueled by population growth, to speed it along would require higher population growth rates, sometimes unnaturally high rates.[55]

Bold is the band that moves independent of population size. They come into the country and occupy a habitat, but the moment return rates begin to decline, they move on, without waiting until crowding forces their hand. And so they move often. In rich and unoccupied Pleistocene North America, when hunting returns declined, there was little incentive or need to stay put, not when there was a better living to be had several valleys over, down the coast, or over the mountains. This strategy favors rapid dispersal over a large area, with much of the continent ultimately passed over lightly, as colonizers leapfrogged from one high-ranked patch to another. They can make a good living, these transient explorers, but it's not without risk: small groups that go too far too fast soon find themselves geographically isolated, alone on a vast landscape. That can put a severe strain on their ability to maintain long-distance contacts and an accessible source of mates for their children, raising the specter of inbreeding and drift (cultural and genetic), and making the group as a whole vulnerable to disappearing altogether in the face of drought, blizzard, disease, or accident.[56] Calamities happen.

Thus, colonizers on a landscape with few other people—and this applies to both the Cautious and the Bold, but especially the latter—not only had to be able to find distant mates and marriage partners (and incidentally exchange information and resources), but they also had to be able to get along with any groups they encountered, kin or no. Having large and open social networks, flexible kin relations, and the ability to move easily between and integrate new individuals and groups, as well as maintain long-distance exchange and alliance networks, all combine to diminish differences among peoples who need to stick together under geographic circumstances that might otherwise keep them far apart for years at a time.[57] Strongly territorial behavior, or hostility toward strangers, would be decidedly disadvantageous to people at this place and time.

On a diverse and unfamiliar landscape, the risk of extinctions is greatest soon after dispersal, when population numbers and growth rates tend to be low. The Cautious hedge their bets and stay close enough to readily maintain overall numbers, but the price is slow colonizing progress. The Bold run greater demographic risks, but the rewards are higher: "with a generalized tool kit, a naïve fauna, an unconstrained social/political environment, and a bit of luck, a lineage might see the Northern Lights, note the transit of the Equatorial sun, and feel the chill winds of the southern oceans in the space of ten or fewer generations. This is its own form of success."[58]

Arguably, then, colonization of a new landscape likely involved trade-offs between a series of competing demands, including

1. *maintaining resource returns,* or keeping food on the table, particularly as preferred or high-ranked resources declined, and in the face of limited knowledge of the landscape;

2. *maximizing mobility,* in order to learn as much as possible, as quickly as possible about the landscape and its resources (in order to reduce environmental uncertainty in space and time);

3. *maximizing residence time in resource-rich habitats,* in order to enhance knowledge of specific changes in resource abundance and distribution (you learn more by staying longer);

4. *minimizing group size,* in order to buffer environmental uncertainty or risk on an unknown landscape; and, finally and perhaps most critically,

5. *maintaining contact between dispersed groups,* in order to exchange information, resources, and especially mates.

Colonizers had to balance the equation between *moving to explore* and *staying to observe.*

So were the first peoples Cautious or Bold? It's a trick question, since these are just the extreme possibilities, and there are many variants in between. But I lean toward the Bold and for a very simple reason. The archaeological record we have, whether of Clovis of pre-Clovis, is not dense enough to match what we would expect to see of Cautious estate settlers (even granting all the caveats expressed in Chapter 4 about the scarcity of sites of this age). That does not rule out a role for population growth in the New World colonization;

it only means this was not the apparent driving force. If colonizers were more Bold than Cautious, what might we expect to see in the archaeological record they left behind?

For one, their traces on the landscape ought to be broad and not deep. If they were wide-ranging explorers who were leapfrogging over large areas, their sites would be few, scattered, smaller, and more ephemeral, marking the brief time spent in any one place— although particular localities, including sources of high-quality stone or other spots where permanent resources were available, should see more repeated occupations. Their toolkits ought to look generalized and ready to be pressed into service for a variety of tasks and contingencies (later, as groups settled in to new habitats and encountered new challenges, we would expect to see new tool forms invented to meet those challenges).

The scale of movement (those leaps) ought to vary in part based on the richness and complexity of the megapatches the group occupied. For example, groups on the open, treeless western plains would be expected to move farther and faster than those in the complex environment of the eastern woodlands. In either case, their movements can be tracked by looking at where they obtained the stone to make their tools. Rock types can often be "fingerprinted" to particular geological formations, and this provides a rough measure of the distance and bearing from source to site (but a rough measure only, because it fails to account for the meanderings of the group or individuals, as explained in the next chapter).

The type of stone, and the distance and direction it was moved, also provides a potential key to unlocking an aspect of landscape learning: namely, how the first Americans may have staged their wayfinding on the unknown landscape. Hunter-gatherers aim to minimize the number and weight of tools they have to carry (only they and possibly their dogs were available as pack animals). But venturing into unknown terrain poses a particular challenge since the colonists would not know where suitable stone might be found when they needed to replace tools that broke (as they inevitably do), nor could they haul months' worth of stone with them. One possible solution would be to scatter "supply depots" of stone on the landscape on their outbound exploring trips, so at the very least they need not return all the way to the outcrop to refurbish their tool kit. These caches of stone would be there if needed, and could be abandoned if not, but either way become fixed and predictable points on an otherwise unknown landscape for the individual who leaves them in place, and for all who are told about them.

Although the movement across the continent might have seemed slow on a human time scale—double-digit generations, say—it will likely look astonishingly fast to us in archaeological time, for those generations are easily encompassed in the "wink of a radiocarbon eye" (to use Beaton's charmingly mixed metaphor[59]). Of course, coupling their speed with the late Pleistocene radiocarbon plateaus (Chapter 1) may make it impossible for us to develop a high-precision chronology of colonization, or track movement within and between megapatches. But if we ever do, we might expect to see a stutter-step pattern to the colonization process as groups spent relatively longer periods of residence within a megapatch, separated by rapid movements between them.

We also expect, as Bird and O'Connell suggest (above), for sites to be located initially in the highest-ranked patches, but that, too, will not always be easy to ascertain, demanding as it does evidence of past animal and plant availability that is more detailed than we can usually muster.[60] Likewise, it might be difficult to see patterns and trends in the diet, such as resources being added to the diet as their rankings change by virtue of knowledge gained about them, given the vagaries of preservation of animal bones and plant remains. Nonetheless, high-ranked, big-game resources ought to be taken, and there should also be evidence of the exploitation of other lower-ranked resources that might be in an area. If the archaeological record is sufficiently detailed, we should be able to spot resource depression setting in as prey and predators wised up to one another, as human populations grew, and as animals became more wary and/or fled an area.

The first peoples, to offset their high settlement mobility and dispersal across their new landscape, had to maintain contact with distant groups and keep open social systems to help recognize strangers as friends. Long-distance contacts can be detected archaeologically in stone exchanged among groups, or in rendezvous sites that brought together people widely dispersed, which are not as hard to spot archaeologically as one might imagine (Chapter 9 points to one such site).

Open social systems are a bit more difficult to detect, but may be seen in proxy form in what are known as *style zones*, the widespread distribution, use, and exchange of instantly recognizable, and sometimes highly symbolic, artifacts such as Clovis or other distinctive styles of projectile points. These forms may have had functions, but could also serve as a *currency* (a term not to be taken too literally) for social and ritual functions, and over long spans and large areas could serve to maintain recognition and alliances—not unlike, say, the custom of athletes exchanging country pins at the Olympics. In a similar colonizing situation but very different setting, Patrick Kirch argues that the exchange of Lapita-style ceramics and other items between widely separated Pacific island populations served as a social glue that helped cement long-distance alliances and a supply of suitable marriage partners in the face of "unpredictable environmental hazards . . . [and] demographically small and unstable groups." Having friends and relations (real or fictive) might have meant the difference between survival and extinction on these remote islands.[61]

Early on in the Paleoindian period, Clovis points should be broadly similar in style and technology across a vast area of North America, more so than any artifact forms in any later cultural period, Paleoindian or otherwise. The extent of the Clovis point distribution is a by-product of the size of the colonizing footprint, but their similarity across that range could reflect the maintenance of a common symbol of an extensive social and mating network, which helped offset the effects of distance.

Still, those effects are inevitable, so we expect over the decades and centuries, as groups spread ever further from another, that there would be cultural drift over time and space as the initial, "founder styles" diverged and diversified. Ultimately, new stylistic forms should appear, and those continent-wide style zones should shrink. As we will see in the next two chapters, we can catch glimpses of some of these hypothesized processes and features. We'll start with Clovis.[62]

PLATE 2

The Meander Glacier flowing out of the
Borchgrevink Mountains toward the Ross
Sea, northern Victoria Land, Antarctica.
Although the glacier is shrouded in snow,
its crevasses appear as a corrugated surface
atop the snow. The prominent peak is
Mount Kinel at 2,160 m above sea level.
(Photograph courtesy of Steve Emslie.)

PLATE 3
Glacial ice calving into the Pacific Ocean off
Patagonia, South America. Such would have
been a common scene along the Pacific
Coast of North America during the Last Glacial Maximum (LGM). (Photograph courtesy
of Anne S. Meltzer.)

PLATE 4
Skeleton of the now-extinct giant ground
sloth, *Paramylodon harlani*. (Photograph
courtesy of Louis L. Jacobs.)

PLATE 5
View across the Pedra Furada site, showing
the great height of the cliff face looming
above the site. The lone figure in the center
distance is James Adovasio. For scale, he
stands just over 6 ft tall. (Photograph by
David J. Meltzer.)

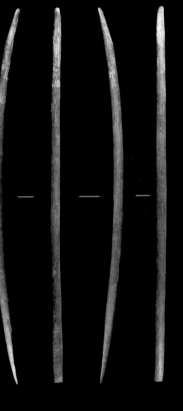

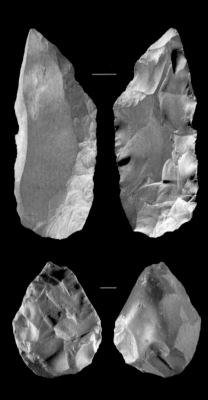

12cm

3cm

PLATE 7

Rhinoceros horn foreshaft (left) and two
of the stone tools (right) from the Yana
RHS site, Siberia. In its form and scored,
beveled ends, the foreshaft resembles
ones found in Clovis sites, though the Yana
specimen is twice as long (but about the
same width) as the average Clovis example.
Although reminiscent of Clovis, the Yana
artifacts are more than 16,000 years older,
making any historical links between the
two occupations tenuous. (From Pitulko
et al. 2004, copyrighted by and used with
permission of *Science*.)

PLATE 8

The Clovis tool kit included unifacial (right)
and bifacial (far left) tools, fluted points
(second and third from the left), and the
occasional stone blade (above the points).
Rarely found but possibly common were
artifacts of bone and ivory, including a
mammoth bone "shaft wrench" (center), perhaps used to straighten spears, and a bone
"foreshaft" (beside it on the right) like that
from Yana RHS. (Photograph by Tom Wolff.)

PLATE 9

Clovis point (in the approximate center
of the image) along with some of the
different types of stone—including chert,
jasper, quartzite, and obsidian—used in
Paleoindian times. The Clovis point has been
reworked as evidenced by the "shouldering"
of the point about midway up both
sides, and was made of Alibates agatized
dolomite, a block of which is in the lower
center of the image. (Photograph by David
J. Meltzer.)

PLATE 10
A quartz crystal biface, a rarity in Clovis
times, from the Simon cache, Idaho.
(Photograph by J. David Kilby, courtesy of
the Herrett Center for Arts and Science,
College of Southern Idaho.)

PLATE 11

The deGraffenried Clovis biface cache, Texas.
Note the broad, thin overshot flakes that
extend across all or most of the face of the
four bifaces. The projectile point preform is
just over 17 cm long. (Photograph courtesy
of J. David Kilby, used with permission of
Mark Mullins.)

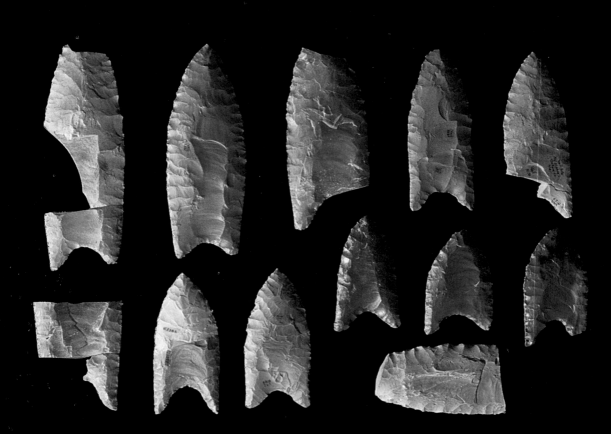

PLATE 12

Projectile points from the Vail site kill and
campground. In the case of all the broken
points, the tips were found in the kill
area, the bases at the camp. (Photograph
courtesy of R. Michael Gramly.)

PLATE 13

Folsom fluted points from the Mountaineer
site, Colorado. From left to right: (1) A
preform that broke in manufacture. The
two large pieces were separated, and one
ended up in a fire, hence its dark red color. A
portion of the channel flake removed for the
flute was found as well, and is shown in its
place on the flute scar. (2) A complete point,
but one that has been used on more than
one occasion, and has been resharpened
while still in the haft, producing a slight
"shouldering" toward the tip. (3) A point that
shattered on impact—perhaps when bone
met stone at high velocity. The impact flakes
were driven off the tip toward the base,
essentially the reverse of the fluting process.
(Photograph by David J. Meltzer.)

PLATE 14

An assortment of Late Paleoindian stone tools from sites in the central Rio Grande Valley, New Mexico, including scrapers (left and top), finely pointed gravers (around the knife), spokeshaves (for trimming wood or bone, above the knife), and cutting tools. The stone for their manufacture was brought in from all corners of the compass. The Swiss Army knife, which fulfills many of these same tool functions today, is just under 6 cm long. (Photograph courtesy of Tony Baker.)

PLATE 15
Some of the bison bone discarded at
Folsom, New Mexico, including a string of
articulated thoracic and lumbar vertebrae
(which form the bison hump and lower
back), as well as a skull, jaws, shoulder,
and leg bones, not all of which came from
the same animal—the skull is from a bison
calf, the backbone from an adult cow.
(Photograph by David J. Meltzer.)

PLATE 16

The Mesa site is located atop this
pronounced, isolated landform that juts
out on the North Slope of Alaska alongside
Iteriak Creek (note the individuals and
equipment atop the rocky ridge). Take
away the helicopter and the field camp
(to the right, below the mesa), and the
landscape would look much the same as at
the time of its occupation 10,000 years ago.
(Photograph by David J. Meltzer.)

8

CLOVIS ADAPTATIONS AND PLEISTOCENE EXTINCTIONS

Edgar B. Howard, of the University of Pennsylvania's museum, had been ambushed before, so this time he moved fast. He caught a westbound train on November 12, 1932, alerted to the news a road crew mining gravel from an old pond on the high plains of eastern New Mexico had struck bones. Lots of them apparently, and big ones, too: mammoth and bison. That summer, Howard had seen scraps of the bones and Folsom-like fluted points ranchers had been finding in wind-scoured "blowouts" in these old ponds. The road crew had hit the mother lode.

Howard knew the stakes. Earlier, his cordial relationship with paleontologist Barnum Brown had soured when Brown stole credit for a fluted point found in Pleistocene bone-bearing sediments in Burnet Cave near Carlsbad, New Mexico. Howard, who had made the find while Brown was far away in Montana, had not been happy. Nor was Howard pleased to learn his idea for a national committee to investigate potential late Pleistocene archaeological sites had been scooped, and the committee formed without him. In the years after the Folsom discovery, everyone wanted in on the action at early sites, Howard most of all. He was, a disapproving Loren Eiseley observed, "addicted to the pursuit of terminal ice-age man."[1]

Four days later, from the small town of Clovis, New Mexico, Howard flashed a telegram back east: "Extensive bone deposit at new site. Mostly bison, also horse & mammoth. Some evidence of hearths along edges. Will tie up permissions for future work & spend a few more days investigating." But only a few more days. Back in

Philadelphia the next week, he announced in the journal *Science* the finding of evidence of humans "at a very remote period," and staked his claim to the Clovis site.[2]

The following spring, Howard and his crew began excavating at Clovis. They had to work quickly. That summer the world's preeminent geologists would descend on Washington, D.C., for the International Geological Congress. There would be field trips out west afterward. Howard wanted Clovis to be one of the stops, but that meant having something to show: a fluted point in undisputed association with mammoth bones would do nicely. Finally, by late July, he had several, "scattered around like a dog buries bones," just what was needed to give his visitors "something substantial to think about."[3]

On August 3, in a scene that harkened back to the landmark visit to Folsom six years earlier, American and European scientists descended on Clovis, including the eminent Sir Arthur Smith Woodward of the British Museum (whose keynote address to the congress triumphantly reviewed the proof "Piltdown Man" was an early Pleistocene human). When they arrived, the visitors saw spear points, fluted but larger and less refined than those at Folsom, amid mammoth skeletons (Figure 45). That evening a

FIGURE 45.

E. B. Howard removed some of the more significant finds from the Clovis site as large blocks, in order to bring them back to Philadelphia for more careful excavation and cleaning. In the midst of that work, he posed, pointing to a beveled mammoth bone rod next to the ulna (lower forelimb) of a mammoth. (Photograph courtesy of The Academy of Natural Sciences, Ewell Sale Stewart Library, Philadelphia.)

jubilant Howard telegraphed the museum to say "Drs Merriam Stock Woodward and Vanstraelen have examined excavations here today [and] they agree evidence obtained indicates association of artifacts with extinct elephant and bison."[4] Howard couldn't sleep that night, happily playing over the events of the day.

Back in Philadelphia the next week, Howard issued another statement in *Science:* Clovis, with its artifacts and "matted masses of bones of mammoth,"[5] showed America was inhabited some 15,000 years ago. Not to be outdone, two weeks later Jesse Figgins (of Folsom fame) announced in *Science* he, too, could report on a recently completed excavation of mammoth bones near Dent, Colorado, which had also yielded two large, fluted spear points like those at Clovis. But he was too late to snag the naming rights that came with priority of discovery. We speak of Clovis points and the Clovis culture— not Dent points and the Dent culture.

Coincidentally, that same summer delegates to the Fifth Pacific Science Congress in Vancouver, Canada, were presented a volume of essays on the origin and antiquity of American Indians. The lead chapter, by W. A. Johnston of the Canadian Geological Survey, for the first time suggested that during the Wisconsin glacial period, lower sea levels meant a "land bridge probably existed" between Siberia and Alaska, and that a migration route south from Alaska through the Mackenzie River Valley had opened in the wake of ice retreat.[6]

Within just a few months, Howard and geologist Ernst Antevs put all the pieces together: 20,000 to 15,000 years ago, the Bering Strait was dry land, and a corridor was open along the east side of the Rockies, removing all obstacles to migrations from Central Asia to the Great Plains. Given how neatly this corresponded to their estimated age of the Clovis site, it seemed a route and a timeline for the peopling of the Americas was emerging. The precise ages on that timeline, of course, would not be nailed down until the advent of radiocarbon dating, still several decades off (Chapter 3).

No matter. The geological evidence of a Pleistocene human antiquity was secure, and Howard, pleased by what he'd accomplished at Clovis but anxious to follow the first Americans' trail back to Asia, soon handed over the reins of the Clovis site to a graduate student and began planning an ambitious field program. He would search for sites along an arc from New Mexico, up along the flanks of the Rocky Mountains, through the ice-free corridor region, and he hoped, ultimately back to Lake Baikal in Siberia.

Howard's grandiose plans flopped. But at Clovis, his successor began to fill in the details of a culture that, within a decade, others would trace across the continent. Unlike Folsom Paleoindians, who mostly kept to the plains, Clovis groups were seemingly everywhere. Their sites have since been found across the continent and in a variety of environments, from the coniferous forests of the Pacific Northwest to the desert Southwest, through the rich grasslands of the western plains to the complex mixed forests of the American Southeast and near tundra of the Northeast. No subsequent North American occupation, Paleoindian or otherwise, was so widespread or occupied such diverse habitats, let alone amid the kind of climatic, ecological, and geological upheaval that was happening in the wake of continental deglaciation. And it wasn't long before Clovis peoples' sudden and widespread appearance was attributed to fast-moving big-game hunters—and blamed for the extinction of the mammoth, mastodon, and nearly three dozen other genera of Pleistocene mammals (Chapter 2). It would take another fifty years to realize matters were not as simple as they appeared. But that's getting ahead of the story.

MADE IN AMERICA

In 1935 Howard traveled to the Soviet Union via Europe, but got nowhere near Siberia. Soviet authorities forbid it. Forced to cool his heels in Leningrad, Howard visited museums, where he examined archaeological collections from across the country amassed since the time of Czar Peter the Great. None yielded any Clovis fluted points. That told Howard something. Fluting may just be an American invention—the *first* American invention. Knowing fluting's place of origin and how it spread would be vital to tracking Clovis movement. But where was that, and when?

The obvious place to start searching was between Alaska and the northern plains, and many archaeologists have done just that in the decades since Howard. All together, about fifty fluted points have been found in Alaska, but mostly in the state's far northern reaches (at or north of about 66°N latitude, about the latitude of the Arctic Circle), and only on the surface or in deposits with younger artifacts. About 60 fluted points also have been found in the one-time ice-free corridor.[7] Yet, few of these have secure radiocarbon dates, and those that do are fewer than 10,500 years old, as they ought not to be if they were left behind by Clovis colonizers moving south. Stylistically, the Alaskan and Canadian fluted points don't look much like Clovis points from the Lower Forty-eight either, but appear more akin to later fluted points. This gives reason to suspect fluting was invented *south* of the Laurentide and Cordilleran ice sheets, and spread from there—including to the north (a migratory backwash, as it were, which may have carried other types north as well, as discussed in Chapter 9).

The Great Plains and Southwest are likely places of origin; the oldest known fluted points occur there. Another possibility is eastern North America. Since more and more varied fluted points occur there than anywhere else, Ronald Mason long ago argued they were invented here, on the principle that diversity reflects age. (Sound familiar? It's the principle Greenberg invoked to fix the relative antiquity of his three language families.) Finding eastern fluted points that clearly predate those to the west would settle the issue, but so far none have been. In the meantime, Adovasio reminds us the unfluted Miller lanceolate from Stratum IIa at Meadowcroft would make a fine fluted point precursor.

Once invented, fluting spread over North America, and persisted through Clovis and subsequent Paleoindian cultures (Folsom points are fluted, as are a variety of eastern North American points). But it's hard to understand why. Fluting is a tricky technology to master, in many ways one of the most difficult steps in the sequence of making a point. It is estimated Clovis points broke while being fluted 10%–20% of the time (failure may have occurred nearly 40% of the time when fluting Folsom points[8]). Mistakes were costly, especially when stone supplies were scarce. And for what? Fluting thins and potentially weakens the point, and served no purpose . . . that we know of.

Early on, it was thought fluting enhanced bloodletting of the speared prey, not surprising since in the 1920s, when fluted points were first found, the memory of World War I bayonets was still fresh. But that idea fell from favor when archaeologists realized the flutes were sandwiched between bone (or wood or ivory) foreshafts, or embedded in a socket and *hafted* (attached) to the spear. Flutes on hafted points, wrapped with sinew and possibly coated in tree resin or tar to firmly anchor them, were too deeply buried to enhance bleeding. Others suggested fluting helped strengthen the bond of point to haft, or perhaps by thinning allowed the points to penetrate deeper and become more lethal. Yet, over the next 10,000 years of post-Paleoindian time, hunters managed to muddle through using unfluted points that were hafted and presumably just as firmly anchored to a spear or arrow shaft, and equally effective as penetrating weapons. Maybe, George Frison and Bruce Bradley argue, the reason we cannot show that fluting directly enhanced the performance of the point is because it didn't.

If that's so, why did the technique endure for at least half a millennium? Part of the answer may be inertia: tradition is a potent force to dislodge. But for a process so risky, so costly, and by consensus so useless, there must be more to it. Perhaps fluting was style or art, a symbolic representation of hunting prowess (showing off!), or part of a pre-hunt ritual: could it be mere coincidence some fluted points preserve traces of staining by red ocher, a blood-red mineral paint? In the ritual realm, as we know, costs don't matter. Admittedly, these possibilities are not easily put to the test, but are more the sort of explanations we fall back on when we cannot see purpose in prehistory. As Stanley Ahler and Phillip Geib say, we "sometimes characterize the unknown as unknowable, just to put our busy minds at ease."[9]

Art or ritual it may have been, but whatever fluting was, a great many Paleoindians employed it. Still, not all fluted points are alike, save in a most general sense: all share a lanceolate shape, are ground smooth along the base and partway up the sides (which ensured the edges did not cut the sinew binding or socket haft), and have flutes that extend from the base partially up one or (more commonly) both faces of the point (Figure 46). On a "classic" Clovis point, the sides are parallel, the flutes travel usually less than one-half the length of the face, and the points are relatively long and thick, though deliberately made Clovis miniatures are known (children's toys, some suspect). A freshly minted Clovis point, based on a sample from Texas (and, no, the points are not necessarily bigger in Texas), would be slightly over 10 centimeters long, 3 centimeters wide, and 7.5 millimeters thick.[10]

FIGURE 46.
Clovis point from the Clovis type site in
New Mexico, displaying the type's diagnostic
features, including fluting and *outre passé*
(overshot) flaking. This particular specimen
is made of Edwards Formation chert, likely
from an outcrop source near Big Springs,
Texas, approximately 300 km southeast of
the site. It was found near the vertebra and
ribs of a mammoth. (Photograph by David
J. Meltzer; line drawing by Frederic Sellet;
arranged by Judith Cooper.)

There is often surprising uniformity in the size and shape of the lower portion or base of a Clovis point, where it would have been inserted into a haft. It appears the aim was to fit the points to a haft, and not vice versa. That implies the hafts were more difficult or time-consuming to make, or perhaps the haft material was harder to come by. Regardless, broken points had to be unwrapped and ejected when they broke, as they inevitably did, and the haft reloaded.

Beyond that, fluted points vary considerably in their size and shape, and in the length and depth of the flutes, the number of flutes (though three flutes per face is about the limit), the kind of chipping used to fashion the point, the finishing of the edges, and a host of other minutiae that delight archaeologists. This is true with Clovis, and especially between Clovis and immediately post-Clovis fluted points, of which there are many styles (especially in eastern North America [Chapter 9]). That sometimes makes it difficult to decide what's Clovis and what's not, particularly in the absence of accompanying radiocarbon ages. Much of the variation results from the effects of use and resharpening of the points, but also from the divergence of people, knapping styles, and techniques over time and space. Think of it as a kind of "cultural drift," as kin and descendants experimented with and introduced their own variations on the Clovis theme.

PACKING THE TOOLBOX

Fluted points are certainly the most recognizable artifact in the Clovis toolkit, but assuredly not the only important one. It's a generalized toolkit: suitable for many tasks, not built specifically for any one (see Plate 8). It includes large bifaces, which could be readily transported and from which flakes could be struck for use, or from which points could be obtained. There were also end and side scrapers for cleaning hides; gravers or flakes with carefully formed small projections or spurs, possibly used to engrave bone or wood; spokeshaves, or notched pieces, which may have been used to shave wooden or bone shafts; and a variety of knives or flake-cutting tools. There is a certain monotony to the Clovis toolkit, with variation expressed more in the presence or frequency of certain tools (end scrapers, for example), or the kind of stone used, than in the types of tools. That sort of pattern tells us more about what was done than who did what.

However, a few distinctive stone-tool types are known to occur primarily in some regions but not others. Long, curving stone blades (their length twice their width) and the multifaceted (polyhedral) cores from which they were struck, were fashioned by stone workers on the southern plains, and perhaps into the Southeast: a particularly rich haul comes from the Gault site in central Texas. Christopher Ellis and his Great Lakes colleagues show that backed bifaces (perhaps themselves hafted) were common, and fluted drills absent, in the eastern Great Lakes, while the reverse seems true in New England.

Where preservation permits, bone and ivory artifacts are occasionally found. These include ivory hammers or billets, and a mammoth bone shaft "socket" wrench, possibly used for straightening ivory or bone rods, from the Murray Springs site (Arizona). No bone or ivory rods accompanied this particular specimen, but examples have been recovered in a dozen other sites scattered across the West, and from underwater sites in Florida.

These bone and ivory rods are usually about 30 centimeters long but fewer than 2 centimeters wide, and are often beveled at one or both ends. A glimpse at how these were made comes from the Clovis site, which yielded a section of mammoth tusk, V-notched at one end as though felled by a beaver, which at the time it was lost or abandoned was on its way to being further reduced and fashioned into one of those ivory rods. These rods have been variously interpreted as spear points, foreshafts for joining stone points to a spear, tools for fine flaking of stone, pry bars for dismembering carcasses, sled shoes, or wedges for tightening up loose haft bindings.[11] There's no reason to limit their function to just one of these possibilities.

Unfortunately, their scarcity reveals little about how bone and ivory were used by Paleoindians, since these were almost certainly used more frequently and in more ways than we can infer. The spotty distribution, however, does say a great deal about how poorly bone and ivory are preserved in the archaeological record.

Which also explains the virtual absence of what may have been their "perishable" artifacts, those made of plant fibers, or animal hide and fur. Such items vanish quickly, save in unusual circumstances, such as where it's extremely dry or perennially wet (recall the string tied around the tent stakes at boggy Monte Verde). James Adovasio reports Clovis-age twined bags, mats, burden baskets, and trays, mostly from Great Basin dry caves (such as Danger Cave, Utah). No Clovis nets have been found. Yet. Adovasio expects they will, since net hunting has deep roots in prehistory, as well we might expect: it requires relatively little expertise, it's effective and nonconfrontational, and animals trapped in nets (even ones as large as elk) can be killed safely at close range by stabbing or clubbing by individuals without great physical strength. Consequently, it's a method that routinely involved "females, juveniles, and the old."[12]

Not all Clovis tools and artifacts appear in all Clovis sites, and more differences in their toolkits and technology will likely emerge as we learn more of the archaeological record. At present, we lack sites and assemblages from many geographic areas, and have found relatively few of their habitation sites where, by virtue of longer periods of occupation and greater number of activities that took place, a wider range of tool classes would be expected.

BUT IS IT ART?

A number of Clovis bone and ivory rods are scored with zigzags; parallel lines; small, curved incisions; and in one instance, zipper-like markings. The markings are far too systematic to be merely accidental. They could be art for art's sake, some other symbolic decoration, or more prosaically, ownership or use marks.

Engraved stones, too, though no more than a few dozen, have also been found in Clovis-age sites. These are incised with cross-hatched, parallel, intersecting, or converging lines. All but two were recovered at the Gault site, and had been made on the locally occurring limestone (Figure 47). That's not the best artistic canvas to be sure, and what

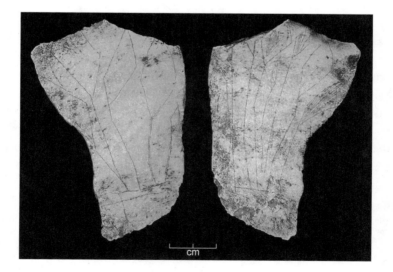

FIGURE 47.

Clovis art: two faces of a piece of engraved limestone from the Gault site, Texas. The meaning of the lines is unknown. Among the interpretations: they represent plants, or perhaps a map of rivers and tributaries. (Photograph courtesy of Michael Collins and the Gault Project, Texas Archeological Research Laboratory, University of Texas–Austin, of a cast made by Pete Bostrom of Lithics Casting Lab, Troy, IL.)

those scratches represent is anyone's guess. But if one applies a bit of imagination, some of the Gault engravings appear to represent stick-figure plants or, letting the imagination go full throttle, maps depicting rivers and their tributaries.

Clovis-age cave or rockshelter art is rarest of all, if indeed we've ever seen any. We're never sure we are looking at what we think we're looking at, as it is often difficult to determine the age of an image on a rock wall. Even so, Larry Agenbroad and colleagues have recorded half a dozen rock panels on the Colorado Plateau in eastern Utah that display paintings and engravings of what appear to be mammoths. Agenbroad worries some of these are modern fakes (shades of Hilborne Cresson!), and none are dated to late Pleistocene times.[13] As noted earlier, the extraordinarily rich and expressive art routinely found in Paleolithic Europe—cave paintings and sculpted figurines of extinct animals and humans appearing in both natural and abstract forms—is absent from late Pleistocene North America. Perhaps by then the animals were already extinct, or possibly people had other things on their minds (or their walls).

STONE FOR THE TAKING

Which is not to say Clovis people lacked appreciation for the finer things. In making their stone tools, they relied almost exclusively on very-high-quality chert, jasper, chalcedony, and obsidian (volcanic glass) (see Plate 9). All are fine grained and lack cleavage planes or other natural fault lines, making them easy to chip into a desired form, and capable of holding razor-sharp edges. Other kinds of stone can be serviceable, but Clovis (and, for that matter, later Paleoindian) knappers only went for the best. Why? Higher-quality stone is less failure prone, which would have been of no small concern to hunters going after large game at a time when hunting was very much an up-close and personal activity: no safe firing from long range here. If a point broke at the wrong moment because of a flaw that weakened the stone's tensile strength, it was not just a bad day at the office. As Frison says, it could mean injury or death.[14] Further, a point made of higher-quality stone lasts longer and can be readily resharpened, thereby prolonging its use. That would be important to wide-ranging hunter-gatherers, who could not predict when they would next be able to resupply at a stone source.

We benefit, too: because these high-quality stone types are often distinctive in color, fossils, chemistry, or composition, we can frequently pinpoint where they were obtained and how. It appears Clovis peoples mostly acquired their stone directly from bedrock outcrops (Figure 48), where it could be quarried in large blocks, rather than from picking through the gravels of a river bed or a glacial moraine in hopes of finding cobbles of suitable size or material. Size was critical to Clovis knappers since some of the bifaces they made were upwards of 20 centimeters in length and width, and more than 600 grams in mass, requiring the starting blocks of raw material—*cores,* they're called—to be of even greater size.[15]

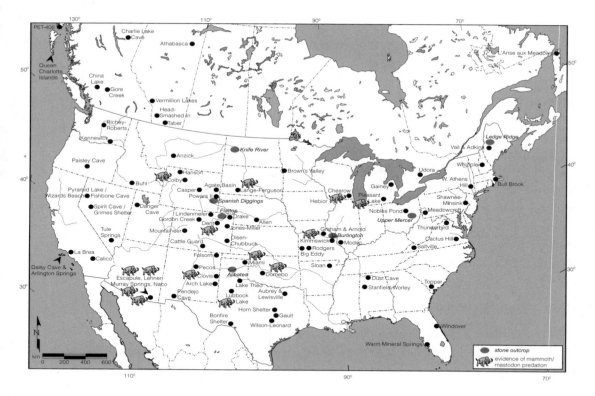

FIGURE 48.

Map of Canada and the United States showing the location of pre-Clovis, Clovis, Late Paleoindian, and more recent sites discussed in the text. Only some of the stone sources used by Paleoindian groups are located on the map; the symbol for the outcrops should not be taken as indicative of their size or shape, but only as indicating an approximate location.

The Clovis knappers' preference for specific stone outcrops reveals the depth of their geological knowledge: they found sources we never have. And they went directly to those sources themselves. No trading was apparently involved, but then what self-respecting hunter-gatherer wants to rely on others to supply a material so essential to survival, or needed in bulk? As open as territories were on this landscape, stone sources were accessible to all, leaving little incentive for a Clovis entrepreneur to schlep double the usual amount of stone, one-half for personal use and the other half for exchange, across great distances on the off chance of finding a trading partner. That said, their sporadic encounters on the landscape likely resulted in an exchange of gifts to mark the occasion and the bonds forged, which may explain why sites occasionally yield a Clovis point or two made with an unusual style or material.

Because stone can often be identified to the outcrop from which it was obtained, the distances between where it was quarried and where it was abandoned serve as an odometer of Paleoindian travels. Sometimes those distances were great. Alibates agatized dolomite from the Texas Panhandle, just outside of Amarillo, was used to fashion Clovis points left in the Drake cache in northeastern Colorado, 585 kilometers away. Although we routinely use the straight-line distance from quarry to site as a measure of Paleoindian mobility, that takes no account of side excursions, return trips, or any other archaeologically invisible travel. Actual mileage may vary.

It certainly did in Clovis times across the continent, the scale depending on the nature and density of resources being exploited. Distances of 300–400 kilometers are common on the treeless western plains, and on the open parkland and tundra of northeastern North America, where resources were locally clumped but widely separated. In the closed, ecologically richer forests of northeastern North America—where, we suspect, groups did not have to be as wide-ranging (see below)—the sites are dominated by locally acquired stone (picked up within 50 kilometers of the site, say), though occasionally contain specimens of more distant or *exotic* material.

In most sites, exotic material occurs as finished artifacts, or at most, as prepared bifaces or cores, rather than as raw blocks of stone (at the Gainey site in Michigan, for example, cores were hauled in from east-central Ohio's Upper Mercer chert outcrops, some 380 kilometers away). Mobile hunter-gatherers without benefit of beasts of burden (save, perhaps, for dogs) would have had their hands full carrying food, hides, tools, children, and the like. Stone artifacts were heavy enough on belts or slings without the extra weight of blocks of stone, much of which was destined only for discard when later flaked into a point or tool.

How they fashioned their points and tools is captured in the manufacturing debris left behind, particularly at sites situated close to stone outcrops, including Gault, Thunderbird (Virginia), and West Athens Hill (New York). These sites are dominated by the vast numbers of flakes *(debitage)* that fly off as a stone tool is chipped. At Thunderbird, a single, three-inch excavation level of a 10 x 10 foot square excavation unit (this was the 1970s; we were slow to go metric) often yielded more than 10,000 pieces of debitage, along with tools that broke in the process of manufacture, and even glimpses of discrete moments in time: among them, flakes piled up as though they had rained down on the sides of a knapper's crossed legs, and the tip of a nearly finished Clovis point found dozens of feet from its broken base (Figure 49). That point had been just a few flakes away from being perfect, and one can only empathize with the knapper's frustration when it broke *(why did I have to hit it that one last time?)*—and readily imagine seeing the broken pieces flying across the site.

Occasionally, if the stone wasn't of the highest quality, it might be slowly heated in sand beneath a hearth to enhance its "flakeability." Experimental studies by Philip Wilke and colleagues show that the optimal method for heat-treating stone is to slowly raise it to a temperature of 165°C–300°C for several hours, then bring it back down slowly (heat or cool it too quickly, and the stone will craze or shatter). Heat-treated stone gains a waxy luster, and depending on its chemistry, can change color: specimens of the yellow green jasper at Thunderbird turned pink and red from heating.[16]

FIGURE 49.

Two fluted points from the Thunderbird site, Virginia, which failed en route to manufacture. The specimen on the left was nearly complete when it broke during final thinning. Its two parts were found some distance from one another, hinting that a healthy pitch followed the fatal break. (Photograph by David J. Meltzer.)

After groups departed the quarry, they looped around the landscape, and over time, artifacts were used or broken, and the stone supply dwindled. When far from a quarry, the groups conserved their resources: fluted points that dulled or broke were carefully resharpened, but if broken beyond repair or whittled to a useless nubbin (say, 4–5 centimeters in length, barely long enough to peek out from beyond the binding of their haft), the point might see new life by being recycled into other tools like scrapers, drills, or wedges.

Just so, groups venturing far into unknown terrain surely kept a watchful eye for new outcrop sources to replenish their toolkits. And perhaps, hedging their landscape learning bets, they also had the foresight to lay in resupply depots along the way.

CACHING ACROSS THE LANDSCAPE

The Clovis-age archaeological record includes a number of stone artifact caches containing bifaces and blades (see Plate 10). Although these could have been used "as is," more likely they were chipped into those forms for easier transport, and with the idea that from these forms, they could readily be made into one of a variety of useable tools. These caches rarely occur at Clovis camps or kills (though there are exceptions), but instead often turn up in otherwise isolated spots, giving the appearance of having been stashed, presumably with the intent of being retrieved later. The cache pieces are generally fashioned of stone from outcrops several hundred kilometers distant.

We now know of at least twenty Clovis caches scattered throughout North America, but mostly in the West. A prime example is the Richey-Roberts site, discovered in the late 1980s by workers installing a sprinkler system in an apple orchard near East Wenatchee, Washington.[17] Full excavations revealed some forty Clovis artifacts, including projectile points, preforms (unfinished points fashioned into a broadly lanceolate shape), bifaces, scrapers, other stone tools, and a dozen or more bone and antler rods. The Richey-Roberts Clovis points, some of which are 21 centimeters long and 6 centimeters wide, are among the largest ever found. Here, as in other Clovis caches, some artifacts were sprinkled with red ocher. Were they being symbolically infused with life-giving blood, or was this a blood-colored warning: but to whom—or what? Or was it something else? No one can say.

At the Anzick cache in Montana, over 100 Clovis stone artifacts, and bone and ivory rods, accompanied the uncremated, ocher-covered remains of a child (Chapter 5).[18] Here, the term cache is surely inappropriate, since the collection of artifacts is likely grave goods, not intended for later use, at least not in the corporeal world. But for those caches that are not obviously burial offerings, a more utilitarian explanation may be sought.

On its face, an artifact cache represents a handy solution to the logistical problem faced by highly mobile peoples: the gap between where a stone could be acquired (which is predictable), and where it was to be used (which is not always predictable, and in any case may be very far away). Annual moves might bring mobile peoples past quarries occasionally, but they can only haul away a limited amount of stone. And yet that stone is "spent" all over the territory, sometimes exhausting supplies far from the source. One solution to that logistical dilemma would be to cache supplies of stone in convenient places around the landscape. When tools broke, wore down through resharpening, or were lost, as all tools inevitably were, resupply at a cache could save a long trip back to the quarry. Seems self-evident.

Later Paleoindian groups were just as mobile as Clovis groups and just as reliant on high-quality, distant stone sources. By that reasoning, we should also see their caches as well, but we don't. Late Paleoindian caches are extremely rare.[19] There are now enough Clovis caches and few enough from the millennium that followed to suggest this pattern is real and not an accident of sampling. How, then, can we explain the preponderance of Clovis caches, and the scarcity (if not absence) of later ones?

Michael Collins sees an explanation in the predictability of Clovis movement: he suggests they knew in advance where they were headed, knew they would be returning to particular spots with perilously exhausted tools in need of replenishment, and in anticipation, left stone behind for future use. Later Paleoindians, he argues, could not predict their future movements with enough confidence to benefit from caching stone. Perhaps. Yet, the stone used by Folsom groups (Clovis successors on the plains) comes from a greater diversity of sources, sometimes from even greater distances. The more complex pattern of Folsom stone acquisition suggests that however unpredictable their movements, they certainly knew where sources were, and were highly adept at scheduling their visits. Moreover, if Clovis groups knew they were returning to a spot and would need and use the stone, then why were the caches still there 11,000 years later?

I suspect a different strategy might have been in play: Clovis groups were so new to the landscape they didn't know and couldn't predict where or when they were next going to find a suitable stone outcrop. By depositing caches about the landscape as they moved away from known sources, they created artificial resupply depots, and thus anticipated and compensated for that lack of knowledge. If they found no stone in the area where they were headed, they at least would not have to double back completely to the original outcrop to refurbish their current tools when those were used up. And if they did find a new stone source, the cache could be ignored and abandoned.

Is it then a mere accident of sampling that most of the Clovis caches found so far are on the Great Plains and in the Northwest, far from stone outcrops? Or could this be telling us that the challenges of exploring those areas were different? Time and more caches will tell.

Stone is especially suitable for caching, since unlike meat or other food, it won't spoil, won't be attractive to scavengers, and barring massive tectonic activity, won't move before you return. A stone cache has a fixed and predictable location. As such, it serves a useful role in wayfinding (Chapter 7), for now one's "map" of a landscape includes not just the natural resources, but some artificially placed ones as well. This helps make it possible to venture further across otherwise unknown terrain. Over time, as new sources of suitable stone are located, and as groups are better able to predict where and when they will be able to replenish their supplies (presumably in the post-Clovis period), then caches become less critical to movement, and ultimately unnecessary.

The obvious flaws to this theory are the occasional caches found close to a raw material source—like the deGraffenreid cache (see Plate 11), apparently stashed in the vicinity of the stone-rich Gault site[20]—and the likelihood there were multiple purposes for caching stone. And, of course, one cannot use the absence of caches in later Paleo-indian periods as proof landscape learning was complete by then, at least not without getting dizzy spinning in a logical circle. That said, I think there is independent evidence—albeit circumstantial—to indicate that might have been the case, though that's for the next chapter, when other clues to the process become apparent.

MOBILE HOMES

Befitting their mobile lifestyle, the dwellings Clovis people built were temporary shelters made of wood or materials that quickly degraded and disappeared (none were made of stone). Their traces are often little more than hard-to-spot circular stains in the soil marking where wooden posts were once planted *(postmolds)*. William Gardner reports one such structure at Thunderbird, a line of postmolds forming a rough oval 9 × 12 meters in size.[21] Other evidence of structures can be as subtle as slightly hardened earth, or artifacts concentrated around darkened soil or burned areas.

Although such clues reveal little of the habitations once standing there, they do hint at site use and the scale of the occupation. In the Upper Midwest and northeastern North America, for example, a dozen or so Paleoindian sites include discrete artifact concentrations, perhaps 4–6 meters in diameter, yielding a like range of tools. These are thought to be individual family households and the remains of their domestic activities. Occasionally, broken artifacts from different concentrations can be fit together, a strong hint that the households were occupied at the same moment in time. At Nobles Pond (Ohio), the concentrations were even neatly arranged in a semicircle.[22] A Pleistocene mobile home park, as it were, but again with the emphasis on *mobile*. In other sites, the concentrations are less well defined and overlap, which bespeak successive occupations in the same spot.

Evidently, these groups did not stay for long in any one place, nor always return to it: most of their sites were used but once. With a continent to themselves, the only incentive to revisit a spot was if it offered a prized resource, whether permanent (the quarry-related sites of Gault, Thunderbird, and West Athens Hill) or temporary (the Vail site in Maine seems to have been established alongside a migratory game trail). Such is the advantage to colonists on a landscape with few other people and no territorial borders. But as discussed in the previous chapter, there was also strong incentive for separate families to gather and socialize, exchange items and information, maintain old kin ties, and establish new connections through marriage. So far, however, no Clovis "rendezvous" has been found.

POINTS ON LAND

But Clovis people left behind more than sites; they also scattered their distinctive points across North America. There has long been something of a cottage industry among archaeologists devoted to tallying Clovis points by state (we're approaching 550 in Texas). David Anderson and colleagues have compiled those into a continent-wide total: it's currently at over 13,000 points from some 1,500 locations.[23] The great majority of these are *isolates*, unaccompanied by other Clovis tools, bones, or trappings of a site. As David Brose remarked (tongue-in-cheek), from the looks of it, these groups ate nothing and lived alone. Yet, where these isolates are found might just tell us something interesting.

A map of their distribution and density reveals isolates are particularly thick on the ground across much of the eastern United States, especially around the Tennessee, Ohio, and Cumberland river valleys. In contrast, many areas have only sparse and widely isolated occurrences, including much of the Great Basin, the Columbia and Colorado plateaus, the northern Great Plains and northern Rockies, and the uppermost and lowermost reaches of the Mississippi Valley.

Anderson interprets the high concentrations of isolates in the Midwest and Southeast as "staging areas," spots on the landscape where groups slowed down and settled in for a time, before pushing still deeper into the continent.[24] Possibly so. However, these are also regions rich in high-quality stone outcrops. Is the density of Clovis points there merely a reflection of the unusually large amounts of archaeological debris that routinely surround such sources, or were such areas indeed magnets for longer-term human occupations by virtue of their being resource-rich and predicable spots on the landscape? It could be both. And what of the scarcity of Clovis points in places like the lower Mississippi Valley? Does that mean there were few Clovis inhabitants, or is it merely a reflection of the last 10,000 years of the river's floods deeply burying any Pleistocene-age remains?

The distribution of Clovis points is not a map of Clovis people on the landscape (as Anderson well appreciates), but a complicated result of many things, including modern land-use patterns (point densities often correlate with the amount of plowed acreage); deposition, erosion, and surface age; archaeological activity (not all states have had systematic surveys for Clovis points); and the presence of high-quality stone outcrops.[25] Few Clovis points does not necessarily mean few Clovis people.

A more representative sample will surely change the geographic details of the Clovis distribution, but probably not the essential fact that the Clovis presence on the landscape was broad and not deep. It's the archaeological footprint of highly mobile people at low population densities, further testimony that movement was tied to topography and terrain, and likely involved leapfrogging across the landscape, rather than proceeding as a wave of human settlement washing across the land.

IT'S ABOUT TIME

The chronology of the Clovis occupation shifts through time and across the continent. The oldest sites are those on the Great Plains and in the Southwest, which range in age from 11,570 BP (the Aubrey site, Texas) to about 10,800 BP. Significantly, the earliest appearance of Clovis follows the opening of a viable ice-free corridor (Chapter 2).

There are reliable radiocarbon dates on only a dozen or so Clovis and Clovis-like (Gainey type) points in the eastern United States, but despite the long-held suspicion Clovis and fluting originated in this area, the oldest of these (Shawnee-Minisink, Pennsylvania) is only 10,940 years old. Most of the others fall between 10,600 and 10,200 BP, a period contemporaneous with Folsom on the Great Plains. Clovis-like materials occur in the Great Basin and far West, although their ages are even less certain, and the cultural chronology is confused by the contemporaneous occurrence of the large, unfluted Western Stemmed points.

A recent radiocarbon study by Michael Waters and Thomas Stafford shows that nearly a dozen Clovis sites from Montana (Anzick) to Arizona (Lehner) to Pennsylvania (Shawnee-Minisink) date to 10,900 BP, give or take a century. That makes them essentially contemporaneous, at least within the bounds of radiocarbon dating.[26] The styles of Clovis points at these sites are quite different (though all are still recognizably Clovis). When the ages of more sites are pinned down as tightly, we'll finally be able to get a precise gauge of the timing of the stylistic change and, by extension, the cultural drift (Chapter 7) that occurred as Clovis people spread across the continent.

In the meantime, the radiocarbon record supports the long-held suspicion Clovis people radiated rapidly across North America, the process taking perhaps no more than 500 radiocarbon years. It may have taken longer in real time. The Clovis period overlaps the early portion of the Younger Dryas, with its radiocarbon plateaus that defy easy calibration (Chapter 2). Although more and tighter-calibrated radiocarbon ages may ultimately change the apparent speed of the Clovis dispersal, perhaps "slowing" it to, say, 1,000 calendar years, it nonetheless will remain one of the fastest expansions of any culture known in prehistory (Chapter 7).

MURDER IN THE PLEISTOCENE?

The traditional explanation (Chapter 7) for how or why they moved so far so fast has been that Clovis people were specialized hunters who pursued wide-ranging big game, including the now-extinct Pleistocene megafauna. Latching on to prey that paid little mind to ecological boundaries ostensibly enabled Clovis groups to do likewise, and thereby speed across the continent without having to learn new adaptive tricks to survive in different environments. After all, a mammoth was a mammoth, no matter where it lived, but if the mammoth happened to be a mastodon, horse, or camel instead, that only required minor adjustments to the hunting strategy. Or so the argument went.

Gary Haynes (no relation to Vance Haynes) proclaims that nowadays "very few if any rational archaeologists want to argue that Clovis people exclusively specialized in megafauna everywhere and at all times."[27] I wholeheartedly agree (and consider myself rational). Besides, specialized hunting was rare enough among known foraging groups. A specialized hunting adaptation that transcended a continent as large and diverse as Late Glacial North America in less than a thousand years would be unique. And yet continent-wide hunting is said to have driven thirty-five genera of mammals to extinction. Specialized, or not, that certainly makes pressing demands on human hunters. Are Clovis people guilty as charged?

There is no doubt they co-existed with some of these now-extinct mammals. Their overlapping distributions place people at the late Pleistocene crime scene, and at least in some instances, Clovis points have been found embedded in large mammal skeletons. We have motive, method, even the occasional smoking gun. Best of all, if Clovis groups were big-time big-game hunters, that accounts for how they moved so far and so fast, and why all these animals went extinct.

The idea humans caused late Pleistocene extinctions is not new, but goes back more than a century. The modern version is due to ecologist Paul Martin, who attributes extinction solely to human hunting: Pleistocene overkill, he calls it. That the animals lost were mostly megafauna (with an adult weight of at least 45 kilograms [about 100 pounds]) he thinks no coincidence: "Large mammals disappeared not because they lost their food supply but because they became one." Their massive size was their death sentence. Yet, what of the small Aztlan rabbit and the diminutive pronghorn? Martin supposes these were "large enough" to have been attractive to Clovis hunters.[28] Otherwise, he believes smaller prey were of no interest to Clovis hunters and were tossed back (Clovis catch-and-release?).

As Martin envisions the overkill process,[29] animal fates were sealed when a band of 100 highly skilled Clovis hunters emerged out the southern exit of the ice-free corridor just over 11,500 years ago. Before them was a vast expanse of territory empty of people but teeming with immense game. All that meat on the hoof, Martin says, was irresistible to Clovis hunters, who liked megafauna so much they ate them to extinction.

Martin crafted an elaborate and imaginative scenario to show how this could have been accomplished (whether it actually happened that way is another matter, of course). The way he figures it, overkill was inevitable: these animals had never before peered down the shaft of a spear, and would have been fatally ignorant of the danger posed by humans. Killing such naïve and vulnerable prey in this target-rich environment was easy, he figures, and if just one hunter in four bagged a trophy a week, it would decimate local animal populations. Slow breeders, as most of the Pleistocene megafauna likely were, would not have been able to reproduce fast enough to replace the animals killed, even if only a third of them were slaughtered by hunters. The hunters would make short work of the animals in an area, then turn their attention to the targets in front of them, and ignore the growing pile of meat and carcasses behind. Martin supposes this gluttonous human population doubled in size every twenty years, an annual growth rate of 3.4%, and made fast time, fanning outward in a killing wave from which no animal could escape, and which washed across the continent at a rate of 16 kilometers a year.

At that rate, within 350 years, Clovis people would have numbered 600,000 and would have reached the Gulf of Mexico. Just 1,200 years after entering North America, they got to Tierra del Fuego. In their hemisphere-long wake were tens of thousands of fluted points (smoking guns?), the rotting carcasses of more than 100 million herbivores, and carnivore populations that would have boomed with the sudden fortune in freshly killed meat on the landscape—and then the wave went bust when the supply ran out (Martin doesn't think the Clovis hunters targeted carnivores, or needed to in order to cause their extinction: killing off their prey would do the trick).

A chilling scenario, Pleistocene overkill, and one that has achieved a celebrity few academic theories enjoy, perhaps not least because this grim homily of wanton slaughter can be (and often is) preached as a sermon of our species' sins. Granted, the human rap sheet is a long one, but this may be one crime against nature we didn't commit. Overkill seems unlikely for many reasons, among them:

- It took modern humans over 100,000 years to reach Alaska. Could it have taken only 1,000 more years to get to Tierra del Fuego? There were vast stretches of unexplored and highly variable environments to be traversed, each of which posed radically different adaptive challenges to groups that were there to make a living and raise families; lacked maps, convenience stores, and transportation; and had to find water, other food, shelter, stone, and other resources critical to their survival. That takes time.

- Essential pieces of the overkill scenario ring hollow: most hunter-gatherers reproduce at relatively low rates, generally less than about 1.5% per year. Is it reasonable to suppose Clovis groups multiplied at 3.4% in order to reach a population size of 600,000 in just a few centuries? That's an unusually high rate that Martin plucked from the *Bounty* mutineers on Pitcairn Island, who apparently shed their inhibitions along with Captain Bligh.

- Likewise, it's doubtful Clovis spread in a wave across the continent, ignoring environment and topography. Armies, not hunter-gatherers, come in waves. And indiscriminate hunting of nearly three dozen genera of megafauna continent-wide demands a comprehensive knowledge of places and animal behavior few hunters—especially hunters new to a landscape—could possess. Besides, as Frison explained, no one hunting strategy would work on all mammoths, let alone all megafauna (Chapter 7).

- For that matter, when a foraging group's return rate begins to decline, they move, doing so long before they've burned through all the available resources in their original habitat (Chapter 7). They stay put only under unusual circumstances, notably, being hemmed in on all sides. That was hardly a problem on a late Pleistocene landscape devoid of other people, with a choice of productive habitats in which to hunt and gather, and where one could leave a habitat long before its animals were gone. Its highly unlikely Clovis hunters behaved like Sherman's army in its scorched-earth march across Georgia.

- The Pleistocene fauna may have been naïve, but studies have shown that even naïve prey populations "process information about predators swiftly," and their populations quickly rebound from an onslaught of predators.[30] The Clovis hunters' advantage would not have lasted long. Besides, even if unfamiliar with humans, the Pleistocene fauna were already well familiar with the likes of saber-toothed cats, short-faced bears, and dire wolves, and could shift their defensive behaviors to respond to their new threat (humans) just as did African elephants.

- Finally, overkill requires there were no humans in North America prior to Clovis times, for then the megafauna surely would have become accustomed to and developed defensive behaviors to respond to human hunters, or perhaps would have gone extinct before Clovis times. And as we now know, there is evidence for a human presence in the Americas (Monte Verde) that predates 11,500 BP.

Reasonable points all, but by themselves they are not sufficient to reject overkill. After all, it is possible people could have moved that fast through the hemisphere. They could have reproduced at near rabbit-high rates and spread in an expanding, earth-scorching wave. And, they might have chosen to hunt animals well beyond the point foraging theory suggests they should have abandoned the patch and prey.[31] America's colonizers may simply have been unlike any hunter-gatherers we know (admittedly, I'm skeptical, but it's not completely outside the realm of possibility). And perhaps the Pleistocene fauna were asleep at the evolutionary switch, and failed to learn how to cope with predation. That's highly unlikely too, as any zoologist will tell you, but let's grant the point for the sake of discussion. Nor can we reject overkill on the grounds people were here in pre-Clovis times. Martin observes that Clovis "is the time of unmistakable appearance of Paleoindian hunters using distinctive projectile points,"[32] and he's right. It is only then that we have evidence people did, in fact, prey upon now-extinct large mammals.

For that matter, there is ample evidence that the arrival of humans can signal the death knell for native animals. David Steadman and others have shown that the grim reaper of extinctions followed the landings of the peoples who colonized the islands of the Pacific. The most famous (or infamous) case comes from New Zealand, where eleven species of moas, large, flightless birds that ranged in weight up to 500 pounds, went extinct soon after humans arrived. In fact, at least twenty-five other species—lizards, frogs, birds, and bats—were also lost. Yet, these losses are not the result of hunting alone, as Donald Grayson argues, because in this and in all other island extinctions, multiple causes were in play. The first New Zealanders hunted, but also set forest-destroying fires, introduced competitors and predators (rats and dogs), and even brought in diseases (such as European bird viruses). It was some combination of these that caused extinctions.[33]

Besides, island life is highly susceptible to extinction. Most Pacific islands are tiny (often no larger than Manhattan), their ecosystems are easily destroyed (forests can be felled in a matter of years or decades by humans needing fuel or construction material), and their native animal populations are small, vulnerable, and cannot be replenished from distant continents because they had evolved into new species since their arrival. This was especially true of birds, which on remote, predator-free islands commonly lost their ability to fly, an evolutionary change that proved fatal when predators arrived.[34]

Stuart Fiedel and Gary Haynes proclaim that a continent is "ultimately just a gigantic island," but that's true only in the narrow sense that it's also surrounded by water.[35] None of the factors that make island life vulnerable to extinction apply to North

America, which supported vast animal populations (100 million animals by Martin's own estimate) and a formidable array of predators (no species could afford to let down its evolutionary guard), and for which there is absolutely no evidence Clovis people caused massive (or even minimal) environmental destruction. No New Zealand moa had to contend with a *Smilodon;* no North American mastodon ever had its forest habitat defoliated. If humans drove North American animals to extinction, it had to have been by their spears, as Martin said. Is there evidence of that?

CROSS-EXAMINING THE FACTS

For many years, those of us who worked with the archaeology of late Pleistocene North America confidently asserted there was no widespread evidence of Clovis megafaunal hunting. Yet, none of us ever demonstrated that was so. There was no systematic scrutiny of the archaeological and fossil records to see which—if any—of the thirty-five genera of now-extinct mammals had been found in Clovis-age sites in such a way as to suggest they were killed by hunters. Recognizing it was time to stop arm waving and start analyzing, Grayson and I did just that.[36] The results surprised us.

We began by amassing as comprehensive a list as possible of all alleged Clovis kills, even ones that offered little more than a fluted point found in the same vicinity as some fossil bone. There were seventy-six such sites. However, the mere occurrence of artifacts and animal bones in the same site or even on the same surface is not by itself testimony of a predator-prey relationship. Therefore, we evaluated each case to see if the fluted points or other tools were found in such a way as to suggest they caused the death of the animal, or if there was compelling evidence of butchery (cutmarks or pry marks on bone) (Figure 50), or if the remains were stacked or otherwise arranged in ways that could not be explained by natural causes. That's the sort of evidence we routinely find in kill and butchery sites from other times and places in the past.

FIGURE 50.

A telltale sign of human butchering: cutmarks (the two roughly parallel lines just left of center, and the nearly straight line above and to their right; the irregular lines to their right are from root etching of the bone) on a portion of a bison jaw from the Folsom site, presumably made while removing the tongue. Not all butchering activities result in cutmarks, but when present, they provide secure evidence of human activity. (Photograph by David J. Meltzer.)

Vetted in this manner, humans proved to be the agents of death or dismemberment in only fourteen of those seventy-six sites. And several of those were unsuccessful kills (animals that were speared, but which managed to escape), while others appeared to be the result of opportunistic scavenging of animals perhaps already near death at a water hole (more on these sites below). Humans undeniably hunted big game on occasion, but they didn't make a habit of it. In fact, a single band of hunters may have disproportionate influence on what we see archaeologically: four of those fourteen sites (Escapule, Lehner, Murray Springs, and Naco) are within 30 kilometers of one another in Arizona's San Pedro Valley, and the artifacts found in them are so similar as to appear to be the work of the same group.

Measure it any way you like, fourteen sites is a tiny number, especially set against the wanton slaughter of 100 million animals envisioned under Pleistocene overkill.

But it's not just that it's a pitiful number. In those fourteen sites, only mammoth and mastodon remains occur. What of the other thirty-three genera of extinct Pleistocene mammals? There are no Clovis horse kills, no camel kills, no sloth kills, no *Hemiauchinia* kills, no tapir kills, no giant beaver kills, no kills of any of the other genera of megafauna that went extinct. And it's not because their bones are rare: horses are the most abundant late Pleistocene fossil we have. That a few horse and camel bones, or dermal ossicles of the ground sloth *Paramylodon*, occasionally end up in archaeological sites is intriguing, but does little more than show they co-existed with Clovis people.[37] If there was a continent-wide slaughter, someone did a superb job of hiding the evidence.

WHEN ABSENCE OF EVIDENCE BECOMES EVIDENCE OF ABSENCE

In a brazen bit of rope-a-dope reasoning, Martin insists the scarcity of Clovis big-game kills *proves* overkill. This is so, he claims, because overkill ostensibly occurred in just a few centuries in any one area, and hence the odds are against a kill site being preserved in the geological record. And, secondly, because he believes Clovis people were tidy: after they made a kill, they cleaned up after themselves, carefully retrieving any stone points buried in the mass of flesh and gore of the carcasses (having excavated several large animal kills and the artifacts found in them, I can attest that Paleoindian hunters were not that tidy, but grant the point for the moment).

It is a rare hypothesis that predicts a *lack* of evidence, but as Grayson and I observed, we have one here, and we have it only because evidence for this hypothesis is, well, lacking.[38] But perhaps there is reason to accept Martin's point that kill sites might be rare. After all, in areas such as eastern North America, where high rainfall and heavy vegetation conspire to increase soil acidity, bone is quickly destroyed. Maybe poor preservation accounts for the lack of kills. Or it might be for want of looking in the right places: Pleistocene-age surfaces are buried now, deeply in some places—eight meters deep, as Reid Ferring discovered at the Aubrey site. Large mammal carcasses shot full of Clovis points lying on those surfaces are often beyond notice or reach, and are usually encountered only by chance. Gary Haynes would add that African elephants have been hunted for centuries, yet their kill sites are rare.

But remember New Zealand's moas? They were slaughtered in far greater numbers than African elephants, and in an even narrower time slice than attributed to Pleistocene overkill. With no hint of hypocrisy, Martin points to the *abundance* of moa kill sites as proof human hunting played a role in the extinction of these birds.

And consider this: bison were hunted on the Great Plains of North America for the last 11,000 years, starting in Clovis times. There is abundant archaeological evidence of planned hunts, bone beds containing hundreds of slaughtered animals, impact-fractured projectile points, and skinning and butchering tools. Bison hunting was often highly wasteful. At Olsen-Chubbuck, 190 animals were stampeded into an arroyo and died, and fully 25% of the animals on the bottom of the carcass pile were only slightly butchered or simply untouched, the meat left to rot in the ground (Chapter 9).[39]

Yet after 11,000 years of this archaeologically well-documented, intense, and often wasteful hunting, culminating in the merciless slaughter of bison by commercial Euroamerican hide hunters in the late nineteenth century armed with deadly, large-bore Sharps rifles ("buffalo guns"), bison are still very much alive today—and even on the bill of fare at many restaurants.

So an animal relentlessly hunted for millennia failed to go extinct, while thirty-five genera of mammals that were never or only rarely hunted did. Perhaps, Martin suggests, bison were not as vulnerable to extinction, because they became "wilier at avoiding hunters."[40] Alas, bison are not that wily: an estimated 10,000–20,000 were trapped and killed over several centuries in Late Prehistoric times by hunters who repeatedly stampeded them (at intervals as close as four years) into the *very same* 30 meter diameter, 15 meter deep sinkhole on the plains of northeastern Wyoming (the Vore site).

The widespread slaughter envisioned by Pleistocene overkill should have left equally unmistakable traces. With 100 million megafaunal victims, kill sites should be more the rule than the exception. And there's no hiding behind poor preservation, even in eastern North America. There and elsewhere thousands of megafauna fossils have been found, yet only two have associated artifacts. Nor, as noted, could Clovis groups possibly be that tidy; after all, they must have had teenagers.[41] Kill sites are genuinely scarce, and not just in North America but in South America as well (where Clovis, strictly speaking, does not occur).

There is no small irony in this. Martin cheerily dismissed pre-Clovis for want of multiple sites; Monte Verde alone wasn't enough. "It is true," he said of pre-Clovis, that "the absence of evidence is not evidence of absence—but there are limits to how long and how strongly one can keep believing when supporting evidence is lacking."[42] Agreed. But doesn't it seem only fair Martin should apply that same standard to finding evidence of Pleistocene overkill?

IS OVERKILL DEAD?

In the early 1980s, at the annual national gathering of the archaeology clan, Jim Mead and I organized a symposium devoted to late Pleistocene environments and extinctions in North America. We invited many of the aficionados of extinction, including, naturally, Paul Martin. It was a well-attended session, and so Martin began his talk by taking a poll of the couple hundred archaeologists in the audience. How many, he asked, thought climate change was to blame for Pleistocene extinctions? About a third raised their hands. He then asked how many believed climate change combined with human hunting was the cause: another two-thirds did. Then he asked how many thought overkill alone was to blame. His own hand shot up. Out in the audience a single hand was partly raised, fluttered briefly, then disappeared—despite Martin's pleading to keep it up. We never figured out whose hand that was.

Science is not a matter of counting votes, of course; it's a matter of evidence. Still, there's nothing wrong with polling a room of individuals at least somewhat familiar with the evidence (though not all in the audience were Paleoindian specialists). Martin was surprised by the results. I was too: overkill almost got one more vote than I expected (I knew Martin would vote for it). I would have guessed more would see climate as the sole cause, and fewer climate plus humans. There's a curious element in this debate: archaeologists and vertebrate paleontologists familiar with the archaeological and fossil record tend to pin extinctions on climate change, while ecologists and zoologists familiar with climate's impact on animals tend to suspect and blame humans. The grass is greener on the other side.

But that's not always the case. Recently, on the heels of our examination of supposed megafaunal kills, Donald Grayson and I published a follow-up paper, "Requiem for North American Overkill," which declared overkill dead. Apparently our report of its death was premature. Stuart Fiedel and Gary Haynes, both archaeologists, shot back with a vigorous defense of overkill.[43] Although they applauded our "scrupulous" vetting of the list of megafaunal kill sites, they snarled we had "grossly misrepresent[ed] the overkill debate," and provided "outmoded data and interpretations and ignore[d] or deliberately omit[ted] the most recent chronological, archaeological and climatic data." In a later paper Haynes went on to denounce our "attack" on overkill as "doubletalk." Lest there be any confusion, here's a dictionary definition of what we're accused of: \'dəb-bəl,-tok\ *n.* deliberately unintelligible gibberish.

Misrepresentation and doubletalk. Harsh words, these. Leaving aside the rhetorical posturing (who isn't guilty of that?), the essential issues at hand are uncomplicated. This dispute is not about speculations about who might have killed what, when, how fast, and how quickly they reloaded, but about kill sites and chronology.

Like Martin, Fiedel and Haynes believe an *absence* of kill sites is to be expected, and consider fourteen kill sites a "phenomenally rich record" of human predation in view of the length of time they think it accumulated—the 300 years Clovis and megafauna ostensibly overlapped on the landscape. In fact, fourteen is indeed a large number, but not for the reasons they suppose: given how many of these sites were found bones first (Chapter 3), fourteen probably inflates their actual importance! That aside, our point was not just that there were only fourteen sites, but that there were only two genera in those fourteen sites, and no horse kills, camel kills, sloth kills, musk ox kills, glyptodont kills, and so on down the list of the other thirty-three extinct genera.

Haynes immodestly proclaimed he could "speak with authority on this matter, having devoted 25 years to neotaphonomic studies[44] of large-mammal skeletons in Africa, Australia, and North America." On his self-appointed authority, he declared that "kill sites of any animal smaller than a mammoth are *never* well preserved." He even tacked a number to his pronouncement: "kill sites of animals that weigh less than 250 kg will not be preserved and fossilized except in special depositional 'traps.'" Ignoring the many kill sites of, say, pronghorn antelope or mountain sheep that are found on the northern plains (animals that tip the scales at 70 kilograms maximum),[45] and not just in the special traps Haynes lays for them, it is still fair to ask again, where are all the horse kills, camel kills, sloth kills, musk ox kills, glyptodont kills, and on down the hefty-animal list? All of them surpass Haynes' minimum body mass requirement.

Fiedel and Haynes shift the ground toward smaller animals. Look at eastern North America, they suggest. There is not a single site with "the remains of any butchered carcasses of elk, deer, bear, or woodland bison of Holocene age." In fact, the remains of deer are extremely abundant in those sites, as has long been recognized. Unless their already-dead carcasses were scraped off the landscape and dragged into archaeological sites (possible, one supposes), they *must* have been killed by hunters. If only the bones of the peccary, tapir, Aztlan rabbit, and other lighter-than-250 kilogram Pleistocene animals were as widely represented in Clovis-age sites, Fiedel and Haynes's point would be well taken.

And consider this: North America once had its own flightless bird—the sea duck, *Chedytes lawi*—which inhabited the islands and mainland of coastal California and Oregon. It was hunted by people, even in Paleoindian times (by the inhabitants of Daisy Cave), and ultimately went extinct. But the sea duck's extinction is a problem for overkill: it took 7,000 years of human predation before this vulnerable prey succumbed—hardly the near-instantaneous demise overkill assumes. As important, human predation of this animal produced an unmistakable archaeological record: its bones are well preserved in sites along the coast.[46] In case there is any question, the sea duck was *much* smaller than a mammoth.

Haynes complains it is unfair for critics of overkill to call for "more and more kill sites." No one is calling for more and more exactly; rather, the request is for more of the genera (of the thirty-five) to have secure evidence of human predation. It's that glaring gap that needs to be closed, and again one cannot entirely blame the fossil record here. Horses, camels, and musk ox (all big beasts) are extremely well represented in the fossil record of North America, far more so than either mammoth or mastodon, and yet there are no secure kill sites of these animals.

As for the chronology, Fiedel and Haynes rightly note there have been great strides in our radiocarbon dating of the extinct animals. Some fifteen years ago, only seven genera were known to have lasted until Clovis time. Nowadays, there's more than twice that number (here's where we stood accused of "deliberately omitting" evidence, ostensibly ignoring radiocarbon ages produced by Russell Graham and colleagues. Since we discussed Graham's chronology in detail and explicitly acknowledged his help, we could only be perplexed by their accusation). More terminal Pleistocene dates may yet come for other genera. Or they may not. The point we were making was straightforward: the chronology of extinctions is yet unsettled. We still don't know whether all animals disappeared simultaneously or not.

Haynes and Fiedel then accuse us of ignoring the Signor-Lipps effect, which is the proposition that rare animals tend to disappear from the fossil record before they go extinct. Thus, the rarer the animal on the late Pleistocene landscape, the harder it will be to find a latest Pleistocene age for it. Haynes draws from Signor-Lipps the message that "the youngest radiocarbon dates on species do not come from the very last members of those species." But of course; we're well aware of that (I use the mirror image of Signor-Lipps in the argument made earlier about why the earliest *dated* site in the Americas is not likely to be the earliest site in the Americas, though it could be, if we get very lucky [Chapter 4]).

We are also aware, though Fiedel and Haynes are not, that if radiocarbon dates are indeed scarce on Pleistocene fauna from the 300 years they shared North America with humans, then that also suggests these animals were already in decline before or by the time Clovis arrived. To test this, we need to more fully understand the extinction process. We need to know when the process began, how long it lasted, and the rate of population decline over that time.

More to think about: if all 100 million animals of those thirty-five extinct genera lasted until the very end of the Pleistocene, only to be suddenly dispatched by human hunters, then why would Signor-Lipps come into play at all? There ought to be *plenty* of terminal Pleistocene ages to go around. There are not.

Haynes portrays critics of overkill as engaging in postmodern meta-narratives, a "constructed reality." In some circles, those are fighting words. Not here. I agree with his point that *all* the arguments in this dispute need to be read carefully and critically, and I would add another: although it's terribly old-fashioned to say, I also think evidence matters.

Paul Martin in his *Twilight of the Mammoths*, a long look back at his and overkill's career, and the vigorous debate over the cause of Pleistocene extinctions he and it engendered, breathed a sigh of relief that at least there is agreement about what was *not* a cause of those extinctions. Perhaps the one thing most specialists can agree upon, he explained, "is that the near-time [Pleistocene] extinctions had nothing to do with a space rock" or asteroid impact, unlike in the case of the dinosaurs' demise 65 million years ago. That was too bad, he mused, since "one would find a highly receptive audience among astronomers and their public had mammoth extinction shared similarities with dinosaur extinction."[47]

As they say, watch what you wish for. (Chapter 2, Sidebar).

TIMING WAS EVERYTHING

Overkill helped Martin solve a vexing problem: what could cause thirty-five very different animal genera living in very different habitats to go extinct simultaneously? Let's briefly turn to the radiocarbon record since the timing of extinctions is often a point of departure in talking about their cause. Without putting too fine a point on it, the reasoning here is that if all the megafauna disappeared at the same time, that suggests a cause that could strike down animals of very different physiology and adaptation across diverse habitats and do so instantaneously. Changes in climate and environment, Martin supposed, varied too much in timing, by area, and in severity to kill off so many animals at once—but he thought voracious, fast-moving hunters could. But if humans were not the cause of extinction, then why did all these animals go extinct at the same time? Or did they?

We often assume as much, but after years of radiocarbon dating fossils of Pleistocene fauna, we cannot say these animals went extinct simultaneously. All were certainly gone by 10,800 BP, but Russell Graham and Thomas Stafford have shown that only sixteen of the thirty-five genera were alive as recently as 12,000 years ago—that is, up to the doorstep of Clovis times. The other nineteen genera lack radiocarbon dates showing they survived until then. In some cases (the Aztlan rabbit, for example), the latest occurrence predates the Last Glacial Maximum. Whether that means these animals were already extinct by then is not clear, since many of those genera are relatively rare in the fossil record, and we know that the number of radiocarbon dates available for a particular animal is strongly determined by how many of its fossils have been found.

Additional radiocarbon dates might bring their terminal dates closer to 12,000 years ago, but then they might not. Until we possess more radiocarbon dates, we cannot assume all genera disappeared simultaneously or gradually, or let assumptions about the timing of extinctions be used to support arguments about its cause.

Perhaps more important, it's not enough to know when the extinctions process ended; we also need to know when it *began*, and what was occurring on the landscape at that time to cause populations to decline.[48] That's no easy task with radiocarbon dating. But here, too, ancient DNA—this time from mammoths, horses, camels, and other fossils—can potentially yield valuable insight: changes in genetic diversity over time can reveal population histories. This work is in its infancy, but it is already yielding results: genetic evidence compiled by Beth Shapiro and colleagues has shown that Alaskan bison populations underwent a significant decline starting about 37,000 BP, neatly coinciding with a warm stretch prior to the LGM, when tree cover peaked, reducing grazing land and serving as a barrier to movement.[49]

If North American extinctions were, in fact, smeared over many millennia, this would be no different than the timing of Pleistocene extinctions elsewhere in the world, and in other New World regions. North of the ice sheets in Alaska, Dale Guthrie has shown that one species of horse disappeared by 31,000 BP, another survived until 12,500 BP, and mammoths lasted until 11,500 BP. For the record: there is no evidence of human hunting of horse or mammoth in Alaska. In fact, and as noted in the previous chapter, everything else seems to have been on the menu at Alaska's Broken Mammoth site *except* mammoth.[50]

CONSIDERING CLIMATE

If different genera disappeared at different times, that opens the possibility extinctions resulted from a more complicated cause: perhaps the complex climate changes taking place at the end of the Pleistocene, which played out across North America over thousands of years in different ways with different consequences for different species in different environments.

Indeed, often unspoken in discussions of Pleistocene overkill is the fact that extinctions were not limited to the very largest, mouth-watering mammals, though those dominated this particular extinction episode, but also included storks, vultures, eagles, jays, and blackbirds, as well as a snake, several genera and species of turtles, and a spruce tree.[51] Likewise ignored is the fact that extinctions were merely one element of much larger, sweeping, end-of-the-Pleistocene changes that, extinctions aside, dramatically altered the ranges of plants and animals.

Against this backdrop of other ecological changes, megafaunal extinction seems less an aberration for which a human cause is necessary, and more a part of larger, natural changes in climate and environment. Paleoecologists are busy exploring how these changes may have impacted habitats, forage, and animal physiology and/or reproduction.

Already, there are several hypotheses, though at the moment (and here Martin is exactly right), these are not well developed. The reason, as Grayson and others have emphasized, is that far more needs to be known about the responses of individual genera and species to the changes that came with the end of the Pleistocene.[52] That's not easy, given the limits of the fossil record and the fact these animals are extinct: our knowledge of what climatic and environmental changes they could (or could not) tolerate is limited. We do not know how changes from equable to seasonal climates, changes in grassland composition, increasing temperature and decreasing rainfall (especially in western regions), or changes in growing-season length (it got much shorter in the far north) would affect each.

We suspect many of the very large mammals had long gestation periods (comparable to the approximately 22 months in modern elephants), as well as low and slow reproductive rates (births came late and infrequently, and rarely produced more than one offspring).[53] If successive offspring died when a long-established birth season suddenly became harsh and inhospitable, the limited reproductive capacity might leave the animals unable to respond (indeed, this is one reason large animals have a high risk of extinction). And it would make them vulnerable to a coup de grâce by hunters who, if they targeted breeding females or young animals (as they did at the Colby mammoth kill in Wyoming), would have sped an already-diminished local herd toward extinction.

Further, biotic communities in place for tens of thousands of years dissolved, and plants dispersed in different directions at different speeds (Chapter 2). How did that upset the delicate balance of animal and plant life, and how did each animal respond to the new forage hand it was dealt? What effect did this have on competition for resources? All of this has to be understood for each animal genus as it changed in numbers, range, and distribution across space and through time.[54]

After the LGM, for example, Guthrie shows that Alaskan horses diminished markedly in body size, shrinking steadily until their extinction around 12,500 BP. Horses are obligate grazers, and as cold, arid, and treeless northern grasslands gave way to a wetter landscape, with more lakes, bogs, and tundra vegetation, the horse suffered from lack of food, and faced increased competition from other large animals such as woolly mammoth. Mammoth was less of an obligate grazer and outlasted the horse by a millennium, but ultimately it, too, succumbed to the disappearance of the northern grassland.

The Alaskan cases are among the few for which we have evidence of animal populations under stress, and good evidence of possible environmental causes for their demise. Even though overkill is wrong, there is still much work to be done to show climate-based explanations are right.

But if Clovis hunters weren't overkilling megafauna, what role did these animals play in their adaptations? And what was the nature of Clovis adaptations? A closer look is in order.

At the time of its initial occupation by people 11,300 years ago, the Clovis site in New Mexico was a spring-fed pond feeding into nearby Blackwater Draw. The springs attracted plants, animals, and people for many thousands of years, even into the twentieth century, when the springs were finally tapped out by irrigation. But just about the time the site's archaeological riches caught the eye of E. B. Howard, a quarry company moved in to mine the site's deeper, commercial-grade gravels. As giant bulldozers dug for the gravel, they ravaged the overlying layers of artifact and bone. In an awkward dance that satisfied neither party, bulldozer operators would temporarily suspend work if they happened to spot bone-rich deposits, and archaeologists would rush in to salvage what they could. Much was saved, but more was lost.[55]

Yet, what was found is remarkable, for the late Pleistocene springs watered frogs, turtles, snakes, mammoth, bison, horse, camel, peccary, deer, and antelope, as well as saber-toothed cat, bear, and dire wolf. Not all of these were food. Ironically, bison, and not mammoth, were one of the most abundant species in the Clovis-age deposits; seven were found, and at least one had been speared, a Clovis point found lodged against its scapula (shoulder blade).

Even so, mammoth remains attracted the greatest attention. Over the years, nearly a dozen were excavated by archaeologists called to the scene; more probably passed unnoticed through the quarry machinery (it was *very* large machinery; one can scarcely imagine how many rabbit or roasted small-animal skeletons were overlooked).

There is no compelling evidence humans killed all or even most of the mammoths found here; we know only that the animals died at this spot. That said, at least three mammoths were found with projectile points and the tools used to butcher them. They died at different times, some on the pond's margins, others in its center. Their skeletons were nearly complete and articulated (the bones lay in proper anatomical position), suggesting the carcasses were carved where they fell, but only partially so. Flesh was stripped from the upper skeleton (forelimbs, shoulders, and ribs) and skulls were smashed to obtain brains, but the remainder of the carcasses were left to rot.

This pattern of partially butchered mammoth bones in marshy, pond, or stream settings recurs in many Clovis sites, from Lange-Ferguson on the northern plains of South Dakota, to Lehner in the San Pedro Valley.[56] The pattern is so persistent it looks purposeful, as though Clovis hunters lurked in such places, looking to drive the mammoths into shallow water, or waiting in ambush for the animal to get stuck in the mud and become an easier target. But appearances may be deceiving. As Frison observed, mired animals that large are difficult to butcher and nearly impossible to extract, as the articulated skeletons at Clovis clearly attest. For that matter, healthy animals (especially full-grown adults) are not easily mired. Skilled hunters, Frison argues, would take mammoth elsewhere and on their terms, perhaps by quietly isolating a juvenile away from the herd's protection, wounding it, then patiently waiting for it to become disabled or die—all the while preventing the animal from staggering to a waterhole to die.

OPPORTUNITIES FOUND

If that's so, why are many Clovis-stabbed mammoths at water holes? In studying African elephants, Gary Haynes saw that a disproportionate number of younger animals frequented water holes when ill (as their body temperature rose) or during drought, often dying there without human help. Such conditions force animals, especially ones with large, water-cooled engines like elephants, to gravitate to water, even if they have to dig to reach it (humans are not the only species that digs water wells; elephants can, too).

Vance Haynes, in fact, sees hints of a brief but severe drought in Clovis times across the American West. His evidence is strongly disputed by others, including Vance Holliday, and at the Clovis site itself: many artifacts, including three Clovis points, were found in spring conduits at Clovis. They were placed there, Haynes believes, as offerings. Were that so, presumably the springs were active at the time or had flowed in recent memory. Regardless, if there had been a drought, it was nowhere as severe as the one that for nearly 2,000 years of the Middle Holocene laid waste to much of the American West. That was a time when humans were forced to dig deep wells for water, and shift their diet to more drought-resistant animals and plants. A pit at the Clovis site resembles such a well, but it was excavated in the 1960s, and observations made at the time (and when it was reopened in 1993) cannot demonstrate its purpose.[57]

Drought or no, sick and enfeebled animals of a species teetering on extinction that wandered into ponds may have become mired, and thereby became risk-free targets for passing (or patiently waiting) hunter-gatherers, who took what meat they could, then moved on. Less danger was involved since the animals could be patiently monitored from a distance. Killing might even be unnecessary if death was inevitable and predictable.

James Judge calls this *opportunistic scavenging,* and thinks it, more so than deliberate big-game hunting, characterizes Clovis.[58] That would explain all those partially butchered animals mired in the mud, and why so many Clovis mammoth "kills" look like animals that died of natural causes, but with a few foreign objects (Clovis points) stuck in their carcasses.

Opportunistic or not, Clovis groups were occasionally quite adept at snaring big game. In the San Pedro Valley, bones of thirteen mammoths, mostly calves or young adults, were excavated on the Lehner Ranch by Emil Haury (Figure 51) and (later) Vance Haynes.[59] Scattered among mammoth ribs, hindquarters, and jaws were thirteen Clovis points made of locally available chert, chalcedony, and quartz crystal, along with stone scrapers, knives, and a chopper. The wide scatter of mammoth bones, their unequal

FIGURE 51.

Excavations at the Lehner mammoth kill site, Arizona, ca. 1959. Lehner is one of a dozen sites for which we have secure evidence of human hunting of mammoth. (Photograph courtesy of Fred Wendorf.)

preservation, their vertical distribution over about 1 meter of deposits, suggest the serial killing of animals isolated from their herds, then dispatched, butchered, and consumed in the centuries around 10,950 BP.

The San Pedro Valley was a ripe field. At Murray Springs, 16 kilometers north of Lehner and at virtually the same moment in time (around 10,900 BP), at least one mammoth and, nearby, eleven bison were killed. The Clovis groups camped close by, hauling in a mammoth drumstick (one of the leg bones) and bison filets, and overhauling broken weaponry. Vance Haynes recovered Clovis points, a variety of tools (including the afore-mentioned mammoth bone shaft wrench), and nearly 15,000 flakes from tool manufacturing and resharpening. Many flakes littered the kills, evidence of the constant resharpening of tools that dull easily when carving animal flesh. As was the case at Clovis, the mammoth carcass was only partially disarticulated (and only the bison bone was burned). The dead mammoth may have had other visitors, too: Haynes found mammoth tracks leading up to the skeleton. Perhaps, as elephants do today, curious mammoths came over to take a look at their dead kin.[60]

The Clovis points used against the bison at Murray Springs were far the worse for wear. Two of them had flakes driven backward off the point tip (impact fractures). One point was found in the kill, and its impact flake in the camp (carried there in the meat, presumably), while the reverse was true of another. Impact fractures occur when stone hits bone at high velocity. Such fractures are rare on Clovis points from mammoth-bearing sites (only one of thirteen points at Lehner was impact fractured, for example), yet common on points in bison kills of all ages. The scarcity of impact damage on Clovis points suggests at the very least that those aimed at mammoths were not penetrating deep enough to hit bone, and those found as isolates may not have been used as points (a possibility for which there is some evidence).

At Clovis, Lehner, and Murray Springs, and localities like Colby and Lange-Ferguson, there were mammoth opportunities found.[61] At Colby, hunters took full advantage: eight mammoths, mostly immature animals (one a fetus), were killed at intervals by hunters some 10,900 years ago, who then carefully stacked their leftovers into two piles, set roughly 33 meters apart. The first pile had remains of three immature mammoths and a few tools. The second yielded the remains of four mammoths, mostly articulated ribs and shoulder bones, along with an innominate (pelvic bone), all topped by a skull. Directly underneath the innominate was a fluted point (an offering, perhaps?). Frison, who excavated Colby, believes these were meat caches, with the one reopened and the meat used, but not the other. Stone caches had been placed on the landscape for use after the hard work; in northern latitudes, meat was cached for the hard times.

THE BIG ONES THAT GOT AWAY

There were mammoth opportunities lost, too. Just down the San Pedro from Lehner and Murray Springs was the Naco mammoth, found eroding out of the sidewalls of Greenbush Creek. Most of the skeleton was there, save the hindquarters, which had eroded and disappeared. Eight Clovis projectile points, variable in size, were found with this one animal (a mammoth pincushion): one lay at the base of the skull, another near the left scapula, two were wedged between ribs, and one against the surface of the atlas vertebra. Yet, no butchering tools were found, nor did any bones show signs of filleting. The Naco mammoth evidently escaped its killers and died quietly of its wounds with hide intact. But escaped from whom? The stone by which the Naco points were fashioned was the same as that used by the hunters who stabbed the Lehner and Murray Springs mammoths.

Nor is Naco alone. Relatively close by (and still in the San Pedro Valley) was the Escapule mammoth, a single animal found with two projectile points. Judging by the stylistic and stone similarities to the points at the other Clovis sites in the valley, it appears to be another escapee of the Lehner and/or Murray Springs carnage.

Out on the Great Plains, there's further evidence of less-than-successful predation. Dent, which sits at the base of a terrace of Colorado's South Platte River, yielded two fluted points, a knife, and a dozen mammoths. Found with them were a great many boulders (some half a meter in circumference), prompting the engaging scenario that a group of hunters came upon the herd and, for want of adequate weaponry, grabbed the closest weapons at hand—cobbles from the terrace—and stoned the animals to death. An engaging scenario, but probably wrong. More recent work at the site revealed that bones, boulders, and fluted points all came washing 10–15 meters downslope from the terrace top about 200 years ago, though the bones themselves date to 10,990 BP.[62] The boulders were probably not late Pleistocene missiles, but cobbles that fortuitously came to rest with the mammoth remains. If there had been a hunt, it was at best unsuccessful, the two points perhaps coming from the carcass of an animal that had carried them around for some time, a thorn in its elephant-thick hide.

So, too, the case of Domebo (pronounced like the name of Disney's big-eared elephant), a mammoth found in south-central Oklahoma. With the bones (dated to 10,960 BP) were two fluted points, and three small flakes. The points, though found near the skeleton, did not cause its death, and there were no signs of butchering. The coroner ruled the cause of death unknown.[63]

And then there's the Miami site in the northern panhandle of Texas, discovered in the 1930s when Dust Bowl drought forced farmers to plow deeper than usual, thereby exposing an ancient pond bed containing remains of five mammoths, three of which were mature or nearly mature individuals. Oddly, there were no bones of other animals in the pond, but there were several projectile points and a scraper. Yet, there's no evidence of butchering, so it may be the Miami mammoths were enfeebled by natural causes or were already dead when a Clovis group came upon the carcasses and began salvaging what they could, losing a few artifacts in the process. Or perhaps it was an unsuccessful kill. Two of the points were found with one of the mature mammoths: could she have staggered to the basin to die and been followed by her young, who lingered there after her death until their own?[64]

EAST MEETS WEST

Mammoth altogether eluded the fluted spear points of eastern Paleoindian groups, but then these open-ground, gregarious herd animals were relatively rare in heavily wooded Pleistocene eastern North America. Their distantly related, forest-dwelling kin, the mastodon, was far more abundant in this region, and yet it was even more successful at avoiding being on the Clovis menu. Except at Kimmswick.

Since 1839, many mastodon skeletons have been mined from this spot, now the Mastodon State Historic Site along the Mississippi River, south of St. Louis, Missouri. There was talk early on that ancient artifacts were there, too. William Henry Holmes himself excavated here in 1901–2, to lay that dangerous talk to rest. It didn't work. In the 1980s, a team directed by Russell Graham reopened Kimmswick, and in ancient pond settings reminiscent of Clovis sites further west, found the remains of two mastodon, a cow and calf, along with a few fluted points, butchering tools, and several thousand tiny flakes. The site is undated, but is assuredly within the Paleoindian age range.

There are hints of mastodon being taken at other eastern Paleoindian sites,[65] but those are strong only at Michigan's Pleasant Lake, where mastodon bones show cut marks, striations, and polishing suggestive of human knife work. Curiously, not a single stone artifact has come from the site (there's that Clovis tidiness again!).

It has not escaped notice that most megafauna kills are found on the Great Plains and in the Southwest. That most of them are of mammoth partly explains the pattern: that's where this animal was especially abundant. Yet, it fails to explain why their forest relatives (the mastodon) were not taken in comparable numbers. It's not because we haven't look for mastodon kills in the east. The search has been ongoing since the 1930s. Nor is it for want of mastodon remains of the proper age. George Quimby and Ronald Mason long ago pointed to the overlap in the distribution of fluted points and mastodons in the Upper Midwest (the Mason-Quimby line, Paul Martin called it).

Rather, the answer lies in the ecological stage on which these adaptations played out. Late Pleistocene environments in the East were more complex and changing more rapidly than those in the West (Chapter 2). In turn, plant and animal species and communities were correspondingly richer in the East, and that made a difference in Clovis-age foraging. As Michael Cannon and I found, the diets of early Americans in eastern North America were broader and less dominated by large mammals than were those of their western contemporaries.[66]

JUST SHOWING OFF

Armed with stone-tipped spears, Clovis hunters could take big game. Occasionally they were successful, other times not. And sometimes they got lucky and took advantage of a bad situation—bad for the prey, at least. That Clovis hunters did not pursue mammoth or mastodon more regularly or ruthlessly comes as no particular surprise.

Hunting big game has more than its share of risks. Like coming home empty handed. Or not coming home at all. And if elephants, their modern, highly intelligent, unpredictable, and vengeful relatives, are any indication, hunting mammoth or mastodon was especially risky. Elephants, as Hadza hunters told James O'Connell, do not behave like animals; they behave like enemies. Yet the Hadza, O'Connell adds, hunt rhinoceros and sometimes even lions. As George Silberbauer observed, it is "admirable common sense not to shoot a flimsy arrow into dangerous prey like elephant and [water] buffalo unless you know exactly what the prey will do next; both species take their annual toll of hunters [even when] armed with high-powered rifles and give rise to innumerable tales of narrow escapes."[67]

Indeed, when elephants are hunted nowadays, it is at a distance with high-powered rifles or up close with metal broad swords, both far superior weapons to Clovis-pointed spears. Spear-toting archaeologists experimenting on recently deceased zoo and circus elephants have shown that bone and stone points can penetrate elephant hide (Frison reports that Siberian freeze-dried woolly mammoth hide was equally thick, if not thicker, and of course there was hair to contend with, too). A spear thrust would be most lethal in those spots where the hide thins or, say, if plunged deep into a lung. Those are easy to target when the animal lies dead before you, but not so easy after you miss your first shot at a live and suddenly very angry elephant. As Frison learned during a cull in Africa, no Clovis point could drop a charging elephant in its tracks. Teddy Roosevelt, after his own frighteningly close encounter with a raging bull elephant while on a post-presidential romp through Africa bagging animals large and small, well understood why smaller prey made up the majority of the diet for African hunter-gatherers.[68]

In fact, big-game hunting can be a very inefficient means of feeding a family. Bird and O'Connell record that despite substantial effort, Hadza men "manage to acquire large carcasses on average only about once every 30 hunter-days,"[69] a daily failure rate of 97%. When big game is bagged, most of the meat is shared, producing more for the group than for the hunter's family. So what's in it for the hunter and his kin? Why would one specialize in the risky and inefficient pursuit of large animal prey?

Kristen Hawkes and Rebecca Bliege Bird suspect big-game hunting and the finely fashioned weaponry that goes with it may be a form of *costly signaling*, or showing off, by which hunters gain prestige among peers and competitors. Hunting success confers status, and marks a hunter as a powerful ally, a dangerous adversary, or an attractive mate. One key piece of evidence in favor of their hypothesis: the children of skilled Hadza hunters do not have better survival odds than other children (indicating they were not eating any better, despite their father's success), but they had more siblings, revealing that skilled hunters had much higher fertility.[70] (It's hard to ignore the coincidence that excavating a big-game kill site confers considerable *archaeological* status; I'll leave to others to judge its rewards.)

It was long ago suggested, tongue-in-cheek, that each Clovis hunter probably killed one mammoth, then spent the rest of his life talking about it.[71] An exaggeration, perhaps. Still, there's a nugget of truth to it. Humans talk. And as Hawkes and Bliege Bird observe, reputations are crafted through storytelling. Roosevelt saw the kernel of one such story emerging just hours after the bull that sideswiped him was killed. "The gunbearers, as they walked ahead of us camp ward . . . began to improvise a song, reciting the success of the hunt, the death of the elephant, and the power of the rifles," one that was soon added to the stories already told. The elephant, like "no other animal, not the lion himself, is so constant a theme of talk, and a subject of such unflagging interest round the campfire . . . " Yet, at the core of each story there had to be a kill, for "as in other domains of male contest," Hawkes and Bliege Bird add, "'trash talk' may have its uses, but reputations for delivering the goods cannot be built upon it."[72]

ROUNDING OUT THE CLOVIS DIET

Smaller game and plants likely do nothing for bragging rights, but they certainly help fill in the bulk of the diet. That is true among modern hunter-gatherers, and was likely true in Clovis times as well. But what else may have been part of the Paleoindian diet has only recently come into view, albeit slowly, as one might expect given how few sites there are, the traditional attention given to large animal kills, and the problems of preservation of smaller animal and plant remains—and perhaps even of the artifacts used to harvest those. A study by Christopher Ellis revealed that when pursuing small prey, hunters the world over commonly used throwing sticks; snares; or bone, wood, or antler points, since those readily stun the animal, do no damage to its skin, nor cause it to sink (if hunting waterfowl or fish).[73] These would leave little archaeological trace (save under circumstances like those at Monte Verde, where wooden lances were preserved).

In the far West for all appearances, Clovis-age groups ate nothing, or at least nothing that's been left behind. There are hints (a few bones here and there) that mountain sheep were taken at higher elevations in the Great Basin. These are animals that can be captured and subdued readily with drop nets, which were available by this time in prehistory. Another hint of their adaptations comes from mapping the distribution of early fluted and stemmed points in the Great Basin: they cluster along the edges of now-dry Pleistocene pluvial lakes, marshes, and springs. Whatever these Clovis and other Paleoindians were doing, Grayson points out, they were doing near shallow water. If later sites in the same area bear witness, what they were doing was exploiting small mammals, birds, fish, and mollusks.

Confirmation that fish were part of a Clovis-age diet in the far West comes from a study of the bone isotopes from the Buhl skeleton, found in Idaho and dated to 10,700 BP. The isotope chemistry of bone can reveal the type and relative proportion of plants and animals in the diet. As the old saying goes, you are what you eat (to which those who study prehistoric coprolites are wont to add, "except what you excrete"). In the Buhl case, her diet consisted of meat and fish—probably salmon, the isotope signature being that of marine rather than freshwater fish. There must have been long, lean periods in her life, too, judging by episodes of arrested bone growth evident in her x-rays.[74]

Moving onto the Great Plains, George Agogino reported finding a turtle roasting pit at the Clovis site: six or seven Pleistocene terrapins stacked one atop another and apparently cooked in their shells. Turtles appear frequently in other Great Plains Clovis-age sites, including Lubbock Lake in far west Texas, and at Aubrey and Lewisville in north-central Texas, where turtles were among the most common vertebrate fossils recovered (at Lewisville they comprised about 90% of the animal remains). So frequent were turtle bones at Lewisville that many "were donated to visitors and to Boy Scout groups as mementos"[75] ("guest goodie bags" are hardly the norm at archaeological sites, but then Lewisville was excavated in the late 1950s by amateurs, innocent of archaeological protocol).

Often larger than their modern relatives and always slow moving, Pleistocene turtles, many now extinct, were the best kind of game. One genus, *Geochelone* (Figure 52), was nearly a meter long, 75 centimeters wide, and 60 centimeters tall. That's a lot of meat under the hood of a prey species that could be pursued (leisurely, of course) without threat to human life or limb. Ultimately, we may discover turtles were a Paleoindian staple, a menu item rivaling faster and bigger game, which may explain their occurrence even at big-game kill sites like Clovis and Kimmswick.

A staple, perhaps, but these were not turtle-hunting specialists. At Aubrey and Lewisville, hunter-gatherers foraged up and down the food chain: bison and deer are present, as well as a variety of small animals, including snakes and lizards, frogs and birds. Some of these

FIGURE 52.

An intact shell of the extinct North American tortoise, *Geochelone*, excavated from the sediments of a Pleistocene pluvial lake bed in west Texas. The putty knife is 18 cm long. (Photograph courtesy of Richard Rose.)

smaller remains could be mere background noise, animals that wandered into the site and died of their own accord. It happens. Yet, many of their bones were also burned, a sign they were consumed by humans. But why would hunters on such a rich landscape stoop so low for forage? Were they that "unfocused," as Gary Haynes put it?[76] To be sure, resource ranking matters to hunter-gatherers, but food is food, and the lower-ranked items could be collected readily while out on the hunt, or gathered by other members of the group.

Archaeologists have long supposed deer or elk were targeted by groups in the complex forests of southeastern North America, as they were by virtually all subsequent hunter-gatherers of that region. The evidence is meager, though deer bones are abundant just across the academic border separating the Paleoindian and subsequent Archaic period (Chapter 9).

Caribou (a member of the deer family) were apparently prey at several northeastern sites, although their bones have only been found at Bull Brook (Massachusetts), Udora (Ontario), and Whipple (New Hampshire); also apparently consumed at Udora were hare and arctic fox. These confirmed and suspected caribou kills date to the centuries around 10,600 BP, and are located in what would have been a swath of more open terrain stretching from Nova Scotia to the Great Lakes. Why caribou in this setting?

These animals historically moved in great herds, wintering in the forest and summering on the tundra. In immediate post-glacial times, tundra was not so extensive (Chapter 2), and caribou ranges and numbers were probably smaller. Even so, however, their migratory habits and herd instincts made them a prime resource in northern forests and parklands.

At Maine's Vail site on the Magalloway River, Michael Gramly found a killing ground, a sandy patch littered with twelve fluted points, some complete and others with shattered tips, but no other tools.[77] Some 250 meters away and across the river (Figure 53) was a camp marked by more projectile points (bases this time), scrapers, knives, gravers, drills, and wedges, the latter for splitting bone and retrieving marrow. Kill and camp literally can be joined: seven point tips from the former re-fit onto bases at the latter, reuniting what had come apart in a split second of impact 10,500 years earlier (see Plate 12).

No bone is preserved in the acidic soils of this site, but Gramly observes the Vail killing ground is at a topographic pinch point in the valley, an ideal spot for waiting hunters to intercept caribou on their seasonal migrations between tundra and forest. He thinks hunters positioned themselves here for several years running, since the camp contained multiple artifact concentrations, possibly from different stints on-site. And they may have cached their leftovers, both in shallow pits on the site and at the Adkins site, just a kilometer away, where Gramly spotted half a dozen boulders, each weighing 100 kilograms (220 pounds) or more. These were arranged around a pit, creating a storage chamber about twice the size of an average household refrigerator (and in Maine at the end of the Pleistocene, easily as cold). Having a food cache is handy when waiting on the unpredictable arrival of a migratory herd, especially given how far the hunters had come.

Their points were made of stone acquired 300 kilometers away in the Hudson River valley. Not surprisingly, the points were intensively used, resharpened, and recycled. The hunters must not have found, or had the time to visit, the outcrop of high-quality Ledge

FIGURE 53.
Aerial view of the Vail camp and kill sites, Magalloway Valley, Maine, ca. 1980. The clusters of artifacts marking the camp were found on the sandy beach of what is now (artificial) Lake Aziscohos, just along the central portion of the tree line; the killing ground was on the prong of land in the approximate center of the image 90 m away from the camp. (Photograph courtesy of R. Michael Gramly.)

Ridge chert just 30 kilometers distant, where they might have refurbished their weaponry. But then, hunter-gatherer mobility is not just a matter of the distance from outcrop to site; it's also about time: the time elapsed since the last visit to the quarry, say, or perhaps the time it might take one away from an ambush spot and chance missing the herd as it passed through. Ultimately, however, they did miss it when, Gramly believes, the caribou chose another valley for their seasonal trek. Caribou can be fickle. With that, Vail was abandoned (though not the valley, as Gramly and his team's subsequent discoveries have shown).

WHEN HUNTERS GATHER

It has been proclaimed plant foods were "neither a provable nor logical part" of Paleoindian diets.[78] That's a bit excessive. It is fair to say the *degree* of plant use among Paleoindians remains unproven. It is both provable and, as Lewis Binford and others have shown, eminently logical. Finding that proof requires good recovery techniques since seeds, berries, and other traces of plants (or even small animals) are not easily detected, even if they are preserved. A little luck helps, too.

Luck and technique came together at Shawnee-Minisink, Pennsylvania. During the excavations (Figure 54), 10% of all excavated sediment was poured into large, water-filled

FIGURE 54.
Excavations in the main block of the Shawnee-Minisink site, Pennsylvania. The three individuals in the deepest part of the excavation are just above the Paleoindian-age stratum. (Photograph courtesy of Richard J. Dent and the Department of Anthropology, American University.)

washtubs; minute organic remains (bits of bone, wood, charcoal, seeds) floated to the surface and were collected with a tea-strainer, while the remainder of the sediment was passed through fine mesh screens. Flotation captured acalypha, hackberry, blackberry, chenopod, hawthorn plum, and grape seeds, as well as tiny fish bones, from meals 10,900 years ago. Shawnee-Minisink, however, is an exception. Most other Paleoindian sites were excavated before flotation became customary, and even today, flotation or some form of water screening is not always incorporated in fieldwork.

Indirect evidence of plant consumption comes from Paleoindian teeth (what few there are), some of which, Gentry Steele reports, are heavily worn—a sign of the grit that gets chewed along with the greens. Wandering into more circumstantial territory, there are possible seed-grinding stones at Clovis, and cutting tools with polished edges at several other sites (including Gault). The polish appears to be "sickle sheen," which develops when tools are repeatedly used to harvest grasses, which are suffused with tiny silica bodies (opal phytoliths) that abrade and burnish the stone.

Although this is a meager record of recognizable plant-processing tools, a bit of context is helpful. First, many plant-processing implements recorded historically (for example, among the Iroquois) are biodegradable. Second, the scarcity of tools might reflect the intensity or, better, the lack of intensity of plant use. Plant foods are routinely lower-ranked resources, since harvesting and processing them is time consuming and difficult (Chapter 7). Consequently, they are added to the diet mostly in times of stress (such as during drought), and become a staple only later in prehistory, when human populations increase and opportunities to move to new territories decrease. At those times, plant use becomes more intensive, and with it comes the heavy-duty tools for large scale plant processing: earth and rock ovens, bedrock mortars, and grinding stones (manos and metates). The absence of these tools from Clovis sites may only indicate plant use was not intensive, not that plants were unused. Collecting nuts, berries, fruits, and green vegetables requires little more than a basket or skin pouch. The chances of building up a visible archaeological record from exploiting such resources is vanishingly small.[79]

Finally, it may be the evidence is right in front of us. Gary Haynes believes Clovis sites show "no attraction to nut-tree-wooded areas or seed rich grasslands."[80] Of course, we cannot say exactly where nut-rich woods or seed-rich grasslands were over 10,000 years after the fact, but we can say that a vast number of Clovis isolates do occur in what were the complex forests of Late Glacial eastern North America, which included nut-bearing trees such as walnut and hickory. Like prehistoric Swiss Army knives, Clovis points were useable for various tasks, whether skinning rabbits, prying open turtle shells, or even digging out stubborn, edible roots. Many isolates examined in eastern North America show wear patterns indicative of use as multipurpose, hafted knives, while very few display impact fractures from hunting damage.

Haynes, in fact, inadvertently (and probably unintentionally) underscores the importance of plant collecting, suggesting it was the trigger to Clovis settlement mobility. He reasons that since the men were off hunting while the women stayed close to camp to gather, the women must have had a strong voice in making the "executive decisions" about when camp should be moved. This was so, he surmises, "because they were the most sensitive to the exhaustion of resources that were within reasonable walking distance of camp."[81] As plant gatherers, they would be.

And move they did. Within a few centuries, Clovis groups were scattered throughout North America. So why did Clovis groups radiate so far and so fast? It's surely not because they were chasing mammoths (or being chased by mammoths!), or because the extinction of those animals left Clovis people with so little to eat they had to constantly press on. More likely, it was because they found themselves on a vast and empty landscape that was unknown and unpredictable, yet rich and untapped, in which there was a great adaptive incentive and benefit to ranging widely.

As they dispersed cross-country, and their knowledge of the landscape and its resources became ever-more detailed, their foraging systems became more stable, their populations increased, and their odds of vanishing diminished. And with that, broad changes were in the offing.

SETTLING IN

*Late Paleoindians and
the Waning Ice Age*

In the 1930s, the United States was staggering from a wicked one-two punch. The Great Depression, fast on the heels of the 1929 stock market collapse, halved the country's industrial output, flung nearly a quarter of its workers from their jobs, and led to the failure of thousands of banks, and with them the disappearance of millions of savings accounts. Homes and businesses were foreclosed, displaced people hit the roads and rails, and even organized crime foundered when Prohibition's repeal dropped the bottom out of the bootleg liquor market. Evidently the distressed nation needed a drink more than it needed temperance.

Then conditions got worse. In early 1932, rain stopped falling across much of the American West, and the consequences of decades of short-sighted agricultural practices hit hard. Having stripped away tens of million of acres of native grassland, and unable because of drought to plant crops to protect the naked land from erosion, a horrified nation watched as massive dust storms carried tons of Great Plains topsoil east across the continent, shading the midday sun in Chicago, and dusting the decks of ships hundreds of miles off the Atlantic coast. Plains residents fled in the opposite direction, their journey to California's farm fields immortalized in John Steinbeck novels, Woody Guthrie ballads, and Dorothea Lange and Arthur Rothstein photographs.

Yet, grim as the combination of Depression and Dust Bowl drought was for the country, it was a boon for archaeology. Large-scale archaeological surveys and extensive excavations were a vital part of President Franklin Roosevelt's New Deal recovery programs. (Why? They cost little in tools and supplies, employed thousands, and produced scientific results.) Moreover, the scouring of land across the West exposed deeply buried late Pleistocene deposits—and Paleoindian sites long hidden from view. Nearly two dozen were found in the span of the Dust Bowl decade.[1]

But the more sites that appeared, the more complicated matters became. Some had projectile points reminiscent of Folsom, others like those at Clovis. Occasionally the new-found points were fluted, but many were not. Clovis and Clovis-like points proved to be widespread, implying the "existence over a wide terrain of a common culture horizon,"[2] but most appeared more geographically restricted. All were obviously Paleoindian in age—they'd been found alongside remains of extinct bison or mammoth (the lesson from Folsom about what and how to look for Paleoindian sites had been well learned). But how old were they, exactly? There was clearly a great deal more to the Paleoindian occupation than anyone had imagined, but how did these forms relate to one another in time and across the continent? Did they represent different adaptations to different environments? Different peoples? Different styles of fashioning stone weaponry? It was hard to say.

Certainly, by the 1930s, American archaeologists appreciated that cultures in the past must have varied, as did those of the present. In 1939 the eminent American anthro-pologist Alfred Kroeber showed them just how complex that variation could be. In his *Cultural and Natural Areas of Native North America,* he brought to fruition an ambi-tious effort to map all the cultures of North America and show their "environmental relations." It was an effort borne of Kroeber's frustration at his colleagues who'd been giving steadily less attention to the role of the environment. As a consequence, much of anthropology had become "virtually a *sociology* of American Indian culture." Kroeber did not intend that as a compliment. Anthropologists could speak at length about native ceremony, social organization, art, and myth, but hardly at all about how groups made a living—a glaring omission, he complained, since "no culture is wholly intelligible without reference to the . . . environmental factors . . . which condition it."[3]

And so Kroeber defined a half-dozen major and over fifty minor culture areas, correlated those with their environment, and attempted to understand the challenges and opportunities in those settings, which might in turn help explain the historical trajectories of different groups and, more broadly, the grand cycles of human history. *Cultural and Natural Areas of Native North America* became an instant classic.

Kroeber rightly appreciated that mapping culture areas provided at best a brief snap-shot of a changing human phenomenon (in his case, cultures as they were constituted at first European contact). Yet, though he was aware of the dynamic nature of culture, he took the stability of the environment for granted, and perhaps reasonably so, since he was only looking back several centuries.

In fact, even though it was generally appreciated in 1939 that the Ice Ages had hap-pened, their effects on ecological communities and how these might have changed over time was poorly known. Knowledge would improve, however, as studies of fossil animals and plants, and glacial geology, painted an ever-more-detailed canvas of past climates and environments. By 1954 archaeologist George Quimby voiced what many were thinking: Kroeber's landmark study might be invaluable for understanding the ecological stage of recent cultures, but "it has become increasingly evident that the cultural and natural areas of the neo-Indian were not necessarily those of the paleo-Indian."[4]

So Quimby tried his hand at describing what might have been the environments of late Pleistocene peoples. "Cultural and Natural Areas before Kroeber" he called his paper, and asked Kroeber for permission to use his name in the title. The great man seemed mildly amused.

With a series of maps, Quimby illustrated the changing North American landscape over the final millennia of the Pleistocene, and suggested where and how Paleoindians may have adapted to that emerging New World. It was, he admitted, a preliminary effort. Even so, fellow archaeologist Homer Aschmann insisted Quimby's hypotheses demanded "immediate examination before they become established dogma." Quimby assured him he needn't worry: "The mortality rates on my previous dogma are too high." Besides, Quimby added, it would be some time before there was a "generally accepted reconstruction of cultural and natural areas in late glacial times."[5]

Quimby was right. We're still working on it.

FILLING IN THE LANDSCAPE

The broad outlines by now are evident, however. As Clovis populations increased in number and expanded into new habitats, cultural drift (Chapter 7) became more pronounced. It's a process that, once begun, continued over the next 10,000 years as environments changed and populations expanded, dispersed, adapted, and evolved. As Clovis descendants settled into different regions of North America, the links between them—so vital when there were few people on the landscape, and they needed to keep in contact with one another—began to break down. It is perhaps no coincidence that a few centuries after the initial Clovis dispersal, there was a shift from the early, broad, and relatively homogeneous projectile-point style to multiple Late Paleoindian point styles in different regions (Figure 55). Although the precise timing varied by area, sometime after 10,900 years ago in the central and western portions of the continent, and after 10,600 years ago in eastern North America, Clovis and its related forms disappeared. By 10,500 years ago, the Clovis fluted form had been supplanted by a variety of regionally distinctive projectile points.

Archaeologists know not to place undue weight on stylistic change in projectile points, for like selectively neutral traits (such as mtDNA mutations), their style can be arbitrary and change randomly, so long as style does not interfere with the workings of the point (which is why invoking a point style change as evidence of the Younger Dryas extraterrestrial event is meaningless [Chapter 2]). That said, point styles are often emblematic of a group, just as boy scouts and street gangs have their emblems. Assuming Late Paleoindian point types mark cultural groups (however defined), and different forms mark different or evolving groups in time and space, we can use them to refer to time periods—the Folsom period, for example—much as historians refer to, say, the Middle Ages.

FIGURE 55.
A sample of Late Paleoindian projectile point types in North America, ca. 10,500–10,000 BP. From left to right, the types (by region) are Lake Mojave (Great Basin), Agate Basin (primarily northern plains), Plainview (primarily southern plains), Barnes (northern Midwest), Dalton (central Mississippi Valley and east), and Simpson (southeast). (Arranged by Katherine Monigal.)

Let's keep with that analogy for a moment. It's easy to put one's finger on the *middle* of the Middle Ages, less so on its beginning or end. But at least with historic periods, one can identify specific tipping points: the Middle Ages ended with the beginning of the Renaissance in fourteenth-century Italy. An arbitrary demarcation to be sure, and hard to pin down tighter than a century, but it makes for a relatively clean break (in Europe, anyway). Not so with archaeological periods based on projectile point styles. They do not have discrete boundaries in time and space, and can overlap on their edges. Moreover, though we often assume a point type is unique to a group—the Folsom folk, or the Agate Basin people, we say—that may not be the case either.[6] A point *style* is just what the term implies: it's the current fashion. And fashions change, even within a population. Perhaps in those Paleoindian sites where more than one point type occurs (and there are many), we are glimpsing the transition from one style to another. The bottom line: styles probably do mark broad periods in time and space, just don't look too closely for well-defined and discrete boundaries.

On the Great Plains and in the Rocky Mountains, Folsom points follow Clovis, a fact made certain when Folsom points were found in deposits stratigraphically above Clovis points at the Clovis site (thank you, Nicholaus Steno), and later bolstered by radiocarbon ages that showed Folsom appeared just after 10,900 BP and lasted until about 10,200 BP. Makers of Folsom points ranged over the North American plains from Canada to Texas, but also ventured west into the Rockies and as far east as Wisconsin. Those who followed Folsom had smaller ranges: some occurred primarily on the southern plains (e.g., Midland, Plainview, Milnesand), others on the northern plains (e.g., Goshen, Agate Basin, Hell Gap), and some in the central plains (Firstview).

Most of these point forms are rare, if not absent, in far western North America where, curiously, Clovis is not followed by multiple styles over time and space, but instead by the Great Basin Stemmed Series, points like Haskett, Parman, Lake Mohave, and Cougar Mountain, which are mostly geographic variations on the same lanceolate theme (with stems for hafting, not flutes). And a long-lived theme it was, persisting over nearly 4,000 years, appearing as early as 11,200 (though most are at least a millennium younger) and lasting as late as 7,000 years ago.[7]

Those relatively few varieties hardly match the explosion of point styles in eastern North America in the millennium after Clovis, though for want of radiocarbon dates their precise ages have in most cases proven difficult to pin down. It appears the early Clovis-like Gainey style is followed in the Upper Midwest and Great Lakes by Barnes points (lanceolates with long, Folsom-like flutes), and in New England by Michaud forms. In turn, these are evidently succeeded (west to east) by Holcombe, Crowfield, and Nicholas points.

South of that region, across northern Alabama, Tennessee, Kentucky, the dominant post-Clovis form appears to be the Cumberland point, which also has long flutes but is much thicker and heavier than Folsom or Barnes. Prevailing in the Carolinas are triangular and elongated Redstone points; Quad points occur in the piney woodlands of Mississippi, Alabama, and the Florida panhandle; and Simpson and Suwannee forms fanned out across much of the rest of Florida and into Georgia. This stylistic proliferation slowed in the centuries after 10,500 BP with the appearance of Dalton points across a broad swath of the Southeast, though there are plenty of variations on the Dalton theme (such as Hardaway points in the Carolina Piedmont, Greenbrier in Alabama-Tennessee, and San Patrice in Louisiana and east Texas).

Lots of types, but no need to fret the details. Instead, let me paint some broad brush strokes. First and most notably, the post-Clovis projectile point sequences west and east do not correspond to one another. After Clovis, Paleoindians go their separate ways stylistically, if not literally. Second, with the exception of the far West and Great Basin, these style zones become progressively more numerous as well as smaller over time. Third, that process seems fastest in eastern North America, where the stylistic landscape is more variegated, and by the end of the Paleoindian period has more style zones in smaller areas than any other region of North America (David Anderson calls the East a "center of technological and social innovation"[8]). Finally, we're now squarely in the midst of the Younger Dryas (11,000–10,000 BP) with its radiocarbon-confounding plateaus (Chapter 2), which means that 900 *radiocarbon* years of Late Paleoindian time from 10,900 to 10,000 BP translates (with current calibrations) to 1,360 *calendar* years. That apparent explosion of styles and the process of settling-in wasn't quite so fast as it appears.

As in Clovis times, later Paleoindians used high-quality stone, and out on the open plains and in the Great Basin went great distances to acquire it, at least initially. The closest of the outcrops used by the occupants of the Folsom site, for example, was 190 kilometers to the southeast, and the most distant was 450 kilometers due north. The oblique triangle formed by just these three stops on their rounds (they made more) has an area of around 90,000 km². That's a substantial territory, but hardly excessive by Paleoindian standards. Comparable ranges are known from the far West and parts of eastern North America (particularly the Upper Midwest). These ranges shrink over Late Paleoindian time, though more slowly on the plains.[9]

Moreover, and also unlike in Clovis times, Late Paleoindians had "home" ranges. As Daniel Amick has shown, Folsom groups on the plains primarily used stone found on the plains and to the east, while those in the Rio Grande Valley obtained their stone mostly from the valley or the mountains to the southwest. Groups in the Great Basin, George Jones and colleagues show, likewise collected obsidian from within clear-cut geographic areas that were roughly oblong in shape and extended some 450 kilometers north to south, and about 150 kilometers west to east. Little, if any, stone moved between those tracts.[10]

Such patterns suggest that territories—if we can use the term without conveying the sense these were jealously guarded—were recognized by this time in prehistory. And, as was apparent with the style zones, the stone "conveyance zones" vary in size across the continent. They are larger on the open terrain of the far West and plains, smaller in eastern North America. Such differences are surely tied to the way in which Late Paleoindian hunter-gatherers used their habitats, and indeed, in some areas they appear to correspond to region-specific or even prey-specific foraging strategies. These were occasionally accompanied by new technologies, presumably in response to the different adaptive opportunities and challenges in these diverse settings. Broadly speaking, Folsom and later Paleoindians of the plains and Rocky Mountains of western North America had a subsistence strategy tied to exploitation of *Bison antiquus*. Dalton groups were thorough-going denizens of the resource-rich forests of eastern North America. Many Great Basin peoples had lifeways focused on animals and plants available in or near marshes and wet meadows. And, in the Pacific Northwest and along the California coast, groups were dining on shellfish.

Taking all this at face value, the increasing stylistic diversification and the ever-shrinking style and stone-conveyance zones, can be read as a relaxation in the pressure to maintain contact with distant kin, a reduction in the spatial scale and openness of the social systems, and a steady settling-in and filling of the landscape. Later Paleoindians no longer spanned the continent as their ancestors had, and their universe had become much smaller.

All this played out in varied ways across the continent. I'll begin (and, admittedly, spend much of my time) on the Great Plains, where Clovis is earliest replaced, and where the subsistence strategy came to center on the largest mammal left standing in the wake of Pleistocene extinctions.[11]

BISON TO THE FORE

Pleistocene extinctions had a profound impact on plains Paleoindians, though it had little to do with depriving them of their favorite foods, but much to do with providing an alternative food source. In the warmer and drier millennia of the late Pleistocene, the plains grassland became less diverse and increasingly dominated by warm-season grasses, which (recalling Chapter 2) grow during the summer and are available well into the fall and winter (in contrast to cool-season grasses, which grow in spring and require cooler and wetter habitats, and were more prevalent in the Pleistocene).

The shift from cool-season- to warm-season-dominated grassland at the end of the Pleistocene would have spelled trouble for most herbivores, since warm-season grasses taste lousy (they produce anti-herbivory enzymes) and for animals that are not ruminants, these grasses have limited nutritional benefit. If mammoth, horse, camel, and other grazers depended on cool-season forms, there would have been less to eat, more competition, and overgrazing of dwindling patches, all of which could spell trouble for individual animals, local populations, and perhaps even species. But what of bison? The answer is *buffalo grass* and *blue-grama grass*. These are both warm-season grasses, were among the main constituents of the newly forming late Pleistocene grassland and, most important of all, are a favored food of bison, which as ruminants have the digestive tools (including a four-chambered stomach) to maximize the nutritional value of these grasses.

Although intriguing, there is no direct evidence that the expansion of the warm-season grassland, triggered by changing climates and benefiting from bison grazing (which helped the grasses spread), led to the extinction of the Pleistocene megafauna. Regardless, with the disappearance of the last of the Pleistocene animals, bison had the plains virtually to themselves. Lacking competitors or significant predators (the megacarnivores were gone, too, replaced by the gray wolf), bison populations on the plains skyrocketed.

It is surely no coincidence that bison hunting began in a significant way in Late Paleoindian times, for here was a high-ranked resource that could be dispatched at *relatively* low risk, and one that was becoming increasingly numerous, predictable, and reliable. Paleoindian bison kills were modest, at least at first. We know of only a dozen or so bison killed by Clovis hunters (and virtually all of those met their death at Murray Springs). Folsom-age bison kills typically netted an average of fifteen animals. But just a few centuries later, bison kills averaged nearly sixty animals, a four-fold increase. Had hunters become more adept, or were bison more abundant and easier pickings? Maybe both. Regardless, bison became a primary resource for Paleoindians, and would continue in that role for plains peoples over much of the next 10,000 years, ending only with the animal's near destruction in the nineteenth century.

We have no reliable numbers of the size of bison herds in the late Pleistocene. In fact, there are few reliable tallies of the size of bison populations in historic times. It has been said that in the 1800s, before commercial hide hunters took to the plains, there were upwards of 60 million bison. The reality was probably half that.[12] That's still a lot of bison, and stories abound of single herds—like one seen in Kansas in 1871 said to contain some 4 million animals—that could blacken the plains well into the distance, and take days to pass by.

FIGURE 56.

Life-size reconstruction of *Bison antiquus* at the Lubbock Lake Landmark, Texas, with Ryan Byerly (who is over 6 ft tall) for scale. (Photograph by David J. Meltzer.)

Bison are wide ranging, traveling hundreds of miles season to season. They are gregarious herd animals, and arrange themselves in groups according to sex, age, season, and habitat.[13] Cow-calf herds are composed of females, males under 3 years of age, and a few older males. Males live either individually or in herds that may be as large as thirty animals.

Modern bison (*Bison bison*) are huge animals: males are upwards of 3.8 meters in length, are over 1.86 meters at the shoulder, and weigh on average 820 kilograms. The largest bull on record weighed about 1,350 kilograms (that's 1.5 tons). Females range up to 3.2 meters in length, are just over 1.6 meters in shoulder height, and weigh on average 450 kilograms. Yet, the species roaming the Paleoindians' landscape, *Bison antiquus* (Figure 56), was 15%–20% larger and boasted horns about 30 centimeters longer.[14]

Despite their massive bulk and deceptively sluggish behavior, bison are powerful runners, capable of reaching speeds of 65 kilometers/hour and turning on a dime. At least on dry land. They don't do as well on ice or in deep snow, but are surprisingly sure-footed in shallow water. They tend to stampede when frightened, a trait often exploited by human hunters, for the bison's speed, combined with its poor eyesight and tendency to run with head low to the ground, makes it very difficult for animals at the front of a stampede to change direction, or for followers to stop suddenly in the event they unexpectedly encounter a deep arroyo or cliff edge. But hunters had to be careful. As Dale Lott observes, the bison's speed, quickness, massive head (a formidable battering ram), hooves like sledgehammers, sharp horns (in both sexes), and, he adds, a "very short fuse," made it a dangerous adversary for people and other animals (wolves, mostly) that intended harm. Almost certainly, if Pleistocene bison were as fast, nimble, and unpredictable as bison are now, they challenged the wits and skills of pedestrian hunters.[15]

HUNTERS AND HUNTED

But Paleoindians met that challenge with sometimes-spectacular success, killing hundreds of animals in a single, deadly slaughter. These large kill sites are the iconic images of Paleoindian times, conjuring visions of stampeding animals crashing chaotically into an arroyo, massive bodies cracking and bones snapping, spears flying in thick clouds of dust, a sound wall of painful bellows of dying animals, and blood and gore everywhere. Symbolic these images may be, but representative they're not. Killing just one or two bison at a time, quietly, away from the herd, and with less fanfare, was probably the more-common strategy—George Frison suggests it was the best strategy. The smaller kills could have been accomplished by small family bands; the large kills likely required a multi-band, communal effort, though what that translates to in actual numbers is anyone's guess. It is certain a lone hunter could not drive 200 animals to their deaths: many hands (and spears) were needed.

Naturally, the smaller-scale kills are not as readily spotted archaeologically as the larger ones, and—I must confess having worked on both—are not nearly as interesting, with the result we know less about them (and the corollary: large, communal kills probably distract us from seeing other aspects of Paleoindian adaptations). Yet, Jason LaBelle and colleagues show that large kills, despite their archaeological visibility and influence on our interpretations of Paleoindian adaptations, represent less than 3% of all known Folsom sites, and perhaps an equal measure of later Paleoindian sites as well.[16]

Whether setting off on a communal hunt or not, Paleoindian hunters surely aimed to reduce their risks—always a good idea when pursuing animals this large—and were highly skilled at trapping, hindering, or otherwise disabling their prey, using the topography to great advantage.

But it's not the topographic feature you think, although that does make an engaging scenario: a herd of bison rushing toward a near-invisible cliff, seeing it too late to change direction, flying over the edge "Wile E. Coyote" style (legs grabbing wildly for firm ground) and smashing on the rocks below. Any broken-bodied survivors would be finished off by armed hunters. This hunting trick can be deadly effective and almost risk free when the topography and the animals cooperate, but it's hardly foolproof. Bison are too nimble to always play their assigned role.

More telling, no Paleoindians appear to have used the tactic. The first jump kills are no more than 5,000 years old (the oldest is Head-Smashed-In, Alberta). The only purported Paleoindian bison jump is Bonfire Rockshelter (Texas), where some twenty-five animals were found at the base of a twenty-three-meter-high cliff. How they died

FIGURE 57.
The base of the arroyo at Olsen-Chubbuck, jammed with bison bones. (Photography by J. B. Wheat, courtesy of University of Colorado Museum of Natural History.)

seemed obvious enough: it was from the plunge over the precipice. But just as in a good murder mystery, the obvious clues are deceiving and the crime scene not as it appears: the bison seem to have been killed elsewhere, and carcass parts moved here afterward. Analysis of the bison remains by Ryan Byerly and colleagues revealed the bones at the site are those meaty parts of a carcass normally transported back to a camp, and not the ones discarded at the spot of a kill. Bonfire may have been an ideal bison-jumping spot (and was so used nearly 7,000 years later), but its protected

BACK TO FOLSOM

The Folsom site in New Mexico is one of the best known sites in North America: it's routinely mentioned in archaeology texts, regularly appears on maps of notable localities, and of course, bestowed its name on the Paleoindian culture.[19] All this because excavations here in 1927 uncovered fluted points between the ribs of what became the type specimen of *Bison antiquus*. That these extinct animals were hunted at Folsom demonstrated for the first time and after decades of controversy that American prehistory began in the Pleistocene (Chapter 3). That distinction earned the site a spot on the very exclusive National Register of Historic Places.

But while Folsom is one of the best-known sites in American archaeology, for a time it was also one of the *least* known—scientifically speaking. The initial goal of the 1920s work was to recover bison skeletons for museum display, and once its significance became apparent, to document the association of the artifacts with the bison and thereby determine the site's age. That was done well. As the decades passed, our knowledge of these Paleoindians grew with the discovery of other sites of this antiquity. Yet, knowledge of Folsom itself lagged behind, owing to the narrow goals of the 1920s work, the methods in place at the time, and the few publications that resulted. Everyone had been so keen to link artifacts to bison that scarcely little more was learned about Folsom than bison were killed here a very long time ago.

Nor was any significant work done at Folsom in the decades afterward. That was Barnum Brown's legacy, for he claimed that after his crews finished excavating in 1928, there was nothing left of the site. Yet, Brown often said such things about sites he excavated to keep others from "poaching" them, and while investigating Folsom's archaeological history, I saw clues Brown just might have been bluffing. Even if he wasn't, we knew so little of Folsom something surely would be learned by working there. Besides, at 2,100 meters above sea level, in the cool and green Dry Cimarron Valley, Folsom is a much nicer place than the blistering hot sand dunes in west Texas where I'd been excavating much of the previous decade. When opportunity knocked with the chance to work at Folsom, it was impossible to resist.

In 1997 I began a multiyear, interdisciplinary project at Folsom, working off the edges of the 1920s excavation where, as it turns out, intact deposits remained (Brown *was* bluffing). With the help of colleagues, the thousands of bison bones and two dozen artifacts found in the 1920s investigations were also analyzed. Although separated by seventy years, the work was complementary. The value of our less-extensive excavations was enhanced by joining our finds with theirs, just as making sense of what was found in the 1920s was improved by interpreting their results in light of what we learned of the site's archaeology, geology, and environment.

And what we learned was this: the site is set within the ancestral Wild Horse Arroyo, where it was joined by a now-buried tributary. The paleovalley and paleotributary had cut deeply into the Smoky Hill Shale bedrock, but by 11,500 BP had begun to fill with a fine-grained silt or loess—wind-blown dust (Chapter 2). By the time the Paleoindian hunters arrived, which we have now pinned down to 10,490 BP (Chapter 4), a meter of loess had built up.

Even so, the paleovalley and paleotributary were still flanked by steep bedrock walls, and almost certainly the Paleoindian hunters put the land form to use. We don't know exactly what their tactics were, but we do know, based in part on reconstructions of the 1920s work, that the greatest density of bison remains occurred at the paleotributary/paleovalley junction. If carcass parts were left where the bison were dropped, it would appear the animals died bunched against the high bedrock walls. Although it is tempting to suggest the paleotributary is the main area of the kill, bison bones were also found in the paleovalley, raising the possibility some bison managed to scramble down and out of the paleotributary, only to be killed by hunters waiting in the paleovalley. Or vice versa.

The hunters' weapons included at least twenty-eight fluted points, the majority made of high-quality stone from the Alibates and Tecovas outcrops in the Texas Panhandle, the remainder of Flattop and Black Forest chert from outcrops in eastern and northeastern Colorado, respectively. No source was closer than 190 kilometers. The hunters' last resupply probably occurred at Alibates, since most of the points were made of this stone. The low number and intensive resharpening of the points made of Flattop chert are more distant echoes of sources visited earlier. By the time the hunters arrived at Folsom, over 25% of their points had been resharpened and reused. Once there, the assemblage suffered badly: most broke on the spot, a third from impacts.

By all measures, this was a successful hunt. A cow-calf herd of thirty-two bison were killed in the fall when the animals were at their peak of body fat. With a complement of flake tools and a quartzite skinning knife, the hunters dismembered the bison, and packed up the meaty parts for transport (they made off with lots of rib racks). Given the time it takes to butcher animals this large, the process must have lasted several days, during which time the hunters must have camped nearby.

And so we searched for their camp, walking the surrounding area, excavating test pits, machine-drilling sediment cores, and found . . . absolutely nothing. Was there no camp, or no camp preserved? Part of the answer came from a test pit dug on the uplands immediately overlooking the bone bed. The excavation bottomed out on Smoky Hill Shale, and charcoal on that bedrock surface yielded an age of 9820 BP. That told us the uplands had been swept clean by heavy rains sometime after the kill, along with any Folsom-age camp debris that might have been there. But perhaps the camp was not up top but down below, in the paleovalley where the hunters would have been less exposed to the wind and elements. No traces of one were found, so if it's there, it's deeply buried, lurking undetected beneath 4–5 meters of sediment.

Yet, I'm not sure there ever was a camp of any significant duration here, given what we learned of the climate at the time. The kill occurred amid the Younger Dryas, when temperatures at Folsom were cooler than they had been a thousand years earlier, or would be just centuries later. The bone bed yielded nearly a dozen snail species that today can only survive at elevations 600–900 meters higher (though it was not cold enough for long enough that alpine trees moved down into the area). It was also drier, judging by that thick drape of loess and the absence of aquatic snails.

In Younger Dryas times, the Folsom landscape was open, grassy parkland with scattered trees, with more heavily wooded areas on the edge of Johnson Mesa to the west, and in the valley of the Dry Cimarron below. Isotope analysis of bison bones and snail shells revealed the grassland was rich in warm-season grasses. It must have been prime bison habitat, at least in summer and early fall, when these grasses covered the landscape. But what of the rest of the year? Today, this is an area of harsh winters and heavy snowfall, as it likely was in the Younger Dryas. Bison don't do well in snow, and there is no isotopic evidence they were grazing on the cool-season grasses of late winter and spring.

This is not an area humans likely overwintered in either. Folsom today is a mix of grassland and woodland scrub, with ample plants for snacking and tea, but almost none that provide the return in fatty acids or carbohydrates that are vital during the winter and spring months. There is no indication foraging conditions were any better in the Younger Dryas. Without the promise of sufficient food, and with winter coming, this would not have been a place for hunters to linger long after their kill.

So why were they here? By the numbers and condition of the stone used to make their points, they'd arrived at Folsom from the Texas Panhandle. It's easy to see how: many of the rivers and streams that flow through the Panhandle have their headwaters in this corner of New Mexico. If the hunters were following those drainages, they would have found themselves in the general vicinity of Folsom. This is not to say all rivers lead to Folsom, only that coming into the area may not have been altogether random, especially if the group left the Panhandle heading north by northwest to areas where we know—based on their stone tools—they'd been before.

Once in the Folsom vicinity, there are only a few places where northbound traffic can easily traverse the high mesas and broken terrain that extend across the boundary land between New Mexico and Colorado. One is Raton Pass, through which the Santa Fe Trail (and now Interstate 25) travel; the other is Trinchera Pass, just 8 kilometers north of the Folsom site, through which Charles Goodnight once trailed cattle to Montana. Perhaps the Folsom hunters were aiming for Trinchera Pass when a herd of bison was spotted, plans were made, and the kill took place. The bison were butchered, meaty parts were readied for transport, and the hunters left before winter set in.

Although it was a highly successful kill, beyond the food these bison provided I suspect the long-term consequences of this episode were insignificant. The hunters' stay was brief, and there is no evidence they ever returned to this spot, though they or others may have come through Trinchera Pass on later occasions. Within a few years, all traces of the kill were buried, and mostly remained so until a town-breaking flood over 10,000 years later brought an astute cowboy out to check his cattle, and spot something unusual in the bottom of the arroyo where he broke wild horses (Chapter 3).

Ironically, however fleeting, ordinary, or even inconsequential this episode may have been in the lives of the Paleoindian hunters who killed those bison some 10,500 years ago, their actions had enormous and lasting impact on American archaeology.

overhang also made it an ideal place to camp, providing a cool respite for a Paleoindian summertime bison kill.[17]

Most commonly, Paleoindians stampeded or maneuvered animals into arroyos, where steep walls, injuries, or both prevented their movement or escape, giving the hunters time and a safe vantage from which to spear the bison. At the Folsom site, 32 animals were killed in an arroyo where it joins one of its tributaries, the former flanked by bedrock walls 5–6 meters high, the latter by 2–3-meter-high walls. Bison can scramble up a slope, but not a nearly vertical one ringed by armed hunters. At Olsen-Chubbuck in Colorado, the trap was hit harder and faster when 190 bison were stampeded into an arroyo over 2 meters deep (Figure 57). The animals in front had no time to stop or change direction when they finally saw the trap before them (remember: they don't see well), and the herd behind pushed forward oblivious of their doom. The result was a thrashing, bellowing pile of bison, with some of those at the bottom wedged head first into the arroyo floor, others violently twisted and contorted from being crushed by the mass and impact of those that followed them in. Survivors were finished off by the hunters.[18]

Elsewhere, other types of landforms or features (natural and artificial) were used, but with the same principle in mind: at the Casper and Horner sites (Wyoming), and Jones-Miller (Colorado), the bison were trapped and killed amid, respectively, sand dunes, an artificial corral, and (possibly) snow drifts.

Trapping the animals was only the first step; keeping them bottled up and speeding their death was the second and sometimes harder task. A wounded wild animal is dangerous and destructive, and bison did not necessarily die immediately on being speared. Modern bow hunters will tell you that even after a direct lung shot, a bison may struggle for life for 30 minutes or more.[20] And from time to time, some animals surely escaped the hunters' traps to live or die elsewhere (like the Naco and Escapule mammoths). All of which meant the weaponry had to be in top working order.

TOOLS OF THE TRADE

The weapons of choice were lanceolate spear points. The earliest of these in the post-Clovis sequence on the plains, Folsom points, like their immediate predecessors, were fluted (see Plate 13). Unlike Clovis, Folsom fluting often extended the full length of the point, which entailed removing a flake upwards of 5 centimeters long yet less than 1.5 millimeters thick, typically from both faces of a point that was itself no more than 4 millimeters thick. That was no mean feat of flintknapping, and it has been the envy and challenge of generations of modern knappers seeking to replicate how it was done. They've tried every technique imaginable, realistic or otherwise (Folsom groups did not have rock-cutting band saws), and there is no consensus on how it was done, or even if there was a single technique. As with all experiments, these reveal how points may have been fluted, but not whether they actually were fluted in the tested manner.

Once fluted, the tip and edges of a newly minted Folsom point were often carefully trimmed with tiny flakes, to give the point symmetry, and the lower edges and base were ground smooth—again, presumably, so they would not cut the sinew used in firmly hafting the point to a shaft.

Folsom points were smaller and more delicate than Clovis, averaging just over 5 centimeters long and 2 centimeters wide (Clovis points are wider, longer, and nearly twice as thick [Chapter 8]). This was so, conventional wisdom had it, because they were aimed at bison, not mammoth, and bison could be taken down with smaller-caliber shot (though as Emil Haury observed, several "small caliber" Clovis points speared some of the Lehner mammoths). Of course, stone projectile points are not bullets, which aim to kill instantly and from far away, the larger caliber being more efficient against big game with tough hide, muscle, and especially, thick-boned skulls.[21] Still, even a small-caliber bullet (or projectile point) propelled at high velocity can be lethal: in Late Prehistoric and Historic times, bison were killed by tiny triangular arrowheads. Fired by a bow, velocity more than compensated for their size and weight (remember Newton's second law: force = mass $\times$ acceleration).

There were probably very few circumstances in which a spear thrust or thrown at an animal would kill it instantly, although these could be propelled with accuracy and force. A Folsom point was found embedded deeply in the neural arch (which encloses and protects the spine) of a bison vertebra at the Lindenmeier site (Colorado).[22] Rather, these sharp-edged weapons were likely designed to tear a hole in the hide large enough for the point and foreshaft to penetrate and cause bleeding.[23] Aim, fire, and be patient: the animal would die in time. In this regard, the only "caliber" that might have mattered was point width (critical to enabling the point and foreshaft to enter) and not length. Clovis points are a mere 6 millimeters wider than Folsom points, hardly a ballistically meaningful difference.

Fluting disappears from the Paleoindian repertoire after Folsom times, though later Paleoindian points retain other ancestral vestiges: a lanceolate form, sharp tips and upper blade edges, and ground lower edges, where the point had been nestled within its haft. Paleoindian point aficionados argue the merits of the various forms as weapons, but Frison, who has experimented with virtually all point types on game animals large and small, was unable to establish that any one was significantly more lethal than another. Like all hunters, however, he has his preferences: he'd arm himself with an Agate Basin point.[24]

Whatever the point type, breaking on impact was common. The broadside of a bison is a picket fence of wide, thick ribs. Weapons failure could be dangerous if the hunter thrust the spear from close by and was still holding on when it broke, leaving no chance for a second stab—and it could take multiple stabs to kill a large animal (which requires quickly pulling the spear out of a moving animal, increasing the chances it will break). Not surprisingly, Christopher Ellis found that among the majority of spear-using hunter-gatherers, their weapons were more often used as javelins (thrown) than as lances (thrust), and when used as lances, mostly served to deliver the close-in coup de grâce to a dying animal.[25]

Throwing might have been accomplished by hand, or perhaps with the aid of an *atlatl*, a throwing stick with a hook at the end that temporarily holds the base of the spear. An atlatl and spear are hurled in an arc that starts behind the thrower's head and comes overhand (much like the motion of a baseball pitcher). As the spear travels the arc, it flexes and stores energy, which is then explosively released, flying from the atlatl at velocities that can average 135 kilometers/hour and exceed 200 kilometers/hour (in contrast, a hand-thrown spear averages about 90 kilometers/hour). A native hunter with an atlatl, one sixteenth-century Spanish chronicler wrote (with a mixture of awe and fear), could throw a spear completely "through a man armed with a coat of mail."[26] With velocity comes distance, which if the modern record for atlatl-aided spear throwing—nearly 300 meters—is any indication, could have been considerable.

Maximizing velocity and distance might require the hunter's entire body be brought into play by standing to throw. That's not always easy or desired if one is crouched behind rocks or brush, or trying to keep an animal or herd from spooking and scattering, but of less concern when animals are trapped.[27] Regardless, the hunter better keep back-up spears close at hand, because once launched, a spear is not easily retrieved, especially from an injured and dangerous animal. In some sites, and Folsom is a good example, one can spot sets of points that are so alike in material and workmanship it appears they were made by the same hand, and possibly launched from the same quiver. But launched by hand or atlatl? Because atlatls were made of wood or bone, they are rare archaeologically, and unknown from Paleoindian times. Evidence of their presence is strictly circumstantial. Studies by Karl Hutchings of fracture velocities on Paleoindian points show some shattered from being propelled at velocities only attainable with an atlatl.

After the kill was made, points were sometimes pressed into service for butchering, but only if tools better suited to the task were unavailable. Often used were skinning knives customized for the task. At Folsom-age sites, these include palm-sized, exquisitely made, ultra-thin bifaces (often no more than 5 millimeters thick), which must have been highly prized, judging by the quality of their material, workmanship, and scarcity. At later Paleoindian sites, the oddly asymmetrical Cody knife served a similar function.

The toolkit found at a kill site can be relatively uncomplicated: points, knives, and perhaps bone butchering tools, the latter often fashioned on the spot for use in dismembering the carcass. But then, kill sites offer only a small window into the full range of activities engaged in by Paleoindians. In their longer-term residential camps (of which, more below), the tools found include scrapers, knives, cores, flakes, wedges, spokeshaves (tools with a concave edge for trimming spear shafts), gravers, drills, and awls, as well as a complement of bone and antler artifacts (see Plate 14).[28]

PRODUCTS OF THEIR LABOR

Paleoindian hunters pursued bison throughout the year, but as Matthew E. Hill shows, there are seasonal trends to the large kills, which partly sort by geography. On the southern and central plains, the peak period for bison hunting was in the late summer and early fall; it was later in the year on the northern plains.[29] Why the difference? Winters are longer and harsher on the northern plains, and meat from kills made at the onset of the cold months could be stored frozen. (We can ascertain when a kill was made because bison calve between mid-April and late May, and their teeth have a regular eruption and wear sequence. Calves found in a site that have only early stage lower first molars, say, were about 4–5 months old at death, so the kill likely took place in late summer or fall, assuming calving and tooth eruption schedules of Pleistocene bison were roughly the same as those of their modern counterparts.)

There's a related fact as well: kills made later in the year often targeted cow-calf herds. Why them, then? In the fall, body fat and nutritional value of cows and calves are at the peak from a summer of grazing; bulls, in contrast, burn off much of their body fat in the fall rut. At the same time, cow-calf herds are less dangerous and more easily manipulated than bull herds. Cow-calf herds are otherwise difficult to hunt during and immediately after spring calving: when frightened, very young calves commonly bolt from the herd, and their mothers follow, creating a chaotic situation that reduces the chances of success.[30]

Although bison was a prime food, not all animals that were killed were butchered, especially at large communal kills, and not all parts of all animals were used. On occasion, Paleoindians practiced *gourmet butchering,* as Lewis Binford calls it, with select cuts of meat removed and eaten on the spot, or dried or frozen (depending on the season), and then taken off-site. In any case, it's unusual to find all the animals at a Paleoindian kill to have been completely butchered, or to have had their bones smashed for marrow, or rendered for grease in hearths and boiling pits. Bone grease is a key ingredient in making pemmican, a highly nutritious, long-lasting, and easily transported food of pounded, dried bison meat mixed with fat, berries, and bone grease in skin bags (which historically was a valuable food for the lean winter and spring months).[31] That Paleoindians were not making pemmican, nor apparently transporting or storing of quantities of meat (practices common among later bison hunters), indicates their food needs were more easily met, perhaps in no small measure because they were able to move freely over great distances, and could exploit bison when encountered.

And sometimes they encountered and killed more animals than they could possibly eat, especially in the near chaos of a communal hunt in which a large herd was being stampeded into a trap. This was not like rounding up slow, dopey cattle. Naturally, there was waste. Olsen-Chubbuck's excavator, Joe Ben Wheat, estimated that 16% of the 190 animals killed here were only partially butchered, and at least 10% of the animals at the bottom of the pile were untouched. Even so, there was still an enormous payoff. Based on the number, sex, and age of the bison in the kill, and using meat weights of modern bison, Wheat calculated the hunters butchered over 21,700 kilograms (23.8 tons) of meat. If Pleistocene bison were 25% more massive than their modern counterparts (his estimate), the harvest would have been 27,100 kilograms (29.8 tons).[32]

If one assumes a single person could butcher an animal every 1–2 hours (the figure varies depending on a host of factors), and that the Olsen-Chubbuck hunters worked steadily at the task, then the butchering process would have taken anywhere from half a day (if, say, 100 individuals were on the job) to several days (if there were only 10 people at work). Whatever the timing, Paleoindians needed to move quickly, particularly in warm weather. Animals had to be disemboweled to prevent internal fluids from spoiling the meat, and the meat filleted and dried before blowflies came to lay their eggs. In cold weather, spoilage would have been less an issue, but butchering more difficult as carcasses froze and hands grew numb. And if there were carnivores or scavengers within smell of the kill, they would have soon become unwelcome guests.[33]

If the hunters were en route elsewhere, the carcasses had to be cut into segments for transport along with the tools, water containers, fire starters, and other gear already being toted. Drying the meat surely helped make it easier to transport, since it would shrink by about 30% and be as much as 75% lighter.[34] Parts hauled away commonly included the high-meat-yielding bones such as ribs, upper forelimbs and hind limbs (bison drumsticks—easily hefted!), and shoulder meat, possibly still attached to the shoulder blade for easy carrying. Often taken, too, were the tail bones (caudal vertebrae), which at first glance makes little sense. There's hardly any meat on these bones. But then it probably wasn't meat the hunters were after, but what was attached to the tail: the hide, which could be made into robes and shelter covers. As the old saying goes, the "tail follows the hide." Routinely discarded at these kills were the low-meat-yielding parts such as skulls, lower limbs and feet, and jaws (see Plate 15), but the jaws were usually not discarded until after the tongues were cut out, these being delicacies often eaten raw on the spot.

If the group was in less of a hurry, or the season and place was right, instead of moving meat to camp, they might move camp to meat. There are several Paleoindian kills with adjoining camps, such as Cattle Guard (Colorado), excavated by Margaret Jodry. Nearly fifty bison had been killed amid sand dunes (Great Sand Dunes National Monument is nearby) in one late summer or early fall. At an adjacent camp, Jodry found that the carcasses had been thoroughly butchered and roasted. Unusual for a Paleoindian site, but perhaps indicative of the longer time spent here, the bones had been smashed for marrow. Over 90% of the 200 projectile points at Cattle Guard were broken and showed impact damage, or had been used and reused so often they were worn down to slugs barely 3 centimeters in length. While enjoying the fruits of their labor, the Cattle Guard hunters refurbished and replaced their broken weaponry, leaving behind more than 18,000 waste flakes, several hundred point preforms, and some 275 channel flakes (created in fluting) indicative of point manufacture.[35]

Cattle Guard was occupied longer than many Paleoindian kill sites, but "longer" is strictly a relative measure based on the density and diversity of archaeological remains. All things being equal, both increase with a longer stay. Unfortunately, we have no means of gauging the actual time that a longer stay represents, though it was probably measurable in weeks. As mobile as these groups were, they likely did not dwell in one spot for very long, except perhaps in winter when movement across the plains and through the mountains was far more difficult, and groups may have slowed or for a time stopped all together.

COPING WITH PLEISTOCENE WINTERS

At Agate Basin, on the high plains of eastern Wyoming, a Folsom group camped on the floodplain of an arroyo some 10,800 years ago, near where at least eleven bison and five pronghorn were trapped and butchered. The people were here through the harsh, cold late winter until early spring, judging by the presence of fetal and neonatal bison remains. They pitched hide-covered structures (with one "tent peg"—a bison rib—found still stuck in the ground), around which was a dense litter of projectile point manufacturing debris, end and side scrapers, cutting tools, gravers, knives and choppers, a crystal-studded abrader (for hide softening?), bone and antler artifacts (including eyed needles), and the slab used to pulverize the red ocher that covered many of the bones and artifacts.

The site's inhabitants had plenty of time on their hands, which they devoted to preparing bison robes—the scrapers are telling us that—and gearing up for hunts that would take place when spring came. Frederic Sellet estimates they fashioned a minimum of thirty-eight new points, more than needed to replace the three broken ones they left behind. That they intended to make camp for a long period is apparent in the stone they used to make those points: it was hauled in from the Knife River chert outcrops about 400 kilometers distant, a trip that would have been extraordinarily difficult once winter set in. And it was a harsh winter: Matthew G. Hill's analysis of the bison remains revealed that the near-starving residents smashed open bison foot and toe bones to suck out the meager marrow they contained.[36]

Folsom groups stayed as long and weathered even harsher climes at the Mountaineer site in the Colorado Rockies, where Mark Stiger uncovered the foundation stones of a couple of Folsom-age structures (Figure 58), the walls of which were made of aspen poles, smeared over with clay mud.[37] To archaeology's good fortune, one of the houses burned after it was abandoned, which fired the mud brick hard, preserving aspen bark imprints on the clay. Within the houses are hearths and artifacts indicative of various domestic chores, such as cleaning hides, fixing tools, and making points. Bison bones

FIGURE 58.
The "floor plan" of one of the Folsom-age house structures at the Mountaineer site. The larger dashed oval represents the approximate outside edge of the structure, based on the distribution of artifacts and features; the smaller dashed circle indicates the location of an interior hearth. (Photograph by Mark Stiger.)

from one of their meals yielded an age of 10,400 BP, making Mountaineer one of the oldest home places in America.

At first glance, it's in an odd spot for one. Mountaineer sits atop Tenderfoot Mountain, a tabletop mesa that looms nearly 330 meters above the surrounding valley, which is itself already 2,400 meters above sea level. There is no permanent water atop the mountain, and most of the game to hunt or plants to gather would have been in the valley below, where the Gunnison River and several large streams flow. But two aspects of the mesa might have made it attractive: It's got a stunning and unobstructed view of the landscape below, making it a great vantage point to spot animals grazing in the valley and plan a hunt (even this fifty-something college professor can make it down the rocky slopes to the valley below in under 20 minutes). Second, in winter, the top of Tenderfoot Mountain is warmer and sunnier than the valley floor, since no mountains shade it and the cold air sinks into the lower-lying basin. And it almost surely was a winter occupation. Why else take the trouble to build insulated, mud-walled huts atop a waterless mesa? In winter, of course, melting snow would provide ample freshwater.

Perhaps not surprisingly, the majority of the artifacts from Mountaineer were made of locally available, fine-grained quartzite, not the usual chert, and bespeak movement restricted to this basin, at least for the period they were at Mountaineer. Judging by the investment they made in their housing and the amount of debris left behind, it was a lengthy stay: weeks or even months (year-round settlement only occurs much later in North American prehistory).

In fact, recent investigations in the Gunnison Valley, as well as similar intermountain basins such as the central Rio Grande Valley (New Mexico), as well as the San Luis Valley and Middle Park (both in Colorado), suggest Paleoindian groups overwintered in these more protected places, rather than spending the cold months on the open plains. "Protected" is something of a loose term. The Gunnison Basin is one of North America's coldest spots today, and likely was in the late Pleistocene, though at least one would not have to contend with the bone-chilling winds of the plains.

A BALANCED MENU

Staying put for long periods has its drawbacks, resource depression most obviously. Food runs away or runs out, as the smashed-for-marrow bison feet and toes at Agate Basin testify. A band of twenty-five families living on bison alone, Frison estimates, would need to kill five or six bison a month for their sustenance. Those might be hard to come by in the dead of winter in a high-mountain basin.[38]

But then, their diets were not dependent on bison alone, however highly these were ranked and prized—especially on the open, treeless high plains, where there are fewer food choices. Most large, communal bison kills made in Late Paleoindian times were made in that setting, but that's not necessarily what the majority of Late Paleoindian sites look like or where they are found. As Jason LaBelle shows, most Paleoindian sites are kills of smaller game or of fewer bison, quarry localities, small artifact scatters (overnight bivouacs, perhaps?), or the occasional isolated point. More telling, many of these occur not on the open plains, but instead in mountain and foothill settings, which provide a greater variety of food resources than the plains proper. The high-ranked food (bison) is there to be sure, but also deer, pronghorn antelope, other smaller animals, and plants.[39]

The conclusion LaBelle draws is that Paleoindians engaged in large, communal hunts on occasion, but for most of the year, the monitoring and hunting of bison was a task left to hunting parties, which would kill a few animals, then transport the meat back to camps. Otherwise, camps were established around less mobile and more predictable resources like water, small game, plants, and stone. These, as Douglas Bamforth observes, often were used repeatedly over time (unlike bison kill spots), and show a far greater reliance on locally available stone.

At these non-bison kill sites, we get a glimpse of the wide range of the Paleoindian diet. One of the more striking examples comes from the Allen site along Medicine Creek in southwestern Nebraska, where Bamforth and colleagues have shown that Paleoindians consumed bison, deer, pronghorn, jackrabbit, cottontail, and prairie dogs, as well as freshwater mussels and fish. Likewise, on the far side of the plains in Wyoming's Big Horn Basin, inhabitants of the Hanson site exploited mountain sheep, deer, marmot, and cottontail rabbit. Even on the plains proper, at Lubbock Lake (Texas), a permanent freshwater spring and pond attracted hunter-gatherers, who took bison, deer, pronghorn, jackrabbit, muskrat, and a variety of ducks that were also drawn to this setting.[40]

GATHERING THE CLAN

During the Late Paleoindian period, we get our first archaeological look at a phenomenon that must have occurred in earlier times and certainly did later: the rendezvous. Such events would have been especially important in the Paleoindian era, since groups likely spent much of their time widely scattered in small family bands. From time to time, they must have gathered to exchange information, resources, and mates; initiate adolescents into adulthood; provide proper ceremonial attention to the sick or recently dead; or just enjoy the company of other people (Chapter 7). How often? No one knows for sure, but among mobile hunter-gatherers of the Arctic, Lewis Binford reports aggregations were timed to children's coming of age: when they were ready to be initiated or married, it was time to gather the clan.[41]

Given the distances bands may have had to travel to the rendezvous, and all they had to carry under normal circumstances, it would have been difficult to come bearing large stores of food (no pot luck dinners these). So we suppose gathering spots were selected for the richness of the local food resources rather than, say, the presence of abundant stone. The location also had to be well known or easily found, so no one missed the event.

The Lindenmeier site in northern Colorado fits the bill. As LaBelle observes, it is tucked into a topographic corner near where a set of low hills comes in perpendicular to the Front Range, and is immediately adjacent to a multistory geological exposure of red and white rocks—a gigantic barber pole that can be spotted from 30 kilometers away.[42] The area was well watered, and there were stands of trees nearby to fuel hearth fires and locally available stone to replace broken tools. Because the site is at the intersection of two biotic communities (the plains and the Front Range), it likely provided ample and diverse plant and animal food for those who gathered here. At the very least, we know Lindenmeier's occupants fed on bison, pronghorn, deer, jackrabbit, wolf, coyote, fox, and turtle.[43]

As Edwin Wilmsen showed, several areas of the site were simultaneously occupied by groups with stylistically distinctive Folsom points, made of different types of stone evidently carried in from different points on the compass, all at least 150 kilometers distant.[44] While here, the groups kept busy making projectile points: several hundred point preforms and nearly 1,000 channel flakes were found. Looking at some of the failed preforms, it's hard not to suspect they were made by novices learning to fashion a Folsom point, or mimicking the activities of the elders.

There was plenty of hide working, with many of the scrapers heavily worn and resharpened, as well as wood, bone, and stone working, judging by the notched stone tools and coarse sandstone blocks for carving and sanding spear shafts and other tools. Delicately made bone needles hint at the manufacture or repair of tailored clothing, which may have been dyed with red ocher (imported from the Laramie Mountains to the north) and decorated with bone "sequins" or stone beads (Figure 59), which were frequently incised on their faces, notched on their edges, and occasionally stained with ocher. But were those bone and stone items more than ornaments? As Wilmsen admits, there may be more to these pieces than we can see or even imagine.

Indeed, more surely occurred at Lindenmeier than we will know from the archaeology, for most of the activities that would take place at a rendezvous may have had little material expression—like conversation—but were a vital part of their lives all the same. So, too, was this particular rendezvous spot. Wilmsen reports Folsom bands periodically returned here to complete the important ritual cycles of their lives. Cycles and movement. Paleoindians were highly mobile some of the time, less so at other times.

And one time they traveled out of archaeological sight.

BACKWASH

The Late Paleoindian period witnessed one final pulse of movement between Alaska and unglaciated North America, only this time populations were not just headed south, and the movement didn't just involve humans. By 10,500 BP, the ice-free corridor was wide open and blanketed by a rich grassland that formed a virtually uninterrupted greenbelt from the Yukon to the Rio Grande.

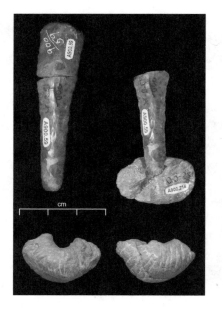

FIGURE 59.

A stone drill and engraved stone bead from the Lindenmeier site, Colorado, shown separately and showing how the drill may have been used to fashion the bead. This particular drill may not have been the one used to craft this bead—though it fits precisely in the center hole of the bead—but the images show how it might have been. (Photographs courtesy of Jason M. LaBelle; arranged by Katherine Monigal.)

Ancient bison DNA reveals what happened next. As Beth Shapiro and colleagues have shown, bison populations in eastern Beringia and unglaciated North America, which had been kept apart since the onset of the Late Wisconsin, began to move. Those in the north headed south, getting as far as the Peace River in northwestern British Columbia by 10,200 BP. In turn, those in the south moved north, hitting their high-water mark near Athabasca, northern Alberta, by 10,400 BP. For a brief period, the two populations of bison overlapped in the center of the ice-free corridor region. Oddly, there was virtually no gene flow between the two, and the southern bison never reached Alaska, nor did any Alaskan bison ever graze in Montana.[45]

However, the people who hunted them may have gone the full distance, judging by the sites of the Mesa Complex, which are found primarily on Alaska's North Slope (the region between the Brooks Range and the Arctic Ocean). These sites were occupied in the several centuries leading up to 10,000 BP, putting their occupation at the tail end of the Younger Dryas and making them essentially contemporary with, but slightly later than, the Paleoarctic sites of interior Alaska (those microblade and microcore sites described in Chapter 6). That also makes them the same age as Late Paleoindian Agate Basin complexes on the northern high plains of the Lower Forty-eight, to which, as it happens, the Mesa projectile points bears more than a passing resemblance.[46]

Michael Kunz discovered the original Mesa site on a large, isolated ridge some 225 kilometers north of the Arctic Circle (but still 290 kilometers south of the Arctic Ocean), which rises 60 meters above the surrounding plain (see Plate 16). Most of the other Mesa Complex sites occupy similar high spots on the landscape. They are in an area of harsh climate, with a mean annual temperature of −9.5°C (15°F) and annual precipitation of just over 20 centimeters. Winds can whistle in excess of 65 kilometers/hour, sometimes for days on end, which has the virtue of keeping the ever-present mosquitoes at bay, but makes for chilly conditions in summer and especially in winter. During the Younger Dryas, it was even drier and colder, which likely explains the nearly twenty small hearths found at Mesa. These were small hearths, for this was not a habitation site but a hunting lookout. The hearths would have warmed hands and feet as hunters sat atop Mesa and took advantage of its 360° view of nearly 100 km² of the surrounding terrain, watching for prey.

The artifacts from Mesa are mostly projectile points (over 125) in various stages of their life histories. As Michael Bever has shown, the hunters sitting up here busied themselves making new points from preforms of locally available cherts, sharpening still useable ones, or discarding those too exhausted to salvage (sometimes first heating them in the fire, likely to loosen the haft that bound them to a spear), and otherwise retooling their weaponry and making ready for the hunt (and incidentally littering the site with more than 20,000 pieces of debitage). There were also some seventy gravers, indicative of working bone. That not much else was going atop this hill is indicated by the scarcity of habitation debris. The rest of the artifact inventory includes a meager collection of flakes pressed into service as cutting tools, scrapers, some hammers and anvils, a few quartz crystals, and small bits of red ocher.

A similar toolkit appears in the other Mesa Complex sites of the far north, the similarity likely reflecting the fact that they were all the by-products of related, wide-ranging, rapidly moving hunters. Hunting what is uncertain. Kunz and colleagues suspect the prey was bison or perhaps musk ox, but not caribou, the denizen of the modern tussock-tundra of the North Slope. The reason? In the late Pleistocene, this was dry grassland better suited to bison, the bones of which are abundant in nearby fossil localities of this age.

Watching for bison and other game may have been something these groups had been doing for some time, and over a large area, for as mentioned, Mesa projectile points bear a striking resemblance to points found on the northern high plains, notably Agate Basin forms, which are approximately the same age. Agate Basin and Mesa appear related, but how? That's unclear, since there are but a couple of sites in the intervening areas of Canada that look like either. Whether Mesa represents Agate Basin hunters moving north up the recently deglaciated ice-free corridor following bison herds in a still-expanding grassland, or the reverse, we cannot say. Yet. But the betting money is on the former.

As such, this may be the most contrarian population movement in human prehistory. During cold intervals, and it was surely cold in the Arctic during the latter part of the Younger Dryas, humans routinely *abandoned* the far north, not migrated into it.[47] Indeed, while these Paleoindians were following open grasslands to the north, there was southbound Paleoarctic traffic down the Pacific coast and through the adjacent forested interior of British Columbia (Chapter 6). But over on the southbound lanes, the traffic moved more slowly: descendants of Paleoarctic groups would not approach the Lower Forty-eight until many thousands of years later.

Over on the northbound lane, Mesa groups traveled much faster (centuries, not millennia), but it was a brief and short-lived movement of people (and bison). Soon after 10,000 years ago, wetter and warmer conditions came to the far north, and in their wake tundra took over the North Slope, spruce forest and peat bogs spread across western and northwestern Canada (Chapter 2), and the grassland corridor between Alaska and the Lower Forty-eight was closed. With that, northern bison populations declined, caribou numbers surged, and the Mesa Complex disappeared.

That disappearance neatly correlates with climate changes that mark the end of the Younger Dryas. As such, it may be one of the few instances of a causal link between Younger Dryas climates and Paleoindian adaptations. Not that we haven't sought others.

LOOKING FOR THE YOUNGER DRYAS

Archaeologists often invoke changes in climate and environment to explain changes in culture. It's a tricky exercise: not all climate change causes culture change, nor does all culture change have a climatic trigger. But if there was a moment when people might have been caught by changing climates, the Younger Dryas certainly seems a likely one. Archaeologists have most assuredly noticed the possibility: the "unexpected onset" of the Younger Dryas is blamed for Pleistocene extinctions (remember over*chill?*), the "fragmentation of Clovis culture and the emergence of more localized regional . . . cultural traditions," and evidently required Florida Paleoindians to dress in furs.[48] But did Paleoindians notice the Younger Dryas? There are reasons to be skeptical.

The most rapid and severe plunge in Younger Dryas temperature occurred at high latitudes, which for the most part were still lacking human occupants. Elsewhere across North America, climates varied, and in places were relatively benign (Chapter 2). Vegetation change was not as abrupt as climate change (there is that century or more lag), and though those changes could be significant, they were most pronounced in more northern regions.

We must be mindful, too, that Paleoindians were highly mobile, still relatively new to the landscape when the Younger Dryas began, had yet to experience relatively stable climates or environments, and weren't fully settled in or dependent on particular resources in specific environments. Adapting to changing environmental conditions was nothing new to them. It was what they did. Only in this case, as Metin Eren aptly notes, instead of people moving to different environments, if anything the Younger Dryas brought different environments to people.[49]

So the notion that the Younger Dryas somehow caught Paleoindians flat-footed seems overly dramatic if not simplistic (especially the stereotype of a suddenly devastating global Big Chill). But even if Paleoindians were not constantly scrambling to keep up with Younger Dryas changes, which mostly occurred on a scale of multiple decades or centuries (still possibly not fast enough to be noticed by people, who directly respond to daily, weekly, and seasonal conditions), the cumulative effects of the Younger Dryas nonetheless may have required a response. The best place to look is on the northern margins of the Paleoindian world: perhaps Mesa in Alaska, or in northeastern North America.

Paige Newby and colleagues observe that during the Younger Dryas, the northeast was largely open tundra, as spruce woodlands (and the northern tree line) had shifted south in response to colder and drier climates. But as the Younger Dryas came to an end and climates warmed, both tundra and spruce woodland retreated north, replaced by closed, mixed pine forests. They "guarantee" that vegetation change had an impact on herd size and migratory behavior of large mammals, namely caribou.[50]

Guarantees aside (there's no fossil record to back it up), that's probably right. Caribou live in arctic and subarctic environments, summering on open tundra and wintering in sheltered forests. The Younger Dryas would have suited them well. When those habitats shifted northward as the Younger Dryas came to an end, caribou likely went along with them, ending up where they were found historically: in far northern Labrador. Moving into now-forested New England and the Canadian Maritimes were non-herd animals like moose and deer.[51]

Newby and colleagues see a correlation between this ecological change (caribou to moose and deer) and the near-simultaneous transition from fluted to non-fluted points (the radiocarbon dating of that transition is a bit wobbly, but we'll grant the point). They think that's no coincidence, but reflects hunters adapting to new prey (that point's harder to grant).

Bear in mind, however, that projectile points were but one element of a much larger toolkit in use. Newby and colleagues admit the changes in the rest of the Paleoindian toolkit over that transition were "more incremental," which is another way of saying there were virtually none, despite apparently changing prey types. Nor is there compelling reason why the method of hafting a point to a spear (whether fluted or not) would change if the target was caribou as opposed to moose or deer. All the points in question were lanceolate spear points, so their killing potential was essentially the same.[52]

More likely, Newby and colleagues are correct in their suspicion that the correlation in the changes in hafting and landscape resources may simply be fortuitous. After all, point styles were changing throughout eastern North America at this time. It is no surprise one landed on this moment of ecological change, especially when that "moment" extends over several centuries.

Still, let's not give up on a possible Younger Dryas influence just yet, for there is no denying the end of this period brought significant ecological change to the northeast and New England, and there was an apparent human response: it's not in the hafting technology, but instead in the number and types of sites. In northeastern North America, tundra and spruce forest gave way to mixed pine forests about 10,000 years ago (depending on location). When that happened, the archaeological record across much of the region essentially went blank. Although there's debate as to why, the most compelling explanation is that the emerging pine-dominated woods, a notoriously unproductive habitat, so limited food options that there was little incentive to stay—and few did.[53]

Those who remained situated themselves close to the area's lakes and rivers, which would have attracted game, migratory birds, and possibly fish (depending on the timing of natural, postglacial re-stocking). Others of their sites are on the floors of what were recently drained glacial lakes, open meadows where moose, deer, and elk would have grazed. These sites are much smaller than those from earlier times related to caribou hunting. The reason, Christopher Ellis and Brian Deller suggest, is that these people were not participating in communal hunts of migratory herds, but rather were targeting individual animals roaming in smaller numbers in these open settings.[54] All the while, the groups were using the same basic toolkit they had in earlier times, though the points themselves changed hafts.

THE DALTON GANG

The biotic communities of northeastern North America were in flux as the Pleistocene came to an end, but that was far less the case in southeastern North America, where the forests were essentially already modern looking, and which provided a much richer variety of foods. The Late Paleoindian presence in this region was correspondingly much more abundant, and characterized by a plethora of point style zones in different regions. In general, all these groups exploited a wide range of animals and plants, though across southeastern North America, there were subtle differences in adaptation, depending on local resources and environment.

The most complete record we have of Late Paleoindian forest adaptations comes from Dalton period sites, which are concentrated in the central Mississippi Valley, but extended from the Ozarks into the Carolinas and date between 10,500 and 10,000 BP.[55] These sites are readily recognized in their Mississippi Valley heartland by the presence of Dalton points, and elsewhere by related types: Colbert, Greenbrier, Hardaway, Nuckolls, and San Patrice. These generally have a lanceolate shape, a slight to deeply concave base, which is thinned in what appears to be a last-gasp effort at fluting, and are often finely serrated (saw-toothed). Over their use lives, Dalton points were frequently resharpened in a way that led to strongly *beveled* edges, giving them a trapezoidal cross-section. Studies of wear patterns indicate these were multifunctional, serving as projectile points as well as knives and saws for woodworking—which in the late Pleistocene southeastern forests would have had no small importance.

Nor are the points the only woodworking tools in use. Newly invented at this time was the Dalton adze, a large, thick, triangular, and relatively heavy biface, presumably used to rough out wooden artifacts from food bowls to dugout canoes (in Historic times, canoes were fashioned by carefully burning the surface of a log, then chipping out the charred portions with an adze; microscopic bits of charcoal have been found adhering to the bits of some Dalton adzes).[56] Given Dalton sites are often located along rivers and streams, and in the bayous of the Mississippi bottomlands, canoes would have been handy transport.

Dalton groups made a good living in the forest. By now all the Pleistocene mammals going extinct had gone, but there was still plenty of food to be had. Dalton peoples exploited mammals from small (chipmunk and cottontail) to large (deer and elk), as well as birds, including turkey and prairie chicken, quail and passenger pigeon, ducks and swans. They pulled muskrat, fish (catfish, gar), aquatic turtles, and shellfish (mussels) from the rivers and streams. They collected walnuts, chestnuts, hickory nuts, persimmons, and acorns from the forest to balance out the diet. And balanced it was: as Joseph Caldwell observed of the remains at one Illinois Dalton site, with only slight exaggeration, "Everything that walked, flew, or crawled" was swallowed by the inhabitants of Modoc Rockshelter.[57]

The Dalton diet speaks, if indirectly, to the matter of landscape learning. Acorns contain high concentrations of tannic acid, and when eaten raw have a bitter taste. To make them palatable, the acid must be removed by leaching, which in prehistory involved placing mashed acorns in a basket and running water through them. No baskets preserve, but Dalton sites have grinding tools that would have served the mashing purpose. Evidently, experimentation by Dalton times had shown how to make acorn meal suitable as food, perhaps as bread flour.

Processing acorns and, for that matter, eating from top to bottom on the food chain indicate diet breadth was expanding in these post-Clovis times. This dietary expansion was traditionally credited to eastern Paleoindians' gradual and increasingly successful adaptations, enabling them to plan more intricate movements across the landscape so as to catch plants and animals at their seasonal peaks of availability or productivity. Archaeologists termed this "primary forest efficiency."

But that's a misnomer. As more food types are added to a diet, efficiency actually drops. Think about all that time that must be invested to make acorns edible, or to juggle all the tactics for pursuing, killing, butchering, and processing very different prey. "Primary forest *inefficiency*" may be a more apt term. Furthermore, expanding the diet is not necessarily a sign of good times.

Unlike their Clovis predecessors, Late Paleoindians in eastern North America were not ranging so widely across the landscape, judging by the fact their stone was mostly acquired from local sources, not exotic ones. And we suspect—based on a greater number of sites—that population density was higher in the later period than it had been earlier. When larger populations and smaller territorial ranges collide, hunter-gatherers have a problem. Mobility is their first order of response to cope with food shortages (Chapter 7): if one hunts out all the local deer or gathers up the available plants, the best option is to pack up and move. That option was readily available when the landscape was still mostly empty and territories had yet to emerge. However, when mobility is no longer possible because someone else is already occupying the adjacent valley, the next option is to stay put but expand the diet, eating more lower-ranked foods.

It still was a good idea to stay on friendly terms with the neighbors, and Dalton groups apparently did. Among the sites in the central Mississippi Valley there occur a small number of points that are longer, wider, thinner, and more finely made than most garden-variety Dalton points, and are routinely fashioned of high-quality Burlington chert, obtainable only at a small outcrop just west of St. Louis. So unusual are these the Dalton label doesn't quite fit, so they are called Sloan points. Points that long and thin break too easily to be useful, and befitting their impractical nature, Sloan points are rarely found in everyday Dalton sites. Instead, they occur as isolates or in caches, a half dozen of which are known from the central Mississippi Valley. Don't think of these as being comparable to Clovis-age caches, for Sloan points could not readily be pressed into service as points or raw material sources, as could a cache of Clovis points, bifaces, and blades. So why the investment in making these fancy, if unusable artifacts?

John Walthall and Brad Koldehoff suspect these had a role in ceremonial exchange, as indicators of wealth or status, or perhaps as symbolic representations of hunting prowess. If the anthropological record is any indication, the Sloan points were cached to stay hidden from view, only to be displayed during public rituals or a when ceremony required their appearance: for example, when rendezvousing with another group and looking to cement an alliance with the exchange of a ritually valuable item.[58]

A SHELTERED EXISTENCE

Just when this process of decreasing mobility and increasing diet breadth began is uncertain, though it was well underway by Dalton times. But a caution here: Dalton sites yield a wide range of food remains in part because many are domestic sites occurring in settings that preserve bones and plant parts: namely, in rockshelters and caves (for archaeological purposes, a rockshelter is a cliff overhang or hollow into which sunlight penetrates; caves go deeper underground and have chambers of permanent darkness).

Curiously, it is only in this period that Paleoindians took up regular residence in nature's indoors. And in the grand scheme of human evolutionary history, it's about time. The world over and deep into the human past, people have occupied caves and shelters, and for good reason. They provide protection from the elements so residents do not have to contend with rain or snow or high winds; in summer they are cool in daytime and warm at night (rock walls absorb heat, then radiate it back); they can be good settings to stash food from the elements, rodents, and vermin; and if one has reason to worry about such things, they can provide a measure of protection from predators. With all those advantages, it is reasonable to ask why, with the obvious exceptions of the Meadowcroft and Paisley caves, no cave or rockshelter in eastern or western North America was inhabited earlier than 10,500 years ago? Why are the oldest and most common artifacts at the bottom of these natural shelters so consistently Late Paleoindian in age?

One school of thought (Chapter 4) attributes the absence of earlier remains to the newcomers' unfamiliarity with the landscape. They did not know where the caves were, and were moving so rapidly they did not linger in an area long enough to find out.[59] Leaving aside the fact these fast-moving and wide-ranging groups found many isolated and obscure stone sources, it is certainly true that where the Clovis presence was thin and shelters are rare, one would not find the other. But what of eastern North America, where Clovis was abundant, and thanks to the porous limestone that rules much of the bedrock geology, the region is pockmarked with caves and shelters? Clovis people may have had reasons for not moving indoors, even when they knew where the indoors were. What those reasons might have been, we cannot say.

Or perhaps there's another explanation. Caves and rockshelters are geologically active features: they weather and erode, fill with sediment, collapse, and retreat. That activity is driven by water and sharply fluctuating temperature, especially freezing and thawing. In the cooler and wetter Pleistocene, there was plenty of that; less so in warmer and drier Holocene climates.

The consequences for archaeology? Michael Collins suggests only a fraction of caves open for occupation by Clovis people are still around today, and those that have not entirely collapsed or filled have nonetheless experienced repeated episodes of erosion and roof fall that have deeply buried their Pleistocene surfaces. Recall that at Meadowcroft Rockshelter, the oldest artifacts were found only after looking under large blocks of sandstone that had crashed from the ceiling (Chapter 4). In other instances, shelters were unopened prior to Late Paleoindian times. Sarah Sherwood and colleagues showed that the Tennessee River near Dust Cave flowed 6 meters higher during the Pleistocene than at present, and its flood sediments choked the cave. It was only after the Tennessee River fell that springs flowing out of Dust Cave flushed it of sediment, and only after the springs dried (after 10,590 BP) that the shelter was open for human occupation (Figure 60).[60]

Archaeological field techniques may be partly to blame here as well, since excavations often take place where the shelter is today, and not where it was in Pleistocene times, which can be somewhere in front of the present entryway. Or, the excavations were unable to penetrate below massive fallen roof blocks. In any case, the result is that the "basement" culture found in the caves and rockshelters is almost always Late Paleoindian in age, and in eastern North America that means Dalton.

There are, by John Walthall's count, at least forty-five rockshelters in eastern North America—from the Ozarks to the southern Appalachians—that have yielded Dalton remains. Most of these are relatively small (say, less than 100 m² of floor area), but a dozen or so are quite large. The largest, Stanfield-Worley (Alabama), is 55 meters wide

FIGURE 60.
Excavations in the Paleoindian levels in Dust Cave, Alabama. (Photograph courtesy of Boyce Driskell.)

and 17 meters deep, with a 9-meter-high ceiling, and with more than 800 m² of floor area suitable for occupation—not bad for a Paleoindian starter home.

These shelters were used repeatedly, for they were fixed and predictable spots on the landscape. The smaller ones served as brief overnight bivouacs for hunting parties, the larger ones accommodated bands for weeks at a time, and both appear to have been preferred residences in the cold and rainy months of fall and winter when it was better to be indoors. The inhabitants slept inside and stayed warm and dry behind slow-burning hearths, but went out front for most other activities, presumably so as not to carpet their floors in sharp-edged debitage and other debris.[61]

The larger shelters have yielded a rich collection of food remains (as earlier noted) and a wider range of artifacts, including scrapers, gravers, flake knives, blades, and wedges used in the activities with which people passed the time. As is true elsewhere, artifact production was high on that list. At Rodgers Shelter (Missouri), the inhabitants stockpiled trumpeter swan bones and stone cobbles from the nearby Pomme de Terre River, in order to fashion bone and stone tools. Found as well were hammer stones and anvil flakers for making the tools, and the worn and broken projectile points that had been brought back from the hunt, unwrapped from their hafts, and discarded.

But it wasn't just tool manufacture and maintenance taking place here: a large number of end scrapers, side scrapers, and awls hint at extensive hide working. Similar tools, along with bone needles, have been recovered at other Dalton shelters, such as the Graham and Arnold caves in Missouri, revealing that clothing manufacture or repair was a regular occurrence. It was colorful clothing, too, if the red ocher that was ground in pitted anvil stones at these sites had been applied as a dye.

PALEOINDIAN OR PALEOARCHAIC?

On the other side of the continent, caves and rockshelters in the Great Basin, although rare, are equally rich sources of information on Late Paleoindian occupations, which in this region are marked by Great Basin Stemmed points. These, as earlier noted, appear in terminal Pleistocene times, but last well into the Early Holocene.

Over that time, the Great Basin experienced far-reaching climatic and environmental changes (Chapter 2). After the jet stream retreated to the north in the wake of the retreating ice sheet, rainfall was no longer funneled into this region, and by late Pleistocene times, the once-deep pluvial lakes had been replaced by shallow ponds and marshes, and occasional springs and streams. The vegetation was changing as well: alpine trees had begun their ascent to higher elevations and cooler temperatures, while the lowlands—first in the south then toward the north—increasingly took on a more scrub-steppe cast of grasses and bushy plants, including sagebrush, saltbrush, creosote, rabbitbrush, and shadscale.

As Charlotte Beck and George Jones have observed,[62] fluted point groups in the Great Basin spent most of their time in the lowland marshes and wet meadows, and on or near the terraces of now-extinct Pleistocene pluvial lakes (Chapter 8). Not so the makers of the Stemmed points, who ranged widely into upland settings and, presumably, exploited a broader range of resources and environments. By virtue of the extraordinary preservation of organic remains in caves and rockshelters in this very dry environment, archaeologists have gotten a rather personal glimpse into what individuals were eating, for these sites yield what they were excreting. In the human coprolites in the late Pleistocene levels in Danger Cave, Utah (Figure 61), for example, were the traces of well-rounded meals that included pronghorn (identified by bits of hair and crushed bone), pickleweed seeds, prickly pear fruits, bulrush, pine nuts, and many insects. This was hardly an atypical case.

Of course, these are just scatological snapshots of individual meals, and not representative of the full range of foods exploited by people who, from the animal bones found in their sites, were also hunting elk, deer, bighorn sheep, the occasional bison, a variety of smaller mammals (especially rabbits), waterfowl, and fish, as well as eating plants. It was a thoroughly mixed diet, the ingredients depending on what was locally available. That diet varied little in the centuries before or the millennia after 10,000 BP (it changed appreciably

FIGURE 61.

Danger Cave, Utah, taken from the floor of the basin, which would have been a salt marsh at the time of the earliest occupation. (Photograph by George T. Jones, courtesy of Charlotte Beck.)

FIGURE 62.

A serrated bone harpoon or leister made of mammoth rib dated to 10,340 BP, from the site of Pyramid Lake, Nevada. (Photograph courtesy of Eugene Hattori and the Nevada State Museum, Carson City, Nevada Department of Cultural Affairs.)

only later, when climatic conditions deteriorated in Middle Holocene times between 7500 and 5000 BP, and the role of seeds—drought and starvation food—jumped dramatically). So seamless is this record of generalized subsistence across the Pleistocene-Holocene boundary that Great Basin archaeologists have given up any pretense that Paleoindians in this setting were big-game hunters, or somehow distinct in adaptation from later Archaic groups. They refer to all peoples from this span of time as *Paleoarchaic*.

Mammoth appear, if briefly, on the Paleoarchaic stage. A point fashioned of mammoth ivory and a serrated mammoth bone harpoon, or leister (Figure 62), were found on the shores of Pyramid Lake, Nevada, a lake that historically was home to a giant species of cutthroat trout (which weighed upward of 28 kilograms). Both artifacts date to a statistically identical 10,350 BP, opening the possibility they came from the same animal.[63] Since the elaborate carving of the leister is the sort of thing best done when bone—mammoth rib, in this case—is relatively fresh or well preserved (bone becomes less malleable as it dries and cracks), these were likely not pieces scavenged from a fossil skeleton, but instead obtained from a recently dead mammoth. There's no small irony in the fact that mammoth survived in the Great Basin until well after Clovis (and overkill) times, and that in the Paleoarchaic at Pyramid Lake, mammoth fishing tackle was used to go after a mammoth-sized fish.

Great Basin fluted points were made with the usual preference for high-quality chert or obsidian. Yet, in fashioning their Stemmed points, knappers often used crystalline volcanic rocks such as basalt and andesite. Projectile points fashioned of such stone can be deadly, though are too coarse grained to allow the razor-sharp edges possible with chert or obsidian. More significant, perhaps, many Great Basin Stemmed points have rounded edges and tips, which Beck and Jones attribute to their use as projectile points as well as tools for cutting and scraping.

Other elements of the Paleoarchaic toolkit included knives, gravers, scrapers, spokeshaves, a few milling and grinding stones (signaling the use of plant foods), and crescents, a tool largely restricted to the Great Basin. As the name implies, these are shaped like a crescent moon, flaked on both faces, often heavily ground along both convex and concave sides, and crushed and abraded on their ends. When first found, they were interpreted as surgical instruments, prompting William Clewlow to wryly observe that if that were so, there must have been "a vast pre-Columbian medical center on the shores of the Black Rock Desert." More likely, he supposed, crescents were used to cut or carve animal hide and flesh, or slice plants such as tule reed, which was used to make fibers for fashioning burden baskets, fish nets, sandals, and the other rare, perishable items found in many of these dry caves and shelters.[64]

It is by virtue of that extraordinary preservation we also glimpse some of the rarest of the perishable artifacts, those related to burial practices. A partially mummified man found in Spirit Cave (Nevada), was laid to rest some 9,500 years ago in a sagebrush-lined pit, wearing a rabbit-skin robe and leather moccasins, and had been carefully wrapped and sewn into a death shroud of woven fiber mats.

THINKING ABOUT LIFE AND DEATH

In fact, in the Late Paleoindian period across North America, we are able to peer further into ritual and ceremony than was possible in earlier times. This is not to say such matters were previously unimportant, only that our knowledge of them is limited to a very few sites, exemplified by the ocher-covered bones of the child buried at Anzick, or the Buhl skeleton with its small collection of grave goods (including stone and bone tools, and a badger baculum).

Part of ritual and ceremony in Late Paleoindian times was related to the food quest, particularly the hunt. At the Folsom-age Lake Theo site (Texas), for example, a small "shrine" (for want of a better term), consisting of several bison jaws, two tibia (the lower and larger rear leg bone), a femur (upper leg), and the spine of a thoracic vertebra (the spine serves as the cantilever for the bison hump), were found set standing vertically in a carefully dug, small, round hole in the midst of a bison kill. It was likely created following the successful hunt.[65]

Some 10,000 years ago at the Jones-Miller site, a wooden post (identified archaeologically by a circular stain in the ground) was set in the earth, and placed alongside it was a miniature Hell Gap point, a fragment of a bone flute or whistle, the jaw of a wolf, and leg parts of what might have been a dog. Soon thereafter, a great bison kill was made over this spot. Coincidence? In historic times on the plains, when a hunt was planned to trap bison in a surround of brush, snow, or some other material, bets were hedged by a shaman planting a "medicine post" in the center of the surround, where offerings were made for a successful kill. If the Jones-Miller feature had the same intent, this was a ritual with very deep roots, and it worked like a charm: some 300 bison were killed at Jones-Miller, possibly on two separate occasions.[66]

At other sites, ritual was less obviously related to empty stomachs. At Powars II in southeastern Wyoming (just 16 kilometers from the Hell Gap site, which was occupied throughout Paleoindian times), a rich outcrop of red ocher was mined by Paleoindians (and only Paleoindians), starting in Clovis times.[67] Recovered at Powars II were many of the usual Paleoindian tools, but they were not necessarily there for the usual reasons. Half a dozen fragments of bison forelimbs and ribs had been pressed into service as digging implements to wrest blocks of red ocher out of the ground. Several dozen flakes had been employed as scrapers to shave red ocher blocks, possibly to put it into a powdered form to be mixed with water, sap, or some other liquid to make a paint (though ocher is an effective coloring agent even as powder). Also found at the site: a few seashell beads, which had been carried (or traded?) a great distance, since the Pacific Ocean is at least 1,600 kilometers away (there used to be a saltwater sea in this area, but that was at least 570 million years earlier).

As intriguing are the forty-two projectile points at Powars II. This was no kill site, yet all the points had been heavily used, worn, and repeatedly reworked; more than half were broken. Virtually all were made of locally available stone. Evidently they'd traveled far and wide, been used often, and were even rendered useless; but instead of being abandoned where they had broken, they were kept and ultimately returned to the red ocher mine near where the stone was first quarried. But why? Michael Stafford and colleagues, puzzling over the matter, drew the reasonable (if hard-to-verify) conclusion the points and beads were left as offerings. Ocher is a potent symbol of life and death, and offerings made at its source a correspondingly powerful ritual act. But why *used* points? Stafford and colleagues wonder if these broken bits had once been the hunters' most successful weapons. Although their hunting value had long since been spent, perhaps some of their power and accuracy could be conveyed to newly-minted ones by the act of their ritual offering.

From sustaining life to accompanying those after death: at the Sloan site, on the summit of an old sand dune in northeast Arkansas, excavations by Dan Morse recovered some 200 small, badly eroded pieces of bone, spread over an area of about 144 m². On careful analysis, the bone appeared to be human, from both adults and children. With the bones were nearly 450 Dalton artifacts, arranged in twenty-nine discrete clusters, perhaps having been placed into pits dug into the dune sand. In some instances, the artifacts were so close together it appears they were buried in leather bags or some other containers.[68] The pattern of artifacts and bone suggested individual graves, but demonstrating that (given the poor preservation) is no easy task, as Morse admits. But if one assumes they were, then about thirty people had been interred here, representing at least a couple of generations.

Fully one-third of the artifacts were Dalton points, the remainder bifaces, adzes, scrapers, abraders, and a variety of other cutting and scraping tools. And then there were a dozen or so Sloan points made of Burlington chert, quarried about 300 kilometers away. Most of those were newly minted or refurbished. In so far as these represent grave goods that accompanied in death what individuals might have used in life, they reveal the toolkit carried by a well-equipped Dalton person. The adzes and bifaces with traces of red ocher on them symbolize something more.

Among societies that are sharply differentiated by social rank, one's standing in life is often carried into death. High-status individuals are buried with expensive grave goods marking their standing (think King Tut). Yet, the grave goods at Sloan—the finely crafted points and tools made of high-quality and hard-to-obtain stone—were evenly spread across individuals of different ages or sexes. Morse concludes there was only minimal social differentiation among the Sloan populace.

Sloan is North America's earliest known cemetery. Cemeteries, of course, are sacred spaces, and this one is no exception: it is located well away from other Dalton sites. There were no distractions or profane acts to take away from any solemn rituals enacted here. Cemeteries were also places to which people repeatedly returned. Of course, that was true of other spots on the landscape, such as stone quarries, for example. But a cemetery was a place of meaningful return as no quarry could ever be. And so it is that after a millennium or more of wide-ranging and untethered mobility, the first peoples have at last become firmly attached to place. They could and, in the end, did come home.

THE END OF THE BEGINNING . . .

Geologists set the date for the end of the Pleistocene at 10,000 BP. Doing so was arbitrary (Chapter 2), but as arbitrary numbers go, it's not bad, coinciding as it does with the end of the Younger Dryas, as discrete a climatic boundary as one could find. Once the Younger Dryas was finished, there was no turning back toward the Ice Age, though the Pleistocene was still in the process of wrapping up. In the far north 10,000 years ago, the once-vast ice sheets continued shrinking; sea levels worldwide were still rising. In the Upper Midwest, the Great Lakes were just reaching their present configuration, and throughout the continent, plants and animals continued shuffling toward their modern, postglacial positions on the landscape. North America as we recognize it would finally appear around 5,000 years ago.

Likewise, archaeologists have traditionally divided the Paleoindian from the subsequent Archaic period around 10,000 BP as well, using as a benchmark the separation of highly mobile Paleoindian "big-game hunters" from the "broad-based foragers" of the Archaic. As should now be apparent, that's an equally arbitrary division. Paleoindians were not strictly big-game hunters, and it was they who first established some of the adaptive themes of the Archaic: use of a wider variety of food resources, smaller territories and mobility, and the advent of new technologies.

In fact, the most significant differences between Paleoindian and Archaic adaptations emerge only much later in the Holocene, when new communities began living in semi-permanent villages, manufacturing pottery, burying their dead in artificially constructed earthen mounds, and domesticating a wide range of plants. The timing of these developments varies by region, and in some the vestiges of ancestral adaptations linger for a very long time.

Although the narrative of North America's first peoples stops here with the end of the Pleistocene and the dawn of the Holocene (doing more goes far beyond the scope of this volume), prehistory in North America does not stop. The descendants of North American Paleoindians weathered severe drought on the Great Plains; bartered Wyoming obsidian for Great Lakes copper and Gulf Coast shells in Midwest commodities exchanges; teased wild plants until they bore seeds and fruits fit to eat, drink, or smoke; dug thousands of kilometers of canals across the arid Southwest and made desert gardens bloom; settled down from their ancestor's mobile ways, and built towns of stone, earth, and wood; and much, much more. Over those 10,000 years, populations boomed, people moved, put down roots, then moved again; environments and adaptations changed, sometimes dramatically; new languages, genetic lineages, and cultures emerged. All independent of any contact with Old World peoples.

Around AD 1000, the long separation of Old World and New came to an end, though only in one far corner of North America. Thule Eskimo and Norse Vikings, moving east and west (respectively), finally closed a circle that had been open since the Pleistocene.

The first signs of their encounters appear in the eleventh century in Thule sites around Baffin Bay, which yield iron, copper, and bronze tools, and even (from a Thule site on Ellesmere Island) chain mail armor, woolen cloth, smelted iron rivets and tools, a carpenter's plane, and a bronze bowl and balance scale fragment. Whether these items moved via peaceful trade (metal for ivory), mutual plunder, or both, we cannot say. But then an exchange of goods was not vital to either party. Norse and Native Americans were evenly matched in technology. The artifacts and scrap metal the Thule got from the Norse merely supplemented their own supplies of native copper and meteoritic iron.

Indeed, it's strongly suspected the contacts between the groups were brief, misunderstandings common, and violence occasionally the end result.[69] A stone arrowhead, like those made by Newfoundland's natives in the centuries around AD 1000, was found having fallen out of a grave in a Norse cemetery in Greenland, possibly carried there stuck in someone's flesh or bone. At the very least, Norse and Native Americans were unsure what to make of one another. The Norse puzzled over Vinland's "Skraelings" in their sagas,[70] and a Thule on southern Baffin Island tried to capture the image of a Norse figure in wood.

Perhaps their only chance for extended interaction was on the northern tip of Newfoundland, where at the site of L'Anse aux Meadows, the Norse—under the leadership of Leif Eriksson according to the sagas—built three timber- and sod-walled long houses (up to 25 meters in length) and several smaller outbuildings overlooking a shallow bay, sometime between AD 980 and 1020. Excavations by Helge Ingstad and others also uncovered remains of a smithy for smelting and forging, a carpentry shop, and a boat shed. Norse ringed pins, loom weights, iron tools and debris, wooden artifacts, hearths, and cooking pits were found amidst the structures.

The very same spot had been used before and after by Native Americans, although there is no evidence they and the Norse were there simultaneously. But then the Norse were not on the ground for very long, as both the sagas and the archaeology attest. The Norse did not plant the farms they had in Greenland, but instead used the site as a winter camp and summer base of exploration.[71] It appears to have never housed more than 100 people, was not kept in good repair, and burned soon after it was abandoned.[72]

It's the only settlement the Norse apparently tried to establish in North America. Their exploration, which possibly reached as far south as Quebec City (judging by butternuts found at L'Anse aux Meadows), never became colonization. The sagas hint they put down no deeper roots in North America because of the Natives, for when Leif

Eriksson, his shipmates, and kin landed, and the Old World and New came face to face, the Old World contingent got the worst of it. The Norse were hamstrung by having smaller forces (their ships could not carry large numbers), tenuous lines of support, and no advantage in weaponry—and they faced Natives possessing superior homeland knowledge and survival skills.

Losses at the hands of the "Skraelings" were a fatal blow to Norse intentions: their plans for western expansion died with the Vinland settlement. Later visits to the continent—likely to satisfy the demand for timber back in Greenland—were fleeting and sporadic. As Leif's brother Thorvald, fatally shot in a skirmish with "Skraelings" on the Labrador coast, gasped in the end, "We've found a land of fine resources, though we'll hardly enjoy much of them."[73]

This initial contact loomed large in the lives and deaths of the players who were immediately involved. But for the vast majority of American Indians living far from the front, those brief encounters in the distant North Atlantic were completely unknown and utterly inconsequential. All that would change with the next round of contact, beginning that October night in 1492 when Old World sailors once again caught sight of the New. For this time, the Europeans brought with them something far deadlier than mere weapons.

10

WHEN PAST AND PRESENT COLLIDE

On March 18, 1835, Charles Darwin disembarked the H.M.S. *Beagle* at Valparaiso, Chile, and set off with "two Peons & 10 mules" on a three-week round trip over the Andes to Mendoza, Argentina. Mules and weather cooperated, the mountains enchanted. He wrote his sister, Susan, that it was all so new and spectacular he "could hardly sleep at nights for thinking over [his] day's work." Best of all, what he saw of Andean geology promised "a very important fact in the theory of the formation of the world" (and, incidentally, inspired his explanation of the Parallel Roads of Glen Roy). Little wonder he ranked this Andean excursion his most successful since leaving England three and a half years earlier.[1]

But in the end, a night spent in Luxan, a small village south of that "stupid, forlorn" town of Mendoza, had far more lasting consequences. There, on the night of March 26, 1835, Darwin was attacked by the giant, blood-sucking Vinchuca bug, *Triatoma infestans,* the "great black bug of the Pampas." It was, he said, "most disgusting to feel soft wingless insects, about an inch long, crawling over one's body. Before sucking they are quite thin, but afterwards they become round and bloated with blood, and in this state are easily crushed." Still, no pain was caused by the wound, and though Darwin and his fellow shipmates were mildly amused by the insect (one of them allowed it a meal), he thought little of the attack then, and not at all later. He left Mendoza the next morning and re-crossed the Andes.[2] Six months later, Darwin was on the Galapagos Islands, where he glimpsed, but could not yet appreciate, the meaning of its wonderfully varied life forms.

No matter. The seeds of the theory of evolution by natural selection were planted, and when they bore fruit a few years later, Darwin could look back at the Galapagos Islands and see what a marvelous laboratory of evolutionary divergence these were. But it was a laboratory to which he could never return. Crippling headaches, dizziness, nausea, stomach pains, and fatigue began just a few years after his return to England, and dogged him the rest of his life. His night in Luxan had caught up with him.

The bug that bit Darwin is the primary vector for *Trypanasoma cruzi*, a microorganism causing Chagas' disease, which attacks the heart, stomach, and intestinal system. Those bitten rarely escape Chagas. Even today, despite concerted efforts to eradicate it (using insecticides, in heavy doses, to kill off Vinchuca populations), Chagas causes more deaths in the Americas than any other parasitic disease, and more than 18 million people are affected by it.[3] The Vinchuca does its work at night, and those bitten often hardly notice the bite or its effects—at least at first. Chagas is a time bomb, Sir Peter Medawar explains: years afterward, the victim suddenly drops dead or, like Darwin, becomes chronically ill with diagnosis-defying symptoms. It was worse for Darwin, since Chagas was then unknown, and his physicians suspected he suffered from hypochondria. And, as Medawar adds, "ill people suspected of hypochondria or malingering have to pretend to be iller than they really are and may then get taken in by their own deception."[4]

Darwin fell prey to one of America's pathogens, a victim of the exchange when an Old World host met a New World parasite. But he was an exception. Old World populations were not overrun by Chagas, nor any other New World pathogen. Through it all, using the excuse his illness provided to free him from Victorian society's distractions, Darwin produced the *Origin of Species* and more than a dozen other books that introduced the concept of evolution, and in so doing forever changed our view of life on earth.

Not so fortunate were the untold millions of anonymous American Indians who died from imported Old World diseases in the decades and centuries after the last discovery of America in 1492.[5] The successors of Columbus brought to America an arsenal of lethal pathogens, including smallpox *(Variola major)*—the most horrific killer by far[6]—measles, influenza, bubonic plague, trachoma, diphtheria, malaria, amoebic dysentery, typhoid fever, cholera, chicken pox, yellow fever, scarlet fever, and whooping cough. In many instances there were repeated epidemics of each. Since these diseases were not endemic to Native American populations, each new breakout proved just as deadly to those born in between episodes of epidemics.

But why, asked Antoine Le Page du Pratz, who lived among the lower Mississippi Valley's Natchez in the 1720s, should "distempers that are not very fatal in other parts of the world, make dreadful ravages among them?" American Indians were dying from what were childhood ailments in other parts of the world. Even at its worst in Western Europe during the 1700s, smallpox accounted for just 10%–15% of all deaths, and 70% of those, Alfred Crosby reports, were children under the age of two.[7]

Why were Native Americans of all ages vulnerable to what were in Europe mere childhood diseases? And why had they no epidemic diseases of their own that might have devastated Europeans, and perhaps even checked their advance into America?

The answers to these questions have little to do with events after 1492. They have a great deal to do with events that occurred before 1492. They have everything to do with the peopling of the Americas.

THE BEHEADING OF PREHISTORY

Precisely how many Native Americans died in the decades and centuries after 1492 is impossible to know, since early on, no one was keeping census records. But Thomas Jefferson suspected the worst. After comparing a tally of Virginia's forty tribes from 1607 when Captain John Smith made contact, to a count made in 1669 by the Virginia assembly, there was no mistaking the "melancholy" message. In only sixty-two years, the tribes had been reduced to one-third their former number.[8]

Jefferson believed diseases were not entirely to blame: "spirituous liquors," war, and land grabbing had played their insidious roles. Still, he knew the fearful toll of disease and, already aware of Edward Jenner's newly discovered vaccine for smallpox, instructed Meriwether Lewis and William Clark to carry on their expedition west "some matter of the kinepox [cowpox fluid] [and] inform those of them with whom you may be, of its efficacy as a preservative from the smallpox; & instruct & encourage them in the use of it."[9] Well intentioned to be sure, but inevitably Lewis and Clark discovered smallpox had arrived long before they did. Even along the Columbia River in the far Pacific Northwest in April 1806, Clark came upon a large, ruined village and the few, pox-scarred survivors of a terrible smallpox epidemic that had roared through three decades earlier. Disease had reached this remote corner of North America long before European liquor, war, or land grabs—and well in advance of any Europeans.

That was hardly an unusual occurrence. Nearly three centuries earlier, in the spring of 1540, Hernando de Soto entered the chiefdom of Cofitachequi (in present-day South Carolina). Although his was the first significant European presence in the region, he saw several "large uninhabited towns, choked with vegetation, which looked as though no people had inhabited them for some time." He learned that "two years ago there had been a plague in that land and they had moved to other towns."[10]

Disease, the enemy no one saw, traveled much farther and frighteningly faster than the Europeans who brought it to America. The reason was deadly simple. Take smallpox: it has no known nonhuman host, but it can survive for extended periods outside its human host in a dried state, including on infected blankets or clothing, where it can remain viable for more than a year. It is also readily spread by breath. With an incubation period of 10–14 days, that's plenty of time for an apparently healthy person to flee a stricken area when the first gory symptoms break out, and unwittingly spread the virus to others in neighboring villages.[11]

So it was that the disease spread among American Indians with astonishing speed and deadly efficiency. Just how far and how fast it could burn through Native populations can be seen in the smallpox epidemic of 1775–82 (which took a small share of the colonial population as well). As Elizabeth Fenn has detailed, that bout began in Mexico City, and by the time it exhausted itself of victims seven years later, it had killed American Indians trading at remote Hudson's Bay Company outposts in far northern Canada (probably reaching there via long-established trade networks).[12]

By the time Europeans arrived on a scene, native populations were often already devastated. Under such circumstances, counts of native populations can be highly suspect, and we must rely more on archaeological evidence—of site and regional abandonment, for example—than on any historic records. Even where Europeans showed up before disease hit (Hernán Cortés in the Valley of Mexico in 1519, for example), reliable numbers are elusive. The earliest explorers, conquistadores, and missionaries could be notoriously bad census takers, often overestimating or underestimating native populations, depending on whether they were in the business of converting native souls (report high numbers to testify to one's good work) or capturing native lands (report low numbers to justify a land grab).[13]

We may never know how many Native Americans were here on the morning of October 12, 1492. Indeed, no one knows how many people lived just about anywhere in those days (census records are a phenomenon of more recent centuries). One school, championed by Henry Dobyns, estimates Native America numbers using a "depopulation ratio," the difference between the "known or confidently estimated preconquest population" in an area against the estimated nadir population for that same area. After examining multiple cases, he concluded that a "sound, if conservative" estimate would put the ratio at about 20:1. It would be even higher if the preconquest population had already suffered a wave of disease. From that, he concluded, the New World pre–European contact population numbered 90 to 112 million, of which 18 million people occupied North America north of Mexico.[14]

These astonishing figures are orders of magnitude higher than all estimates before or since, and have been hotly disputed. Too many unsupportable assumptions, critics say, are behind the numbers, not least Dobyns's belief that there were multiple epidemics, each with extremely high mortality rates, and that most populations had already declined before any Europeans arrived to make a count. It is better, the critics say, to tread cautiously and use more direct evidence.

Douglas Ubelaker, drawing on tribe-by-tribe census figures and available archaeological evidence, puts Native North American populations at the time of European contact at a far more conservative 2.4 million.[15] Carefully derived as it is, this estimate is also suspect since it is based on counts that came in, Ubelaker readily admits, too low and too late. Recall that by the time René-Robert de la Salle arrived in southeastern North America in the late 1600s, the region was virtually depopulated. He saw only remnants of once-teeming communities that had greeted de Soto a century earlier (Chapter 6).

The actual number surely lies somewhere between these two extremes, but no one really knows. What we do know is that American Indian populations bottomed out in the late nineteenth century, the timing varying by tribe and area. At the nadir, there were an estimated 514,510 Native North Americans (their numbers have since rebounded).[16] Here's the root of the claim that from European colonization until the twentieth century, more Native Americans died than were born.

Overstatement, maybe. But no matter the actual number, imported pathogens caused Native deaths in fearful numbers. By Dobyns's estimate, it was a decline of 97%; by Ubelaker's lower figures, at least 78%. Those are staggering losses either way. Even in the 1960s and 1970s, more than four centuries after initial contact, mortality rates for remote Amazonian tribes exposed to measles for the first time were upwards of 75%.[17]

The post–European contact population history of Pecos Pueblo is probably typical in microcosm of what was happening continent-wide (although atypical in that we have record of it). By 1927 Pecos was a long-uninhabited archaeological site (Figure 63) where A. V. Kidder was excavating when he was called by Frank Roberts to come see the first Folsom point found in situ (Chapter 3), and from which Earnest Hooton thought he spotted so many varieties of human crania (Chapter 5). Yet, 300 years earlier, Pecos Pueblo had been a thriving community, with a population of more than 2,000 people (as estimated by a Franciscan friar living among them). The numbers might have been higher still if the Pueblo hadn't already experienced one (or perhaps more) epidemics. In the intervening centuries,

FIGURE 63.

Pecos Pueblo. The rooms in the foreground were part of the South Pueblo. The large building in the background is the ruin of the church built by Spanish missionaries in the early 1700s (an earlier church was razed during the Pueblo Revolt of 1680–92). (Photograph by David J. Meltzer.)

Pecos was battered by repeated smallpox epidemics—possibly a dozen altogether—and weakened by raids from mounted Comanche warriors; the loss of labor for planting and harvest, and the famine that followed; out-migration and social disruption (especially as the older leaders of the community, its rituals and ceremonies, who would have been more susceptible to disease, died off); and the other grim collateral consequences of a village reeling from recurrent and cumulative waves of death. By 1838 only 18 survivors remained. That year, they packed up their belongings and abandoned the Pueblo.[18]

Depopulation on a scale of 78%—let alone of 97%—inevitably had far-reaching and devastating effects on people (of which more below), and possibly even on nature itself. As millions died, farm fields returned to forests, bringing about an artificially induced ecological succession (what European colonists in eastern North America described as pristine woodlands were often second-growth forests). Animals whose numbers had previously been reduced by hunting pressure returned; the chroniclers of de Soto's expedition make no mention of bison, but when de la Salle arrived the next century, many villages and farms had been supplanted by large bison herds grazing areas of open parkland.[19]

Human population losses might even have been large enough to alter global climate—a not-uncontroversial suggestion made by William Ruddiman, who plotted changes in atmospheric CO_2 concentrations from ice cores against historical epidemics. The sharpest decline in atmospheric CO_2 over the last 2,000 years occurred between the 1500s and the 1750s. That suggested to him that the decimation of American Indian populations during these centuries, when abandoned farm fields were overtaken by carbon dioxide–consuming forests, temporarily checked rising atmospheric CO_2 levels. He even supposes this was the trigger for the Little Ice Age (Chapter 2).[20] That CO_2 levels rose again in the late eighteenth century is less a testimony to the rebound of Native American populations than it is evidence of the influx of Old World colonists clearing then-vacated New World forest lands.

THE CHILDHOOD OF DISEASE

European colonists thrived in the New World, while around them, American Indians died. Blame that disparity on evolution and the contingencies of history.[21] Many of humanity's deadliest diseases come from species with whom we come in contact, for viruses or pathogens that might be relatively benign in one organism can turn deadly in another (e.g., Ebola or HIV/AIDS). When a pathogen first appears in a human population, neither young nor old are spared a bout with the illness, and the result is either death or lifetime immunity. Early on in many cases, the result is fatal. But as individuals most vulnerable to the pathogen are eliminated from the gene pool through natural selection, it is the genes of the hardier survivors, who mature and reproduce, which over time come to dominate the gene pool. Humans co-evolve with the disease. As a result, diseases once deadly to both adults and children eventually become just childhood ills. The longer a community has lived with a disease, the less destructive its epidemics.

Old World communities did not start out riddled by disease. For the first few million years of human prehistory, our collective ancestors were few, subsisted by hunting and gathering, and were mobile, much like the first Americans. That had to change in order to spark the evolution of epidemic diseases, for these require two essential ingredients: a pathogen source—usually animals living close by—and a large supply of victims to sustain a chain of infection (measles, for example, requires a minimum of 200,000 to 300,000 hosts to ensure its survival). In the Old World, these vital ingredients only appear with the advent of agriculture, pastoralism, and settled village life.

When humans began to settle in place, clearing land for farming, piling up trash, and fouling their water, they created breeding grounds for rats, lice, worms, mosquitoes, and all manner of disease-causing or carrying organisms. And when they began living with their newly domesticated animals, drinking the same water, breathing the same air, and using animal waste to fertilize their crop fields, they unwittingly shared a volatile mix of pathogens from which sprang potent new diseases. Domesticated animals are a major source for many of humankind's most deadly diseases, like smallpox, which is similar genetically to other orthopoxvirus forms hosted by camels, cows, and rodents;[22] measles appears to be derived from bovine rinderpest or canine distemper; influenza seemingly originated in avian influenza or hog diseases. In all, humans inherited dozens of diseases from animals, domesticated and wild.

Lots of people began living in close contact with one another around 9,000–10,000 years ago, earliest in Mesopotamia and then later throughout the Old World from Africa to Asia. Within a few thousand years, they had tamed wild cattle, sheep, pig, and goat (among other animals), and the combination of farming and herding fueled a steep rise in human populations. More mouths demanded more farmland and pasture. Sanitary conditions declined: it was life cheek by jowl with other people and other animals. In involuntary experiments taking place in countless early villages and towns, new diseases were born and spread. Their arrival went unchallenged.

Epidemic diseases make their archaeological debut some 3,000 years ago, in smallpox pustules visible on Egyptian mummies, though this was apparently a milder form of the virus. The earliest historic record of pandemic smallpox comes from fourth-century AD China (suggesting, as has since been shown in a study of the smallpox genome, that two separate strains of this virus made the jump to humans).[23] Diseases, of course, figure as biblical plagues. A wave of typhoid fever struck Athens during the Peloponnesian War, while the Antonine (smallpox) and Justinian (plague) pandemics, nearly 400 years apart, crippled the Roman Empire. These infectious pathogens reached the distant corners of Europe by Medieval times.

By 1492, having been exposed for thousands of years to dozens of pathogens, Old World populations had effectively "adapted." Natural selection had done its work well. Old World populations still suffered occasionally spectacular epidemics, like the Black Death of the fourteenth century (1347–51). Yet, their populations emerged, sometimes deeply scarred but largely intact. Mortality rates during the Black Death were a mere 30%.

Not so the Americas, where neither old nor young were spared. American Indians were so utterly defenseless against imported European diseases that it was taken by some as a sign of higher authority. "The natives," John Winthrop wrote in 1634, "are neere all dead of small Poxe, so as the Lord hathe cleared our title to what we possess."[24] The Natives themselves, discouraged by their losses and the seemingly stunning good health of the invaders, were often convinced their own deities were no match for the European God.

Others blamed native health care: "When a nation is attacked by the smallpox," du Pratz wrote in 1758, "it quickly makes great havoc; for as a whole family is crowded into a small hut, which has no communications with the external air, but by a door about two feet wide and four feet high, the distemper, if it seizes one, is quickly communicated to all. The aged die in consequence of their advanced years and the bad quality of their food; and the young, if they are not strictly watched, destroy themselves, from an abhorrence of the blotches in their skin. If they can but escape from their hut, they run out and bathe themselves in the river, which is certain death in that distemper."[25] Du Pratz could not understand why native physicians, otherwise quite skilled, could not cope with these common ailments.

They were quite skilled, but their skills were honed against long-familiar ills and were no match for alien pathogens. They were understandably baffled by disease transmission. When smallpox broke out among the Piegen Blackfoot, one survivor reported, they had no idea "that one Man could give it to another, any more than a wounded Man could give his wound to another."[26] More to the point, in those centuries, no one on either side of the ocean understood disease etiology, let alone could have stemmed the viral and pathogenic assault on New World peoples.

The obvious answer, then, to why Native Americans suffered so terribly from introduced infectious disease is because they had no previous exposure and hence had built up no immunity. And when a group with adult immunity spreads the disease to one without, mortality is nearly universal. Crosby likens the process to a forest that has not burned for a very long time: when ignited, burning is swift, and destruction is nearly total.[27]

So we know the immediate cause for the catastrophic population loss among American Indians: a lack of immunity. But that begs the larger question: why did they lack immunity? It's not that they didn't have the ability to develop immunity to diseases. It's that they didn't have immunity to these *specific* diseases. For that, they needed exposure time and proximity. They had neither, which brings us back to the peopling of the Americas.

FROM SIBERIA TO SMALLPOX

No matter the haggling over the precise homeland of the first Americans, archaeological, dental, genetic, linguistic, and skeletal evidence indicates American Indians were descendants of Asian populations, not groups emigrating directly from Europe or the

Middle East. Moreover, their ancestors had left Asia and the Old World thousands of years before the first cow, pig, camel, or sheep was ever penned, and well before village life hatched the first epidemic diseases. No ancestral American suffered through the thousands of years of agony and illness as diseases jumped species and then co-evolved with human hosts.

The port of entry for America's first peoples was the Bering Sea region. They could, and likely did, walk across from Siberia to Alaska when expanding continental ice sheets dropped sea levels worldwide and Beringia surfaced. Crossing its Mammoth Steppe, blanketed by parkland and grazed by mammoth, horse, and bison, was possible anytime between 27,000 and 10,000 years ago (during the last millennium, when the breach was still narrow enough to freeze in winter). The recent genetic evidence of a possible Beringian standstill (Chapter 5) suggests the first peoples may have been relatively isolated in this region for much of that time.

From there they traveled south, either between retreating margins of the Laurentide and Cordilleran ice sheets, or down the Pacific coast. Both entry routes were blocked for much of Late Wisconsin time, and may have been biologically impassable when first opened (the coastal route opening sooner). In any case, it appears from the evidence at Monte Verde that the first Americans were here by at least 12,500 BP and possibly earlier still. Certainly by 11,500 BP, Clovis Paleoindians were widespread, possibly representing a second migratory pulse to the New World, one that may have spread across the continent in less than a thousand years.

Beringia and the routes south served as more than just passageways. These were also, in physical anthropologist T. Dale Stewart's words, "germ filters"[28] that would have effectively stripped away any diseases lurking among the colonizing populations. For as Stewart observed, the harsh Arctic cold would have killed off the vectors (carriers) of many infectious diseases (such as snails, worms, or mosquitoes), along with any of their diseased hosts. If a pathogen somehow infected a small band on this thinly settled landscape during its journey, the group would have been naturally quarantined, and would have disappeared with little notice and no widespread effect. All diseases are parasites: they need hosts to sustain their deadly work. Beringia did not then, as the Arctic does not today, support dense populations that would prolong a disease's run.

Once the first Americans made it here, their rear flank was well protected by the vast distance between the source regions of disease and northeast Siberia. Smallpox only reached this far corner of Eurasia in the seventeenth century. Given more time, infectious epidemic diseases might have crossed the Bering Strait and entered the Americas from the west, or perhaps not. The small and scattered populations that lived in far Northeast Asia and the relatively low level of traffic across the Bering Sea would have made it difficult for Old World diseases to sneak up on New World peoples from behind (the Bering Sea would have been just as effective a barrier preventing the backwash of diseases from the New World to the Old, were there any New World diseases to wash back).

The timing, route(s) of entry, and subsequent geographic isolation of the first Americans helped ensure there were no opportunities for them to adapt and co-evolve with Old World diseases.

But there's more to their lack of immunity than just geography—there's genetics, too. Whether Clovis or pre-Clovis, the first Americans came here as a relatively small founding population, possibly having passed through a population bottleneck, the signature of which is apparent in the small number but widespread distribution of mtDNA and NRY founding lineages among Native Americans, and in the frequency differences of those lineages here relative to Asia (Chapter 5). A founder effect is also apparent in other aspects of their genome, most importantly on chromosome 6, which houses the large cluster of genes that comprise the major histocompatibility complex (MHC).

The MHC is one of our main lines of defense against viruses or other pathogens that have found their way into our cells. Once embedded inside the cell walls, the pathogen is invisible to the body's other defenses (lymphocytes, for example). To scan for those, the MHC encodes human leukocyte antigens (HLAs) to bring to the cell's surface bits of peptide antigens lurking within, enabling them to be scanned by the immune system. It's a constant "news scroll" across a molecular screen of cell health. If the antigens detected are the body's own, they are ignored; if they are foreign (say, fragments of a growing virus or traces of a malignancy), the body's T-cells go on the attack, and destroy the cell and all within it.

The susceptibility of New World populations to epidemic disease, as epidemiologist Francis Black carefully explained, "was not [because they] had inappropriate genes or, individually, deficient immune systems," but because the people of the New World were unusually homogeneous in their MHC antigens (by one estimate, Europeans have thirty-five classes of HLAs, Native Americans fewer than seventeen). That genetic homogeneity, of course, is an echo of their "bottlenecked" Pleistocene population prehistory. And because of that smaller diversity of HLA alleles, there was a far greater chance that a deadly virus introduced into a population would encounter the same allele in successive hosts. Having survived one host, it was essentially pre-adapted to the next, and so was able to readily sustain its deadly progress through the population. Black reports, for example, that there is a 32% chance that a virus passing between two Native South Americans will not encounter a new MHC type, but only a 0.5% chance when it passes through a more MHC polymorphic African population.[29]

Population history matters, and though it helps explain why American Indians lacked immunity to Old World diseases, it fails to account for the flip side of the question: why they had no evolutionary experience of their own with epidemic diseases on this continent, nor infectious pathogens that could be passed to Europeans. Those answers, too, are found deep in prehistory.

When the first Americans arrived, the North American landscape was crowded with now-vanished animals. A common feature of all, aside from their mostly enormous size, was that all were extinct by the end of the Pleistocene. Paleoindians were assuredly contemporaries with more than a dozen of the greatest of these beasts, and Paul Martin has argued vigorously, though not persuasively, for a human hand (or Clovis point) in their extinctions (Chapter 8). More likely, the climatic and ecological changes at the end of the Pleistocene—when it became warmer and more seasonal, and summers grew longer—ultimately did in these Pleistocene mammals, just as it caused the extinction and range changes in a wide range of other animals (and plants).

These were the very same changes occurring halfway around the world that set the stage for the domestication of cattle, goat, pig, sheep, camel, ass, elephant, horse, duck, goose, water buffalo, yak, and cat. In sharp contrast, indigenous animal domestication was insignificant in the New World: just three species of South American camel, the turkey, possibly the muscovy duck, and the guinea pig (they'd brought their dogs with them).

There are two reasons for that striking imbalance in the number of animal domesticates. First, the New World fauna was impoverished. Pleistocene extinctions had wiped out thirty-five genera of animals: that was 80% of North America's large mammals. Gone were horses, elephants (mammoth and mastodon), and camels, animals whose Old World relatives survived the end of the Pleistocene, thereby allowing Genghis Kahn to gallop on horseback to global empire, Hannibal's army to ride elephants over the Alps, and camel caravans of the Persian Empire to travel the Silk Road. All that remained of North America's large mammals were bison, moose, elk, deer, musk ox, mountain goat, and pronghorn antelope. This was a narrow pool from which to seek potential animal domesticates.

Second, not all animals, especially not those North American survivors, are easily domesticated. Animals so suited, Jared Diamond argues, must be social and display dominant and submissive behaviors. They should not compete with humans for food and, better still, should be a potential food source. They cannot have an instant flight reflex when startled. They must breed readily in captivity. And, they must be adaptable to a wide range of climates and environments.[30]

Of the surviving New World large mammals, only bison come close to meeting those strict requirements. But penned bison are unruly animals (Chapter 9). Besides, what's the incentive to domesticate? Domestication is partly a response to food shortages. For more than 10,000 years, bison were an abundant, predictable, and reliable food source. Why take the trouble to corral them? No one did.

The combination of Pleistocene extinctions and the difficulty or disincentive for taming surviving mammals produced no significant New World animal domesticates with which to share and hybridize deadly new pathogens. Only in South America, where several camel species were herded, was there a likely animal source for disease strains. These animals and their herders, however, lived high in the Andes in small and dispersed groups, and were too few and too isolated, William McNeill explains, to sustain infections in the wild. Guinea pigs, also domesticated, served as the primary reservoir for Chagas' disease, but Chagas does not have the pandemic power of smallpox or measles.[31]

However, New World peoples did domesticate a cornucopia of plants (more than in the Old World, in fact), from acorn squash to yams, as well as avocado, cocoa, chili pepper, cotton, gourd, lima bean, manioc, potato, peanut, pumpkin, sunflower, tomato, and tobacco (virtually all of which were carried across the ocean, indelibly altering Old World menus). The most important, however, was the triumvirate of corn, beans, and squash. These plants fed the great pre-Columbian towns of North America, and the Mesoamerican civilizations of the Olmec, Maya, Toltec, and Aztec. Potatoes, peanuts, and yams did the same for South America's Chavin, Moche, Wari, Chimu, and Inca.

So successful was food production in the New World that by 1492, there were cities and towns in the Western Hemisphere to rival in size and complexity any in the Old. The Aztec island capital of Tenochtitlán (now Mexico City) had broad avenues lined with palaces, temples, and houses, and was surrounded by a lake filled daily with canoes, bringing to its markets goods traded near and far. It had a pre–European contact population estimated at over 150,000, with perhaps 1 million in the surrounding valley.[32] That's about as many people in one valley as there were in *all* the British colonies on the eve of the American Revolution. A city on the scale of Tenochtitlán testifies that crowds are necessary but not sufficient for the evolution of epidemic disease, though they can spread one with lightning speed and deadly efficiency.

But there were no epidemic diseases for the city's inhabitants to spread. The manner and timing of their ancestors' journey to the New World, the long-term genetic consequences of their having passed through a population bottleneck, the more than 12,500 years of prehistory they spent in isolation from the rest of the world's populations, and their lack of animal domesticates, all rendered their descendants defenseless against the infectious diseases introduced on these shores in the decades and centuries after Columbus.

It was an unfair fight. The New World had its native infections, particularly in the tropics of Central America, including Carrion's disease, tapeworm, hepatitis, encephalitis, yaws, and venereal syphilis perhaps, and of course, Chagas' disease—the bane of Darwin's later life.[33] Although all could be debilitating, none seem to have done well outside the Western Hemisphere (with the possible exception of syphilis) or against arriving Europeans. Certainly none of these ills changed the demographic course of Old World colonial history in America.

For European colonists, the New World was a vast, rich, and so far as they were concerned untapped land (never mind that it was already inhabited) in which they thrived. Time and again, Crosby relates, Europeans in America remarked on the healthiness of their new homes, and their claims are probably true, despite their Chamber of Commerce ring. To be sure, the English colonies of Roanoke Island and Jamestown got off to terrible starts because of their stubborn disinterest in Native adaptations, an inordinate amount of time spent searching for gold, an initial refusal to learn their new landscape but instead attempt to impose their knowledge of English landscapes and land use practices on Tidewater Virginia, and—not least—their arrival in the midst of record drought years.[34] But even in the Tidewater, the colonists' lives soon improved, and indeed, with a bit of learning of their new landscape, and a few concessions to the ecological differences between Europe and America, it was soon apparent that many Old World plants and animals could also thrive in this new setting.

As early as Columbus's second voyage, "the Admiral [Columbus] took on board mares, sheep, cows and the corresponding males for the propagation of their species; nor did he forget vegetables, grain, barley, and similar seeds, not only for provisions but also for sowing; vines and young plants such as were wanting in that country were carefully taken."[35] Later colonists imported crops like wheat, oats, and rye, as well as (perhaps unintentionally) dandelions, ferns, thistles, plantain, clover, stinging nettles, crab grass, and other weeds. These plants exploded in areas now cleared of forest, and overgrazed by the domesticated animals the Europeans brought with them both intentionally (pigs, cattle, sheep, and horses) and not (rats).[36]

Walking larders of pigs and cattle often accompanied Spanish entradas. Horses came too, serving as transportation and beasts of burden and providing a lopsided military advantage. The animals that came with Europeans may also have had another, more insidious role. Fast-multiplying and rapidly spreading pigs may have been equally to blame for the transmission of diseases such as trichinosis, tuberculosis, and leptospirosis—first to the native fauna, then to native peoples.

Within a few centuries, introduced plants and animals, coupled with imported farming and land use practices, utterly changed the American landscape. *Ecological imperialism,* Crosby calls it, and it was almost entirely one-way. Virtually no New World animals gained a toehold in the Old World, although as mentioned many plants domesticated by American Indians were carried to Europe and became successful staples—corn, potatoes, peppers, and tomatoes—or addictive vices, like tobacco (tobacco may be a small measure of retribution for European depredations in the New World, but it was at best a pyrrhic victory: the great demand for labor to tend the New World's tobacco fields was one of the sparks that set off the slave trade[37]).

European colonists and their descendants in America found themselves free from crowds, possessing plenty of farmland and pasture, and enjoying higher living standards than they had back home. They could, and usually did, build far better lives.[38] For the descendants of the first peoples of the New World, it was a very different story.

THE CONSEQUENCES OF CONTACT

Although their numbers were decimated, Native Americans ultimately survived the waves of disease and depopulation, and the collateral blows of warfare, famine, enslavement, and exploitation. But they did not survive unchanged. The consequences of European contact for American Indians were many and deep, often cascading in unpredictable ways through native culture. Consider, however, a few critical aspects that bear directly on our study of the first Americans.

In the aftermath of epidemics, survivors abandoned old towns and regions: sixteenth- and seventeenth-century documents hint of extensive refugee movements throughout North America. Refugees banded together with other survivors, who were often from previously unrelated communities, possibly speakers of different languages, and occasionally even traditional enemies. Naturally, as cultural and linguistic strangers came together, the social and political rules that once governed their lives—often deeply rooted in kinship networks since shaken or destroyed—had to change. Established relationships and alliances as well as ritual and ceremonial systems were abandoned, oftentimes because disease disproportionately killed the older members of a community, who were the repositories of those traditions and lore. The old rules no longer applied.

The Natchez of the lower Mississippi Valley were, in their heyday, one of the most powerful tribes on the river. As de Soto and his entrada approached the seat of their chiefdom (Quigaltam), he sent a Native emissary ahead to announce to the Natchez chief that he (de Soto) was the "son of the sun and that wherever he went all obeyed him and did him service." The chief would have none of that nonsense, since he believed *he* was the brother of the Great Sun, and would not travel to pay homage to anyone. If de Soto was who he said he was, let him prove it by drying up "the great river" (the Mississippi). Unaccustomed to rebuke, a furious de Soto vowed to "lessen that arrogant demeanor" and sent horse-mounted shock troops to slaughter hundreds of innocents in a nearby village.[39] The Natchez managed to survive, but not de Soto, who was already ill from fever and died shortly thereafter.

The Natchez were still a powerful tribe two centuries later when visited by du Pratz, but what he reported of their "peculiar" social organization puzzled generations of anthropologists. According to him, the Natchez had two major social classes, "stinkards" (the unflattering name assigned by du Pratz) on the bottom, and "nobility" on the top. The nobility, in turn, were comprised of three classes: "suns," "nobles," and "honored people," in that order. Du Pratz described elaborate rules of intermarriage, descent, and inheritance among the classes that—if they actually had been followed—would in a matter of generations have caused the extinction of the so-called stinkards. Yet, they were the largest class among the Natchez. Later attempts to explain the numerical instability of the system included the possibility that the stinkards were disproportionately fecund; that the system was demographically unstable, and if allowed to play out, the stinkards would ultimately go extinct (one source estimated within ten generations); or that maybe du Pratz and other observers simply got it wrong.[40]

As Jeffrey Brain and others argue, however, the reports were correct, and the system *was* unstable over the long term, largely because it represented an agglomeration of two separate social systems. As du Pratz observed, the stinkards spoke "a language entirely different from the nobility."[41] It appears that between the time de Soto first traveled through the Southeast and when du Pratz arrived a century and a half later, the Natchez population had collapsed, and their territory had contracted as a consequence of introduced epidemic disease. The Natchez that du Pratz encountered were different from those who had stood up to de Soto, for the latter-day group included a fusion of Tunica-speaking Chitimacha, Koroa, Tiou, and Grigra refugees.[42]

The Natchez had become a haven for refugees, and by du Pratz's time nearly half their villages were made up of Tunican peoples and speakers. Their own population decimated, their territory abridged, and their numbers composed of people who did not speak their language, the Natchez had to adapt. And so they altered their definitions of status and rank, and made up new rules regarding marriage, kinship, and inheritance. Deeming the Tunica-speaking refugees lower-class but allowing intermarriage enabled the Natchez to assimilate these once-strangers, bolster their own sagging numbers, and ensure the continuance of their nobility.[43]

The Natchez are but one example of how a group was re-shaped as a consequence of the ripple effects of European contact, and an illustration of how genetic exchange may well have occurred far more rapidly than linguistic or social assimilation. There is good reason to suspect that the amalgamation that took place in the lower Mississippi Valley was not unique, but repeated often and elsewhere in the Americas as remnants of peoples and villages banded together in order to survive.

The inevitable result of all this was a post-1492 mismatch between language, genes, and culture of a 'group,' one that played out across North America as survivors amalgamated, creating entities very different from those previously extant (and with it, decreasing any possibility of seeing Goldilocks-level consistency in artifacts, genes, and languages).[44]

Indeed, a sharp and deep divide separates American Indian populations before and after 1492. The peoples whose customs and cultures were systematically recorded in the late nineteenth and early twentieth centuries are the biological and cultural descendants of what were once much larger and differently constituted populations now known only archaeologically from pre–European contact times.

THE PRESENT MAY NOT BE KEY TO THE PAST

But tracing those lines of descent is complicated, since cultural, linguistic, physical, and genetic lineages were likewise casualties. Many languages simply died out along with their speakers in the centuries after European contact. Lewis and Clark were explicitly charged to record what they could of the languages they encountered, for Jefferson well realized and deeply lamented the rapid disappearance of Native languages. There had been a "profusion" spoken in the Carolinas when John Lawson traveled through there in 1701. By 1743 these had been reduced to about twenty languages and dialects. By 1760 there was but a single tongue remaining, Catawba, the last speakers of which died out in the 1950s.[45] For many Native languages, particularly in eastern North America, no more than a few words were ever recorded.

Even where speakers survived, languages were altered or had disappeared. To communicate with each other, refugees from different groups had to adopt a single tongue, and over time lost their own. On the heels of the 1837–38 smallpox epidemic, the Arikara living in villages along the Missouri River in what is now South Dakota—remnants of a farming people, who may have once numbered 20,000—were down to just 380 members. Their numbers were so devastated they were "compelled to come together for protection" with Mandan, Hidatsa, and other surviving Missouri Valley peoples, despite the fact that "the Different Villages do not understand all the words of the others." Arikara is in the Caddoan language family; Mandan and Hidatsa were Siouan.[46] A similar scenario played out in many areas, with the end result that overall American Indian linguistic diversity plummeted in the centuries after European contact.

Yet, it is on that record of Native American languages that Greenberg's Amerind family is constructed. By itself, this says nothing of whether the specific words and grammatical forms he used to group languages within Amerind are correct. That is for linguists to decide. But the loss and borrowing of words does give pause to our accepting his Amerind family, given the vulnerability of the method used to construct it and the potential for errors in word lists. There may be much less to "Amerind" than meets the ears.

Cultural changes were no less profound, as native peoples became incorporated into or dependent on European economies, which imposed a sapping demand for labor (now itself a commodity), opened a ravenous market for items not previously valued, such as fur, gold and silver, and tobacco, while it shut down demand for others, such as trade in high-quality stone, feathers, and marine shells. Likewise, changes occurred as Indians adopted Old World plants, animals, artifacts, and technologies; as they were forced into unfamiliar environments, new adaptations, or slavery; and as their rituals, ceremonies, traditions, and bonds of kinship and society were torn apart and reconstituted in the chaos of demographic collapse and the loss of lore and knowledge.

So, too, the rules of conflict and warfare were forever changed, and not just by the shifting alliances and factionalism that accompanied the jockeying of the European powers for New World dominance or the need to offset population losses from disease. Europeans brought new weapons of war—guns and horses—which made it possible to do battle on a much larger geographic scale (horses greatly increased the combat reach), and also made it far more deadly (one might be able to dodge an arrow, but not a lead ball).[47] To be sure, there was continuity in elements of both culture and material culture amid all these changes, but their contexts changed in the centuries after European contact.

If the linguistic and cultural ties before and after 1492 were frayed, no less so were physical and genetic links. Depopulation from epidemic diseases created a population bottleneck in the decades and centuries after contact, quite unlike any experienced by American Indians since they had left northeastern Asia over 12,000 years earlier. Most of the people responsible for the prehistoric record, as Robert Dunnell observed, have no living descendants.[48] Yet, a population bottleneck may not be a genetic bottleneck if losses from diseases were random with respect to the founding haplogroups and haplotypes. However, the lineages harbored among Native American populations surviving contact are in some sense a sample—of indeterminate size—of lineages that were present in the New World in early October 1492. Haplotypes and even haplogroups that may have existed as recently as 1492 could have been lost in the decades and centuries that followed. This is why aDNA has such great potential in the study of the first Americans, and has already begun to add in previously unknown lineages.

What then of modern American Indians, who today contribute their mtDNA and Y chromosome DNA to our efforts to understand their ancestry? Can we trace modern tribes back to Pleistocene ancestors? Or trace Pleistocene artifacts, skeletal remains, or ancient DNA up to modern tribes? These are no longer just academic questions.

TO WHOM DOES THE PAST BELONG?

Under the terms of the Native American Graves Protection and Repatriation Act (NAGPRA), skeletal remains found in the United States must be returned at the request of a "lineal descendant" or, shy of that, the tribe with the "closest cultural affiliation."[49] NAGPRA is a complicated law, but at its core calls for repatriation when two criteria are met: first, the skeletal remains must be demonstrably Native American. That's defined as "of, or relating to, a tribe, people, or culture that *is* indigenous to the United States" (italics mine: and keep an eye on that *is*—it becomes important). Second, and if such remains are deemed Native American, then NAGPRA requires they be returned on request to the descendants of that person, as identified by "geographical, kinship, biological, archeological, anthropological, linguistic, folklore, oral tradition, historical, or other relevant information or expert opinion."[50] Under the terms of NAGPRA, all those lines of evidence are equally weighted. Demonstrating affiliation only requires a preponderance of evidence, *not* scientific certainty.

Where the skeletal remains come from a recently occupied site, demonstrating affiliation is relatively straightforward, and remains have been returned without dispute over their identity. All the skeletons (nearly 2,000) excavated at Pecos Pueblo by Kidder and described by Hooton (Chapter 5) were returned in 1999 for reburial at what is now Pecos National Monument. There was no question they belonged to the descendants of Pecos.

Kennewick, on the other hand, hits NAGPRA right where it is weakest: establishing biological or cultural affiliation across a vast span of time. An individual who lived nearly 10,000 years ago could be an ancestor of *many* modern individuals and tribes and not just one, or related to none at all. Directly linking past to present across that long span is no simple matter. Geneticists have the best chance, if they can identify in ancient mitochondrial or Y chromosome DNA a lineage that can narrow down the potential population(s) of descendants. However, it would have to be a very specific haplotype rather than, say, Haplogroup A, which is found in 40% of all American Indians (of course, if that haplogroup were found in an ancient skeleton, it would fulfill NAGPRA's first requirement of establishing that an individual was Native American). But even if such a specific haplotype is found in a skeleton of that age, it will still prove difficult to identify one (and only one) tribe of descendants—unless that group was largely isolated over that entire time span. And, of course, this assumes the lineage was not lost in the depopulation that followed European contact.

What of the other approaches? Archaeology can trace (and date) the cultural evolution of artifacts, but the trail from past to present is poorly marked, and one can hardly follow a single people through time by their artifacts and tools, or by their skeletal or dental remains. In fact, even where those appear morphologically unlike modern American Indians—as is said to be the case for certain crania—we are still unable to state conclusively that no historical relationship exists, or that physical remains of an individual are not those of an American Indian. Remember: such a conclusion is based on a tiny (and assuredly unrepresentative) sample, from individuals that may have been members of populations that were small and isolated, and that are separated from us by 400 generations of population and evolutionary change. Linguistics fares worst of all. The record of Native American languages is woefully incomplete, and there is a limited time depth over which one can still hear the echoes of a distant language (the radical historical linguists notwithstanding).

In a sense, then, it's a good thing NAGPRA doesn't require scientific certainty, since insisting on that standard for lineal descent at such a span of time is scientifically impossible. For that matter, as David Hurst Thomas reminds us, given what we know of modern population biology, the "idea of tracking fixed, enduring, bounded ethnic groups is outmoded and quaint, if not a little racist."[51]

Yet, NAGPRA demands affiliation be established all the same. And that sets scientific evidence on a potential collision course with tribal traditions, and American Indians against archaeologists and physical anthropologists, for NAGPRA allows "lineal descendants" to be identified not just by physical remains but by equally weighted evidence from folklore, oral traditions, and geographic proximity. To be sure, Native American traditions of origins (and there are many) have rich and deep resonance, but they cannot be uncritically accepted as detailed population histories either.[52] Under the ambiguous terms of NAGPRA, both sides of the Kennewick dispute could be right, even if both were wrong.

And they were. The filing of the *Bonnichsen et al.* lawsuit in October 1996 blocked the Army Corps of Engineers from returning the Kennewick skeleton to the five Pacific Northwest tribes that were claiming him as one of their own. In 1998 the Corps instead handed Kennewick over to the Department of the Interior, which was better prepared to deal with such matters. The Interior Department, in turn, brought in a team of scientists, conducted an investigation, and in 2000 they concluded from their interpretation of the evidence and the law, that Kennewick was indeed Native American (based on the skeleton's great age) and was affiliated with the five tribes (based on their oral histories and geographic proximity).

Although it was recognized there were "both cultural continuities and cultural discontinuities between the modern day claimant tribes and the cultural group that existed during the lifetime of the Kennewick Man," Bruce Babbitt, then secretary of the interior, declared that the "preponderance of evidence" tilted in favor of the tribes.[53] There were hushed, conspiratorial whispers that this decision had been made inside the White House for raw political reasons.

No matter. Bonnichsen and his fellow plaintiffs challenged the ruling, and thanks to the lawsuit Kennewick was by then under court control. Two years later, federal Judge John Jelderks ruled that the secretary of the interior had no grounds for concluding Kennewick was Native American based merely on its antiquity, which simultaneously rendered moot the conclusion that Kennewick was affiliated with the five tribes (moot or not, the judge criticized that conclusion as well, citing the "difficult challenge" of demonstrating affiliation over so many millennia).[54] Jelderks's ruling was appealed by the federal government (as well as by a California man who claimed he was a descendant of Kennewick—this case has never wanted for comic relief), but in 2004, the Ninth Circuit Court of Appeals upheld Jelderks's ruling. The plaintiffs in *Bonnichsen et al.* could proceed with their study of the Kennewick remains. They have, though their results have yet to be published.

The Court of Appeals judgment pivoted on a single word. NAGPRA defines remains as Native American if they are "of, or relating to, a tribe, people, or culture that *is* indigenous to the United States."[55] The Court took the present tense of the verb "is" to be "significant"—being a Native American meant bearing some relationship to "a *presently existing* tribe, people or culture."[56] But does "is" mean a present-day tribe only or, as Babbitt argued, could "is" and "was" be used interchangeably? The Court did not think so. This wasn't the only time during the Clinton administration that a point of law depended on what the definition of "is" is.[57]

The Appeals Court ruled Congress's purposes would not be served by requiring the transfer of human remains to modern tribes without evidence of a clear-cut relationship. But does that mean this ancient skeleton is not a Native American? Or, more generally, that there is no relationship at all between ancient and modern individuals? Hardly: just as we are hard pressed in the case of ancient skeletal remains to link past to present, so, too, we cannot prove there is no affiliation at all.

The battle over Kennewick was waged in federal courts—certainly not the usual venue for debates over skeletal remains—and at a cost of several million dollars. It also sparked plenty of fights within the anthropological community about how, by what methods, or even whether more ancient skeletal remains can be linked with modern peoples; about the relationship between American Indians and those of us who study their past; and about the peopling of the Americas. There's no small irony here, since a skeleton that's only 8,400 years old (as Kennewick is) may in the end say very little about the first peoples of the Americas (Chapter 5).

But it certainly jangled nerves. Alan Schneider and Robson Bonnichsen, lead attorney and plaintiff, respectively, in the Kennewick lawsuit, paint the effort to repatriate ancient skeletal remains as nothing less than an assault on Western Enlightenment thinking by the forces of "ignorance, intolerance, and superstition." They angrily railed against the government officials in Kennewick, accusing them of having agendas but no expertise, of conducting secret studies and dodging peer review, and of imposing a "monopolistic control over research opportunities . . . to promote the theories they favor and discourage investigation of theories they oppose."[58] One would almost think they lost the Kennewick case. But they won, and gained the right to conduct radiocarbon dating, metric analysis of the crania, DNA and isotope analyses, examination of the site sediments—essentially the same studies performed earlier by the government-selected scientists, putting the lie to the claim that the government's "desire to promote personally favored theories of New World human origins appears to have influenced the selection of the studies conducted by the government in the Kennewick Man case."[59]

Yet, their voices were no less strident than those on the other side of the repatriation aisle. Vine Deloria could hardly disguise his anger—and glee—that Kennewick had finally exposed to the world that archaeology had become little more than a discipline in which there is a "triumph of doctrine over facts." Like any other group of "priests and politicians," he snarled, "scientists lie and fudge their conclusions as much as . . . lawyers and car dealers." And there were scientists who agreed with him. Kennewick was "never an issue of Indian beliefs versus science," Jonathan Marks wrote, "but of Indian beliefs versus pseudoscience."[60]

So to whom does Kennewick or, for that matter *any* skeleton of great antiquity belong? Is it possible he/they has no living descendants at all among any modern American Indians? Of course. Does that mean he is not Native American? Of course not.

But that still leaves hanging several questions: In the absence of demonstrable descendant groups, who should take custody when remains surface? Should analysis be allowed? What should be the proper ultimate disposition? How should the wishes of very distantly related or unrelated modern American Indians be measured against the value of knowledge about the human experience that might be gained by scientific study of the remains? At what point, as David Hurst Thomas asks, do ancient bones stop being tribal and become simply human?[61] Put in the most general terms: who owns the past?

There are many answers to these questions, depending not just on whom you ask but how you ask, for there is no clear solution, and there are several positions that can be reasonable (though perhaps not always in the Kennewick case) and right. An answer can be sought in statute and law, but there is enough ambiguity in NAGPRA one can arrive at several different interpretations, all of which can be legally justified. Yet, the legal answer NAGPRA provides may not be the ethical answer, let alone the scientifically defensible one.

Thomas argues, rightly, that judges should be freed from having to carry the burden of defining the future of America's past. We should try to solve our problems ourselves, by having conversations across the divide between American Indians and archaeologists, and seek common ground.[62] On this point even Schneider and Bonnichsen, and Deloria and Marks agree. Grudgingly.[63]

Common ground has been found before. In early July 1996, several weeks before Kennewick turned up on the shores of the Columbia River, human remains were discovered in On Your Knees Cave along the Alaskan coast (Chapter 5). The excavators, who had a longstanding relationship with the local tribes, immediately notified them of the discovery, and within a week, a cooperative agreement for the excavation and analysis of the remains was in place. The investigation proceeded over the next several years without incident, rancor, or attorneys.

Outcomes like Kennewick are not inevitable, but they do show how much the past can matter, and have consequences for those who study it and those whose ancestors lived it. But that's not all.

REPLAYING THE TAPE OF PREHISTORY

Broad principles, such as Darwin's evolution by natural selection, govern our collective human past, but they alone cannot explain why human prehistory unfolded the way it did. That depends on a swarm of factors, large and small, specific to the circumstances in which our ancestors found themselves at a given moment in time, the decisions they made to move forward, and even in no small measure to chance: such as spending the night of March 26, 1835, in Luxan, Argentina.

There is, as a result, a contingency to prehistory. One supposes it was "inevitable" that humans, once evolved (which was itself hardly inevitable), would one day reach the Americas. But who the first peoples were, why they came, when they arrived, where they came from, how large the group was, and how quickly they spread, was not inevitable at all. It only looks that way afterward, after a long and singularly unpredictable string of choices made under a variety of circumstances brought them to the western edge of Beringia at the end of the Pleistocene.

But replay the tape, Stephen Jay Gould argued, change a few of the quirky details, let their effects cascade over time, and the past would have unfolded very differently, producing historical trajectories that would be "equally explicable, though massively different in form and effect."[64]

What if the first Americans hadn't left the Old World when they did. Or if they had brought diseases or an immunity conferred by Old World ancestry with them. Or if they had not come via Beringia, or gone through a population bottleneck. Or if 80% of North America's megafauna had not gone extinct at the end of the Pleistocene. Or if the remainder had been more readily domesticated, or if there had been strong incentive to do so. Or even if all those savants had been right, and the American Indians had been the Lost Tribes of Israel. And what if Native Americans had not been devastated by disease, or had their own to give Europeans in return.

We cannot say. But consider this: in the spring of 1519, a Spanish force under the command of Hernán Cortés landed on the Mexican mainland. Cortés had been sent by Diego Velázquez, the governor of Cuba, to trade and explore, but not to conquer.

The governor wanted that plum for himself. Cortés pretended to care. Trained in law and devious of mind, the moment he touched the mainland he promptly side-stepped the governor's orders by incorporating a town (Veracruz): according to the canons of medieval Castile, that entitled him to shed any obligations to the governor and appoint himself an agent of the Spanish king. The way Cortés figured it, King Charles (recently elected Holy Roman Emperor) would surely approve Cortés extending his ecumenical and imperial reach, especially if it made them both rich.[65]

Cortés learned from native Totonacs, who greeted him, of the great Aztec Empire to the west and, more important, that the Empire's rule over the region was riddled with pockets of violent resistance.[66] He headed inland over the mountains, alternately fighting and gaining key allies among the Aztecs' sworn enemies (like the Tlaxcalan). The Aztec ruler, Montezuma II, heard Cortés was coming. Runners streaming in from outlying areas reported a procession of strange, "chalky faced" men atop bizarre, deer-like beasts (thanks to Pleistocene extinctions, of course, only the Aztec's most remote ancestors had ever seen a horse).[67] Worse, he heard they had machines that could blow up mountains (cannons), kill from great distance (guns), and wore impenetrable clothing (chain-mail armor).

There had been prophecies the feathered-serpent god Quetzalcoatl would return, which the strangers' impending arrival seemed to match. Uncertain whether the Spaniards were gods or enemies, Montezuma hedged his bets: he'd treat them as gods, but try to keep them at a distance in case they were enemies. Couriers bearing exquisite gifts of gold, silver, and precious stones were sent to Cortés, hoping that would be enough, and make him turn away. Montezuma must have soon regretted the gesture, for it had precisely the opposite effect.[68]

On November 8, 1519, as much of the city watched in astonishment, Cortés led his army down one of the causeways across Lake Texcoco into Tenochtitlán. Montezuma and his courtiers were there to greet him. Cortés, knowing little and caring less what he actually heard, took Montezuma's welcoming remarks as an abdication speech.[69] On behalf of King Charles, he accepted the transfer of authority over the Aztec Empire, then quietly put Montezuma under house arrest. For the next six months, Cortés commanded the increasingly restless city, steadily looting its treasuries of gold and silver.

While Cortés was pilfering Aztec riches, back in Cuba it finally dawned on Governor Velázquez that Cortés was long past due, and was likely playing him for a fool—or, worse, getting rich at his expense. Velázquez dispatched a force under Pánfilo de Narváez to bring Cortés to heel. Hearing from his native allies that another Spanish ship had landed on the coast and guessing what was up, Cortés hustled away with most of his army to confront Narváez, leaving behind a small force under the command of an unsavory lieutenant, Pedro de Alvarado, with orders to keep a tight lid on Tenochtitlán.

By now, however, the Aztecs were acutely aware Montezuma was being held against his will, and they were becoming openly defiant. Alvarado, outnumbered and apprehensive, agreed to allow some 600 Aztec priests, noblemen, and worshippers to use one of the great courtyards in the palace complex for the feast of Huitzilopochtli. But while they were inside, Alvarado's men sealed the doorways, trapped the unarmed worshippers, and butchered them with broadswords. If Alvarado thought the slaughter of these Aztec leaders would quell the unrest, he was as foolish as he was cruel.

Word of the bloody massacre flashed through the streets of Tenochtitlán, and the city exploded in open battle. The Spaniards were soon trapped inside the palace when Cortés, having defeated Narváez and receiving desperate bulletins from his men in Tenochtitlán, quick-marched back and reentered the city on June 24, 1520, this time with over 1,000 Spaniards (now including many of Narváez's men), and many thousand Tlaxcalan warriors. Yet, Cortés was still vastly outnumbered, and his attempt to use Montezuma to manipulate his people failed when the native ruler was killed—though by which side, neither agreed, save to blame the other.

Seven days later the Spaniards and their allies were pinned down and on the verge of annihilation. Cortés called for retreat under the cover of darkness. But as his troops tried to sneak off, Native spotters sounded the alarm, and fierce fighting broke out along the dangerously narrow causeways across the lake. The Aztecs speared Spaniards into the water where, weighed down by their chain-mail armor and looted gold and

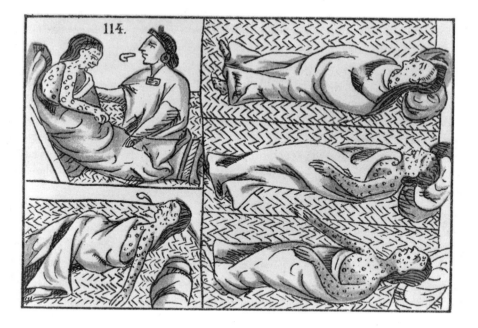

FIGURE 64.

An image from the *Florentine Codex*, which was compiled in the latter half of the sixteenth century, which shows Aztec victims of smallpox. (Image courtesy of the American Museum of Natural History, Negative number 286821.)

silver, they sank and died. Cortés managed to survive, though he'd been badly whipped, losing all but a third of his army that night.

The Spaniards fought their way out of the valley, but the Aztecs didn't pursue their advantage. Instead, they allowed Cortés to escape and take refuge among the Tlaxcalan on the far side of the mountains. The following spring, reinforced by more Spaniards and nearly 150,000 native warriors, Cortés returned to lay siege to Tenochtitlán. After more bloody fighting, he conquered the city in August 1521. It was an astonishing victory against overwhelming numbers, especially since it came so close on the heels of devastating defeat.

Some credit victory to Cortés's superior weaponry and horses, and the terror these created in foes who'd seen neither. Others speak of his missionary zeal, courage, and military genius in exploiting the strife between the Aztecs and their native enemies, and using boats on Lake Texcoco to choke off the canoe traffic that brought supplies to the island capital. Zeal and courage Cortés had plenty of, but so did the Aztecs. They hadn't his weaponry, but could a small number of firearms, cannons, and horses make a difference? The previous June it hadn't. Once the tactical surprise wore off, as it did quickly, the Aztecs began mounting razor-sharp obsidian blades on long pikes to skewer horse-mounted Spaniards.

In the end, what brought Cortés victory at Tenochtitlán was neither arms nor tactics, but smallpox. One of Narváez's men had been infected, and in June 1520, the disease jumped to the Native population. While Cortés and his mauled force were off licking their wounds a horrific, if unintended, and utterly one-sided bout of germ warfare began. Smallpox raged through Tenochtitlán, killing young and old alike. By the time Cortés returned in 1521, the Aztecs were devastated by disease, hunger, and death, and could offer little significant resistance. The siege was the final blow. Spanish chroniclers tell of stepping over piles of smallpox victims (Figure 64), whose rotting bodies choked the streets.

Cortés was badly whipped in June 1520. And though he returned a year later with his horses, firearms, cannons, Tlaxcalan allies, a makeshift navy and a grand strategy, it's doubtful he could have retaken Montezuma's Aztec kingdom with its one million souls. The numbers were against him. But with smallpox having cut a wide swath through the native population, he could—and did—conquer the Empire.

Tenochtitlán's fall was unusual, but only because the devastation by introduced epidemic disease was rarely so militarily decisive.[70] Still, the triumphs of the conquistadores and all who followed were, as Crosby put it, ultimately the triumphs of smallpox.[71] We may not be able to say what would have happened if we could replay the tape of prehistory, but we can say that were it not for introduced diseases and their dreadful effect on Native American populations, American history after 1492 would have been very, very different.

And the triumphs of smallpox and other introduced epidemics had been foreordained over 12,500 years earlier when a group of hunter-gatherers stepped onto the western edge of the vast, grassy plain of Beringia, then headed toward the rising sun.

FURTHER READING

There is a vast literature on the early archaeology of America. Relatively recent overviews, which provide details on many of the sites in question, but are pitched for a broader audience, include James Adovasio and Jake Page's *The First Americans: In Pursuit of Archaeology's Greatest Mystery* (2002, Random House, New York), Tom Dillehay's *The Settlement of the Americas: A New Prehistory* (2000, Basic Books, New York), James Dixon's *Bones, Boats and Bison: Archeology and the First Colonization of Western North America* (1999, University of New Mexico Press, Albuquerque), Danièle Lavallée's *The First South Americans: The Peopling of a Continent from the Earliest Evidence to High Culture* (2000, University of Utah Press, Salt Lake City), and the papers in *The First Americans: The Pleistocene Colonization of the New World* (2002, University of California Press, Berkeley), edited by Nina Jablonski.

Recent scholarly works on the topic include the edited volumes by Michael Barton, Geoffrey Clark, David Yesner, and Georges Pearson, *The Settlement of the American Continents: A Multidisciplinary Approach to Human Biogeography* (2004, University of Arizona Press, Tucson); Robson Bonnichsen, Bradley T. Lepper, Dennis Stanford, and Michael Waters, *Paleoamerican Origins: Beyond Clovis* (2006, Center for the Study of the First Americans, College Station, Texas), David Madsen's *Entering America: Northeast Asia and Beringia before the Last Glacial Maximum* (2004, University of Utah Press, Salt Lake City), and Douglas Ubelaker, *Handbook of North American Indians, Volume 3, Environment, Origins, and Population* (2006, Smithsonian Institution, Washington, D.C.).

For a recent and highly readable discussion of the causes and consequences of ice ages and climate change, see Wallace Broecker and Robert Kunzig's *Fixing Climate: What Past Climate Changes Reveal About the Current Threat—and How to Counter It* (2008, Hill and Wang, New

York). Other fine discussions include Doug Macdougall's *Frozen Earth: The Once and Future Story of Ice Ages* (2004, University of California Press, Berkeley), Richard Alley's *The Two-Mile Time Machine: Ice Cores, Abrupt Climate Change, and Our Future* (2000, Princeton University Press, Princeton, New Jersey) and (though now a bit dated) John and Katherine Imbrie's *Ice Ages: Solving the Mystery* (1976, Enslow, Berkeley Heights, New Jersey).

Dale Guthrie's *Frozen Fauna of the Mammoth Steppe: The Story of Blue Babe* (1990, University of Chicago Press, Chicago) and the recently published *Human Ecology of Beringia* by John Hoffecker and Scott Elias (2007, Columbia University Press, New York) summarize Beringia's geological, climatic, and environmental history. The most recent syntheses of the Pleistocene glaciation, climate, and environment of North America appear in the volume *The Quaternary Period in the United States* (2004, Elsevier Science, New York), edited by Alan Gillespie, Stephen Porter, and Brian Atwater; the articles, however, are written for a more technical audience.

On Native American languages, the best single source is the Smithsonian Institution's *Handbook of North American Indians*, particularly Volume 17, *Languages*, edited by Ives Goddard (1996, Smithsonian Institution, Washington, D.C.). Joseph Greenberg presented the results of his analysis in *Language in the Americas* (1987, Stanford University Press, Stanford). Merrit Ruhlen's *The Origin of Language: Tracing the Evolution of the Mother Tongue* (1994, John Wiley and Sons, New York) is a staunch defense of Greenberg's (and his) method of mass comparison, and a readable discussion of what that method yields for a linguistic history of America and the world. For a thorough history of historical linguistics and a synopsis of Native American languages, including a relentless critique of Greenberg's method and results, see Lyle Campbell's *American Indian Languages: The Historical Linguistics of Native America* (1997, Oxford University Press, Oxford).

Turner has summarized his dental evidence in multiple places, most recently in "Teeth, Needles, Dogs, and Siberia: Bioarchaeological Evidence for the Colonization of the New World," which appeared in Jablonski's *The First Americans: The Pleistocene Colonization of the New World* (cited above). See that same volume for a thoughtful and accessible appraisal of the cranial evidence by Gentry Steele and Joseph Powell, "Facing the Past: A View of the North American Human Fossil Record," as well as Powell's valuable book, *The First Americans: Race, Evolution, and the Origin of Native Americans* (2005, Cambridge University Press, Cambridge). For an entertaining and appropriately skeptical look at how the cranial evidence is sometimes presented, there is Jack Hitt's "Mighty White of You: Racial Preferences Color America's Oldest Skulls and Bones" (2005, *Harper's Magazine*, July, pp. 39–55).

The genetic evidence emerges so quickly that a book devoted to the topic would be obsolete before it ever appeared in print; most discussion on this topic is restricted to technical journals, but a few good summaries for a broader audience are available, including Colin Renfrew's edited volume, *America Past, America Present: Genes and Languages in the Americas and Beyond* (2000, MacDonald Institute for Archaeological Research, Cambridge); Andrew Merriwether's "A Mitochondrial Perspective on the Peopling of the New World," in the Jablonski volume (cited above), and Theodore Schurr's "The Peopling of the New World: Perspectives from Molecular Anthropology" (2004, *Annual Review of Anthropology* 33:551–83).

David Hurst Thomas' *Skull Wars: Kennewick Man, Archeology, and the Battle for Native American Identity* (2000, Basic Books, New York) is a superb and eminently readable treatment of Kennewick, NAGPRA, and the relationship between archaeology and American Indians. Roger Downey takes a journalist's look at the Kennewick controversy in *Riddle of the Bones: Politics, Science, Race, and the Story of Kennewick Man* (2000, Copernicus Books). James Chatters has his say in *Ancient Encounters: Kennewick Man and the First Americans* (2001, Simon & Schuster, New York). The National Park Service maintains an informative Kennewick website with many of the primary documents at www.cr.nps.gov/archeology/kennewick/index.htm; another valuable source is the *Tri-Cities Herald's* Kennewick Man Virtual Interpretive Center at www.kennewick-man.com/index.html.

Questions of landscape learning have only recently come to the fore in discussions of colonization. In the past, we too often made simple assumptions such as "it was just like the Thule," and paid it no more mind. The single best scholarly source is Marcy Rockman and James Steele's edited volume, *Colonization of Unfamiliar Landscapes: The Archaeology of Adaptation* (2003, Routledge, London). It's also a topic of interest for several papers in the volume edited by Barton and others (cited above). "Behavioral ecology and archaeology" by Douglas Bird and James O'Connell (2006) and "Evolutionary foraging models in zooarchaeological analysis: recent applications and future challenges" by Karen Lupo (2007) provide excellent, if technical, discussions of foraging theory.

Useful volumes on Clovis include those edited by Robson Bonnichsen and Karen Turnmire and published by the Center for the Study of the First Americans, including *Clovis: Origins and Adaptations* (1991) and *Ice-Age People of North America: Environments, Origins, and Adaptations* (1999). I find much to disagree with in Gary Haynes's *The Early Settlement of North America: The Clovis Era* (2002, Cambridge University Press, Cambridge) but recognize it as a useful and detailed summary of the Clovis archaeological record. Likewise, however much I disagree with my friend Paul Martin, there is much to recommend in his *Twilight of the Mammoths* (2005, University of California Press, Berkeley)—just not the parts about overkill! The best single source on Pleistocene extinctions is a volume edited by Martin and Richard Klein, *Quaternary Extinctions: A Prehistoric Revolution* (1984, University of Arizona Press, Tucson). Although a bit outdated, the book nonetheless presents a comprehensive look at the issues involved. So, too, does Ross MacPhee and Hans-Dieter Sues's edited volume, *Extinctions in Near Time: Causes, Contexts, and Consequences* (1999, Kluwer Academic/Plenum, New York).

A deeply knowledgeable treatment of Paleoindian hunting is George Frison's *Survival by Hunting: Prehistoric Human Predators and Animal Prey* (2004, University of California Press, Berkeley). I relied heavily on it here. Bonnichsen and Turnmire's edited volume *Ice-Age People of North America: Environments, Origins, and Adaptations* (cited above) provides a valuable, region-by-region look at Paleoindian adaptations, not just Clovis. A handy, one-stop source for many of the Paleoindian papers that appeared in *American Antiquity* is Bruce Huckell and David Kilby's *Readings in Late Pleistocene and Early Holocene Paleoindians: Selections from American Antiquity* (2006, Society for American Archeology, Washington, D.C.).

More region-specific treatments include *Haida Gwaii: Human History and Environment from the Time of Loon to the Time of Iron People* (2005, University of British Columbia Press, Vancouver), edited by Daryl Fedje and Rolf Matthewes, a thorough reporting of their cutting-edge investigation into the human history of the Queen Charlotte Islands, and the islands' place in the colonization of the coast and continent. Still focusing on Beringia and Alaska are the edited volumes by Michael Bever and Mike Kunz, *Between Two Worlds: Late Pleistocene Cultural and Technological Diversity in Beringia* (2001, *Arctic Anthropology* 38), and Frederick West's edited volume *American Beginnings: The Prehistory and Palaeoecology of Beringia* (1996, University of Chicago Press, Chicago). Coming closer to mid-latitude North America is the volume edited by R. Carlson and L. Bona, *Early Human Occupation in British Columbia* (1996, University of British Columbia Press, Vancouver). David Anderson and Kenneth Sassaman's co-edited volume *The Paleoindian and Early Archaic Southeast* (1996, University of Alabama Press, Tuscaloosa) covers much of the Southeast. There's been a great deal written about Folsom and later Paleoindian adaptations and occupations of the plains, but few grand syntheses or even edited compilations (though lots of articles and books on specific sites and Folsom technology). We could use one. The early archaeology of the Great Basin is covered in Donald Grayson's *The Desert's Past: A Natural History of the Great Basin* (1993, Smithsonian Institution Press, Washington, D.C.). The latest word on early Californians can be found in *California Prehistory: Colonization, Culture, and Complexity*, edited by T. Jones and K. Klar (2007, Altamira Press, Lanham, Maryland).

For discussions of the impact of disease among Native Americans, see Alfred Crosby's classic *Ecological Imperialism: The Biological Expansion of Europe, 900–1900* (1986, Cambridge University Press, Cambridge), and Elizabeth Fenn's *Pox Americana: The Great Smallpox Epidemic of 1775–82* (2001, Hill & Wang, New York). For a more technical summary, see the multivolume set edited by David Hurst Thomas, *Columbian Consequences* (Smithsonian Institution Press, Washington, D.C.). A readable treatment of the pre- and post-European contact period is Charles Mann's *1491: New Revelations of the Americas before Columbus* (2005, Knopf, New York). William Ruddiman's *Plows, Plagues, and Petroleum: How Humans Took Control of Climate* (2005, Princeton University Press, Princeton, New Jersey) is a provocative look at the impact of humans on global climates, including the possible connection between disease de-population and atmospheric CO_2 levels.

Finally, to see how the peopling of the Americas fits in the larger context of world prehistory, excellent single sources are Steven Mithen's *After the Ice: A Global Human History 20,000–5000 BC* (2003, Weidenfield & Nicholson, London), and the more technical but nonetheless valuable book, *The Human Past: World Prehistory and the Development of Human Society*, edited by Chris Scarre (2005, Thames & Hudson, London).

NOTES

PREFACE

1. Like this. Some notes will be little more than names, dates, and page numbers. Others will elaborate on matters in the text. All can be ignored without penalty if one is so inclined.

2. Deloria, quoted in D. Thomas 2006:222. Zimmerman (1997:53) considers our suppositions about the Bering Land Bridge to be dogma. For thoughtful discussions of the differences between archaeology and oral history, and relations between archaeology and Native Americans, see Anyon et al. 1997, Ferguson, Watkins, and Pullar 1997, McGhee 2004, and D. Thomas 2000 and 2006. Archambault (2006) summarizes the main themes among the origin traditions of Native American groups, and some of the variations on those themes. Echo-Hawk (1997) identifies some of the elements in oral tradition that he believes might represent distant memories of a Pleistocene world.

3. Elliot 2006:9.

4. The term "New World" appears to have been coined by Peter Martyr, an Italian scholar attached to the court of Ferdinand and Isabella, probably in late 1494, which he subsequently used in the title to his collected letters and essays about Columbus's voyages, *De orbe novo: The Eight Decades of Peter Martyr D'Anghe*, first published in 1516.

CHAPTER 1: OVERTURE

1. Schlederman 2000; Wallace 2000.

2. Unless otherwise noted, all ages given in this book are in radiocarbon years.

3. By using high-precision Uranium/Thorium (^{234}U/^{230}Th) and corresponding ^{14}C ages from cores drilled into coral formations, the calibration curve has now been extended to approximately 26,000 years ago.

4. Broecker 2003; see also Hajdas, Lowe, and Newnham 2006; Hughen et al. 2000; Muscheler et al. 2008; Robinson et al. 2005.

5. Broecker 2006:1146.

6. For the record, in making these calculations I used round numbers and made the arbitrary assumption that the radiocarbon ages had a standard deviation of ± 100 years. I then took the resulting 1 sigma calibrated age range, found the median, and rounded to the nearest 5 years.

7. Kittler, Kayser, and Stoneking 2003.

8. Noonan et al. 2006.

9. Goebel 2004; see also Hoffecker and Elias 2003.

CHAPTER 2: THE LANDSCAPE OF COLONIZATION

1. As it happens, Glen Roy was the scene of the last inter-clan battle fought in Scotland; the MacDonells defeated the Mackintoshes here in 1688.

2. Darwin 1839a.

3. Darwin to Lyell, August 9, 1838, and Lyell to Darwin, September 6 and 8, 1838, in Burkhardt and Smith 1986; Darwin 1839b.

4. Darwin to Lyell, September 6, 1861, in Burkhardt et al. 1994.

5. Ruddiman 2005.

6. Macdougall 2004.

7. Buckland 1823.

8. Agassiz to Buckland, ca. 1838, in Agassiz 1887:289.

9. Krajick 2002.

10. Matters are slightly more complicated, since foraminifera disproportionately fix ^{18}O in colder water. Fortunately, ^{18}O shifts in both cases in the same direction, thereby amplifying the effect. Were it to do otherwise, the matter would be even more complicated. Regardless, the influence of water temperature must be accounted for in the analysis, and can be to a degree. For a discussion, see Macdougall 2004:173.

11. Alley 2000a.

12. The International Commission on Stratigraphy defines the Pleistocene thusly: It begins "just above top of magnetic polarity chronozone C2n (Olduvai) and the extinction level of calcareous nannofossil *Discoaster brouweri* (base Zone CN13). Above are lowest occurrence of calcareous nannofossil medium *Gephyrocapsa* spp. and extinction level of planktonic foraminifer *Globigerinoides extremus*," and its end is defined by "exactly 10,000 Carbon-14 years BP. Near the end of the Younger Dryas cold spell" (www.stratigraphy.org/). The feisty response from the Quaternary community is in Clague 2006.

13. The number of glacial episodes tallied in the marine isotope record met the Milankovitch model's predictions, but the timing proved not quite right. In a now-classic 1976 paper, James Hays, John Imbrie, and Nicholas Shackleton used spectral analysis (which can

tease apart multiple, superimposed cyclical curves) to plot the amplitudes and frequencies of changes in global ice volume occurring at cycles of 100,000 years, 42,000 years, and 23,000 years. These matched the predicted Milankovitch orbital frequencies to within 5% (Hays, Imbrie, and Shackleton 1976). This was too close for coincidence. Yet, Hays and colleagues were puzzled: their spectral analysis only explained about 60% of the variance in the marine oxygen-isotope curve. More troublesome, why were peaks in global ice volume coincident with the relatively weak forcing of the 100,000-year eccentricity cycle, and not the more powerful solar forcing of the obliquity and precession cycles? They were uncertain. Over the next several decades, glacial history was teased out in ever-finer detail in ocean sediment and ice cores, which revealed occasions when deglaciation occurred nearly 10,000 years before an insolation rise (we are justifiably baffled when effects precede their cause). Equally perplexing, peaks in global ice volume proved to be on a 100,000-year eccentricity cycle for only the second half of the Pleistocene. Before that, ice sheets kept pace with the much stronger 41,000-year obliquity cycle. Climate change also proved to have occurred at much faster rates—on the order of decades, centuries, and millennia (the sorts of changes that can have a more direct impact on humans)—than could be explained by changing insolation. But perhaps most puzzling, glaciation occurred in the Northern and Southern hemispheres simultaneously—at least on a long time scale—and yet orbital forcing ought to have opposite effects in the two hemispheres. The earth's climate was proving to be far more complicated than could be explained by a single variable such as insolation. As it turns out, at shorter time scales the northern and southern hemispheres do see-saw in their climates, which reveals that changing the amount of sunshine is vital, although only part of the equation: also important are other factors, including the volume of ice on Earth's surface, its reflectance, Earth's adjustments to ice sheet loading, climatic effects of the ice sheets themselves—both by their sheer mass, as well as what happens as they grow or come apart—and perhaps most important of all, circulation changes in the world's oceans that drive carbon dioxide, which is proving far more influential in climate change over the Pleistocene than previously imagined (Alley 2007; Broecker and Kunzig 2008; see also P. Clark, Alley, and Pollard 1999; Karner and Muller 2000; Marshall et al. 2004; Rind 2002; Schrag 2000).

14. Alley 2000a:118–20, 2000b.

15. In LGM times the Antarctic and Greenland ice sheets were larger, though not by much: there are limits to how far ice from these island-continents could extend over the ocean. Although there were many glaciers that expanded in LGM times, the other major continental glacier on the planet was the Fennoscandian ice sheet, which covered northern Europe and extended into northern Asia.

16. Dyke 2004.

17. Clague, Mathewes, and Ager 2004.

18. Macdougall 2004:165.

19. There is an ongoing and sometimes testy dispute over relative sea level history, driven by disagreements over how to interpret coral terraces in the Bahamas and New Guinea (which 'record' water depth at times past), and incorporate those data into complex models that account for—among other variables—the surface mass of water and ice,

and alteration of the rotational state of Earth as it adjusted its shape to those shifting masses (e.g. Lambeck et al. 2002; Peltier 2002).

20. Brigham-Grette et al. 2004; Keigwin et al. 2006.

21. Guthrie 1990.

22. The battle lines over the Productivity Paradox were drawn in the chapters in Hopkins et al. 1982.

23. P. Anderson, Edwards, and Brubaker 2004.

24. Dyke 2004:393.

25. Fedje et al. 2004; Fedje and Josenhans 2000.

26. D. Mann and Hamilton 1995; D. Mann and Peteet 1994; also Clague, Mathewes, and Ager 2004; Mandryk et al. 2001.

27. The details of Pleistocene climates are revealed in multiple sources: ice and deep sea cores, fossil records of plants and animals, and various geological indicators (sediments, past lake levels, and the like). These inform at different spatial and temporal scales. Take the Greenland ice cores: past temperature can be read from them directly from the ice (surprisingly, it still hasn't warmed from its LGM cooling), or via captured $\delta^{18}O$ (which varies by 1‰ every time temperature changes 1.5°C). Yet, the temperature is that of ice and air over Greenland, not of temperate and tropical regions. Those ice cores also contain salt spray and dust, and though these record climatic phenomena beyond Greenland, these, too, may have little bearing on other, more distant regions. Those cores also track changes in greenhouse gases, and though these reveal global patterns in warming and cooling, precisely how these played out in specific areas of North America depends on local insolation patterns, atmospheric circulation as modified by elevation, snow cover, to mountains, oceans, ice sheets, among other factors. The bottom line: climate reconstruction requires suites of evidence—and careful attention to the time scales at which these are relevant.

28. P. Clark, Alley, and Pollard 1999; Marshall et al. 2004.

29. Denton et al. 2005.

30. Grimm and Jacobson 2004.

31. Benson 2004.

32. The hydrological mechanics of the Lake Missoula flooding are explained in Waitt (1985:1280–1283).

33. Baker and Nummedal 1978; Booth et al. 2004.

34. J. Brown 1971; Grayson 2006a; Lawlor 1998.

35. S. Webb et al. 2004.

36. Grayson 2006b; also Kurten and Anderson 1980.

37. Poinar et al. 2006.

38. Why the lag? Joerg Schaefer et al. (2006) hypothesize that initial warming triggered the return of thermohaline circulation, but so destabilized the Laurentide ice sheet that its outer edges began breaking off, and armadas of icebergs floated into the North Atlantic: deep ocean sediment cores from that time are clogged with ice-rafted rocky debris. That influx of freshwater from melting icebergs may have slowed or stopped the thermohaline conveyor, stymieing Northern Hemisphere warming: further testimony that insolation alone cannot force climate's hand, but is amplified or diminished by other factors.

39. How fast does a surge, surge? In 1986 the Hubbard Glacier in Alaska shot forward at the nearly visible rate of approximately 10 meters per day across the mouth of Russell Fiord; by 2002 it had dammed the fiord and turned it into a lake, trapping seals and porpoises in it.

40. Dyke 2004.

41. Caves dug into the thick loess bluffs of Vicksburg, Mississippi, sheltered the town's inhabitants during Ulysses S. Grant's siege in the Civil War summer of 1863. When Vicksburg fell, a much-relieved Abraham Lincoln was moved to write that "the Father of Waters again goes unvexed to the sea." He had no idea how the "Father of Waters" had in Pleistocene times built those bluffs, making Vicksburg an important strategic spot on the river, and enabling the city to withstand Grant's lengthy siege.

42. During the first of these pulses approximately 16,000 years ago, global seas rose 10–15 meters in just 100–500 years (2–15 meters per century). The second major meltwater pulse—named, incongruously enough for those of us counting since the LGM, Meltwater Pulse 1A—at about 12,500 BP contributed a 20-meter rise in global sea level over a period of 500 years (4 meters per century). See Bassett et al. 2005; P. Clark et al. 2002, 2004; Lambeck, Yokoyama, and Purcell 2002.

43. Isostatic rebound continues to this day. It takes the land a very long time to recover from the former presence of ice.

44. Fedje et al. 2004. As one might imagine, the combination of isostatic rebound, tectonic uplift, and sea level rise make reconstructing the sea level and shoreline history in a particular area quite complicated (see Fedje et al. 2004:102).

45. Broecker and Kunzig 2008; Broecker 1995; Broecker and Denton 1990.

46. In hydrographic parlance, that's a flow of 20 Sverdrups, where 1 Sverdrup equals a flow of 1,000,000 m^3/second. For comparison, the Amazon River moves at 0.2 Sverdrups; the Mississippi River in flood is downright sluggish at a mere 0.016 Sverdrups.

47. Salinity is so critical, in fact, that deep water does not form in the north Pacific because the water there is too fresh. Instead, Pacific cold water moves north at depth, rises and warms, then moves south on the surface.

48. See Alley 2000a and Broecker and Kunzig 2008. Thermohaline circulation is sensitive to freshwater perturbations even as slight as just 0.1 Sverdrup; the hypothesized Younger Dryas flood (at .305 Sverdrups) came in at three times that amount—about 1.5 times the flow of the Amazon (P. Clark et al. 2004:1141; Ellison, Chapman, and Hall 2006).

49. Alley 2007. There was an Older Dryas, but it was a very brief and far less consequential episode of cooling.

50. Alley 2000a:105–6; Kohfeld et al. 2005. Other facts in play include the degree iron or other nutrients that can increase phytoplankton mass and thus the amount of carbon fixation, changes in water temperature (CO_2 is dissolved more readily in cooler water), and sea-ice cover (which can stymie ventilation of CO_2 from the abyss).

51. Raynaud et al. 2003; Siegenthaler et al. 2005. Climatologist William Ruddiman argues that anthropogenic CO_2 first appeared as early as 8,000 years ago when Neolithic farmers began clearing and burning forests and releasing CO_2 (Ruddiman 2005:88–94). Regardless, the precise links between atmosphere, ocean, CO_2, and temperature are yet to be fully understood. Scientists across the globe are working on it, and it is more

than a strictly academic exercise. Gaining insight into what drives climate over geological time will enable us to gauge which ongoing changes are natural, and which are the result of human activity. Doing so will perhaps allow us to predict how the earth's climate will respond to future forcing, both natural and human. That information is coming, and none too soon.

52. Broecker 2003:1519. More specifically, during the first major post-glacial meltwater release, when freshwater was returned to the ocean, there was a rise in atmospheric ^{14}C of as much as approximately 100 parts per million (‰), followed by a decline of nearly 150‰ (after the second major meltwater discharge), and then just plain radiocarbon confusion for a period of about 200 years before 11,100 BP when the clock stalled again (from a decline in atmospheric ^{14}C), followed by about 200 years when the radiocarbon clock ran three times too fast, owing to a sharp rise (approximately 70‰) in atmospheric ^{14}C likely the result of the MOC shutdown and decreased uptake of ^{14}C by the deep ocean. Over the remainder of the Younger Dryas there was a gradual decline in the ratio of $^{14}C/^{12}C$, followed at the end of that period with an abrupt drop in that ratio (Alley 2007; Broecker 2003:1519; P. Clark et al. 2004:1142; Hughen et al. 2000:1951–53; Muscheler et al. 2008).

53. Alley et al. 2005; P. Clark, Mathewes, and Ager 2002, 2004. There is debate over the timing and magnitude of these pulses of sea level rise, as well as which melting ice sheets are to blame, whether in the southern or northern hemisphere—see Peltier 2005.

54. The Bølling-Allerød is also marked in the Greenland ice cores by higher $\delta^{18}O$ levels (Severinghaus and Brook 1999:930; Jouzel 1999:911). However, the timing is so tight it is not clear if the Bølling-Allerød followed the 12,500 BP meltwater pulse or preceded it (Alley et al. 2005).

55. MacDonald et al. 2006; Raynaud et al. 2003. During colder periods, tropical wetlands are the primary source of methane (Alley 2007). Methane (CH_4) is a more-efficient greenhouse gas than CO_2, but thankfully for our modern circumstances, it only has about a ten-year residence time in the atmosphere. Otherwise, our world would be heating up a great deal faster than it already is.

56. Shuman, Bartlein, and Webb 2005; T. Webb, Shuman, and Williams 2004; Williams, Shuman, and Webb 2001.

57. Grimm and Jacobson 2004; T. Webb, Shuman, and Williams 2004.

58. FAUNMAP 1996; D. Mann, personal communication.

59. Curry 2007.

60. Teller 2004.

61. On the bond between thermohaline circulation shutdown and the Younger Dryas, see Alley 2007; Broecker and Kunzig 2008; McManus et al. 2004. Fisher et al. 2008, Lowell et al. 2005, and Teller et al. 2005 summarize the geological and hydrological complications with the eastward-draining Lake Agassiz hypothesis. But if it wasn't Lake Agassiz's water flowing out the St. Lawrence seaway that's to blame, where might the freshwater have come from? Alternative possibilities include a large meltwater discharge from the Keewatin ice dome into the Arctic Ocean via the McKenzie River (Tarasov and Peltier 2006), or perhaps Arctic sea ice flushed out into the North Atlantic

as global sea levels rose (Bradley and England 2008). Perhaps. But Lake Agassiz water remains in the mix, as the timing of its drainage is still far too poorly known to preclude a meltwater pulse down the St. Lawrence, and a fascinating recent study of sediment cores from the St. Lawrence estuary revealed evidence of a Younger Dryas-age spike of chemical elements that could only have come from western Canada, and hence had to have flowed east with Lake Agassiz's water (Carlson et al. 2007).

62. Firestone and Topping 2001.

63. Southon and Taylor 2002.

64. Firestone, West, and Warwick-Smith 2006.

65. Becker has worked on such claims before. She and colleagues claimed there had been an ET impact at the Permian-Triassic boundary that killed off most of life then on earth. Their claim hasn't fared well: "three *Science* papers later," one observer noted, and Becker's group has still "failed to make its case" for an impact (Kerr 2007).

66. Kennett, quoted in Pringle 2007.

67. For a recent review of the geological evidence for that impact and its consequences, see Kring 2007.

68. The sites reported here are mostly the ones in *The Cycle of Cosmic Catastrophes*. There's sleight of hand here, however unintentional: the authors note there are more than fifty sites with a black mat—implying all are evidence of a great fire, contain extraterrestrial elements, and so forth. But black mats are never defined, they examined fewer than ten of those particular sites, and not all the sites they examined even had black mats or radiocarbon ages to indicate they actually dated to this time period. There is also debate about whether there are fifty sites with a black mat; further, it has been shown that not all black mats are late Pleistocene in age, so if they are impact markers, there were a lot more impacts over time.

69. An event of this magnitude ought to have a global signature, so its evidence needs to be found in a great number of sites across the planet, not just a sprinkling of Clovis sites and one in Belgium. On the other hand, since the proposed impact was atop the Great Lakes, there also needs to be a concerted effort to examine sediment cores from the tens of thousands of lakes in that region.

70. Firestone et al. 2007; Ivester et al. 2003; West, quoted in Pringle 2007. Geologists in the Carolinas flatly dismiss the idea the Carolina Bays are related to impacts (Pringle 2007). See also Pintar and Ishman 2008.

71. Koeberl and Gabrielli quoted in Kerr 2008; Kring, quoted in Kerr 2007; Pintar and Ishman 2008.

72. Kennett, quoted in Dalton 2007; Firestone et al. 2007.

73. The post-Clovis so-called decline is nothing more than a small number of post-Clovis Redstone points in South Carolina. Yet, points are not people, and point counts are hardly a valid proxy of population (were it only so simple), but instead a function of many things, not least the availability of stone (Chapter 8). More telling, no Redstone point has ever been radiocarbon dated, so it's not known whether they directly follow Clovis. And as one of the Firestone team admits, few Clovis sites anywhere in eastern North America have been radiocarbon dated, and all of them might postdate the time of the ET impact (Goodyear 1999). Finally, on the Great Plains, there are many Clovis

sites and, more importantly, many more well-dated Folsom sites, which follow Clovis directly in time. If there was massive population collapse, Folsom people failed to get the memo.

74. Firestone et al. 2007.
75. Berger and Loutre 1991; Denton et al. 2005; Yu and Wright 2001.
76. Alley 2000b; MacDonald et al. 2006; Raynaud et al. 2003. The atmosphere of the Younger Dryas had upwards of seven times more dust and three times more salt in the air than in the millennium that followed, which was itself twice as dusty and salty as the atmosphere was in the preindustrial era (Alley 2000a:112–15).
77. Shuman et al. 2002:1786–7; see also Grimm and Jacobson 2004; Yu and Wright 2001.
78. Shuman et al. 2002; Reasoner and Jodry 2000; Fall 1997.
79. Shuman, Bartlein, and Webb. 2005:2197.
80. Alley 2000a:114–16.
81. Broecker 2006:1147.
82. Shuman et al. 2002; Shuman, Bartlein, and Webb 2005; T. Webb, Shuman, and Williams 2004. The Younger Dryas saw the most abrupt and widespread changes in North American biota of any time since the LGM, but it is important to add that it was mostly the pace of change that was so noteworthy; otherwise, the vegetation changes that occurred during the Younger Dryas account for less than 25% of the total changes that took place in North America's biota over the millennia between the LGM and the onset of the Holocene.
83. Alley 2000a:182. By the most recent evidence, the peak of the 8200 BP event spanned only 60–70 years, and evidence of its geographic impact seems largely restricted to the circum-North Atlantic region (the British Isles, Norway, and Newfoundland).
84. Dyke 2004; Alley, quoted in Kerr 2006.

CHAPTER 3: FROM PALEOLITHS TO PALEOINDIANS

1. Jefferson 1788.
2. Haven 1856:3.
3. Jefferson 1788:226.
4. Jefferson 1788:227.
5. The story of the establishment of human antiquity in Europe has been told many times; good general treatments include Grayson 1983 and Van Riper 1993.
6. Haber 1959.
7. Cuvier's demonstration of extinction, and the role played by American fossils as well as information from American Indians, is discussed in Mayor 2005.
8. Cuvier 1796, in Rudwick 1997.
9. Bowler 1976:31.
10. Grayson 1983; Meltzer 2005.
11. Lubbock 1865:269.
12. Buckland 1823:90; Pettitt 2000.
13. Lyell 1863:68.

14. Darwin to Hooker, June 22, 1859, in Burkhardt and Smith 1991:308.
15. Quoted in Evans 1943:103.
16. Lyell 1860:94.
17. Dawkins 1863.
18. Lubbock 1865.
19. Quoted in Grayson 1983:217–18.
20. Darwin 1859:488; the Boucher de Perthes quote is in Darwin 1871:3.
21. There are no book-length treatments of human antiquity in the Americas; the discussions here are drawn mostly from my own published articles, particularly Meltzer 2005 and 2006b. Both of these provide reasonably comprehensive summaries with references, for those who wish to pursue the topic in more detail.
22. Abbott diary, February 14, 1877, Firestone Library, Princeton University.
23. Wright 1912:228.
24. Wright 1890; Abbott 1889.
25. Holmes 1890.
26. Henshaw to Abbott May 21, 1890, Charles Abbott Papers, University Museum, University of Pennsylvania.
27. Undated sketches, William Henry Holmes Papers, Smithsonian Institution Archives, Washington, D.C.
28. Abbott 1892a.
29. Holmes 1893a, 1893b.
30. Abbott 1892a:271.
31. Abbott 1892b:344–45.
32. Holmes 1893a.
33. Chamberlin 1892; McGee 1893; reviewing Wright 1892.
34. The material in this sidebar is drawn from Meltzer and Sturtevant 1983, and Meltzer 1990.
35. Abbott to Wright, April 5, 1899, George Frederick Wright Papers, Oberlin College, Oberlin, Ohio.
36. Meltzer 1983, 2005.
37. Hooton 1930.
38. Upham to Holmes, August 23, 1902, Upham to Chamberlin, August 23, 1902; Warren Upham Papers, Minnesota Historical Society Minneapolis, Minnesota.
39. Chamberlin to Upham, November 12, 1902, Thomas Chamberlin Papers, University of Chicago, Chicago, IL.
40. Abbott to Wright, April 20, 1904, George Frederick Wright Papers, Oberlin College, Oberlin, OH.
41. Hay 1918.
42. Gilder 1911.
43. Kidder 1936; Roberts 1940; N. Nelson to O. Hay, April 5, 1920, Oliver Hay Papers, Smithsonian Institution Archives, Washington, D.C.
44. Sarah Rooke's grave marker is—rightly—one of the largest in the Folsom cemetery, erected by her fellow operators; it reads, in part, "With heroic devotion she

glorified her calling by sacrificing her own life that others might live." It's a fitting monument.

45. Osborn 1922; see Skinner, Skinner, and Goorhis 1977.

46. Cook 1925.

47. W. D. Matthew to J. C. Merriam, November 27, 1925, John C. Merriam Papers, Library of Congress, Washington, D.C.

48. Hrdlička scrawled his opinion across his copy of Cook's paper.

49. Cook 1927:117.

50. Evans to Figgins, October 17, 1929, Figgins Papers, Denver Museum of Nature and Science.

51. Cook 1927:116.

52. Figgins to Howarth, July 22, 1926, Papers of the Director, Denver Museum of Nature and Science.

53. Figgins to Cook, December 28, 1926, J. D. Figgins Papers, Denver Museum of Nature and Science, Denver, Colorado.

54. Figgins to Brown, June 8, 1927, Vertebrate Paleontology Archives, American Museum of Natural History, New York.

55. Carl Schwachheim diary, September 4, 1927, Department of Anthropology, Southern Methodist University, Dallas, TX.

56. Meltzer 1983.

57. Cook to Loomis, November 12, 1928, Harold Cook Papers, Agate Fossil Beds National Monument, Agate, Nebraska.

58. Meltzer 2006b.

59. Hull 1988:31.

60. Abbott to Robins, June 28, 1897, C. C. Abbott Papers, Academy of Natural Sciences, Philadelphia, PA.

61. Rudwick 1985:420.

62. Kidder to Figgins, October 13, 1927, Papers of the Director, Denver Museum of Nature and Science, Denver, CO.

63. Cook to Hay, December 23, 1926, Oliver Hay Papers, Smithsonian Institution Archives, Washington, D.C.

64. Lucas to Figgins, November 18, 1927, Papers of the Director, Denver Museum of Nature and Science, Denver, CO.

65. Figgins to Gregory, December 30, 1927, Papers of the Director, Denver Museum of Nature and Science, Denver, CO.

66. N. Nelson to Figgins, August 16, 1927, J. D. Figgins Papers, Denver Museum of Nature and Science, Denver, CO.

67. Cook to Hay, January 25, 1928, Oliver Hay Papers, Smithsonian Institution Archives, Washington, D.C.

68. Kidder to Figgins, October 13, 1927, Papers of the Director, Denver Museum of Nature and Science, Denver, CO.

69. Boneless Paleoindian sites are more difficult to find, though not impossibly so. Archaeologist Reid Ferring, equally skilled as a geologist, found the Aubrey Clovis site while walking the outlet channel below an Army Corps of Engineers' dam in north Texas. No large bones called him to the locality, just the knowledge that a Pleistocene-aged surface

was exposed there. On that surface, he found the tip of a fluted point, and beneath it an iceberg of a Clovis site. It is all a matter of knowing the right place to look. Ironically, after three years of meticulous excavations resulting in an extraordinary archaeological record, the Aubrey site finally did yield one mammoth rib fragment (Ferring 2001).

70. Simms 1996.
71. Krieger 1953:238–39.
72. Krieger 1964:44.
73. C. Haynes 1964.
74. C. Haynes 1969.
75. C. Haynes 1967.
76. Abbott to Putnam, June 20, 1888, Peabody Museum Papers, Harvard University, Cambridge, MA.

CHAPTER 4: THE PRE-CLOVIS CONTROVERSY AND ITS RESOLUTION

1. Ironically, questions were later raised about Senga's antiquity, stratigraphy, and context.
2. Guidon 2002; Guidon and Arnaud 1991.
3. Krieger 1964; MacNeish 1976; Morlan 1988. The five on Morlan's list were Bluefish Caves, Yukon Territory; Meadowcroft Rockshelter, Pennsylvania; Taima-Taima, Venezuela; Tagua-Tagua, Chile; Los Toldos, Chile. None has yet to be fully accepted.
4. Martin's low opinion of pre-Clovis archaeology, expressed in his *Twilight of the Mammoths* (2005) among other places, is biased—even by his own admission—by his strong opinions about the role the first Americans played in the extinction of the Pleistocene fauna. I'll take up that matter in Chapter 8.
5. Hrdlička 1907; C. Haynes 1969.
6. The dispute over the utility of the criteria played out in several places, most especially in two of the volumes edited by pre-Clovis proponent Alan Bryan (1978, 1986).
7. Leakey et al. 1969.
8. For further reading on Calico, see the chapter "Misadventure at Calico" in Virginia Morrell's *Ancestral Passions* (1995), and also Mary Leakey's *Disclosing the Past: An Autobiography* (1984). The Calico Conference proceedings were published in Schuiling 1972 (from which the Simpson quote is drawn, page 36).
9. Toth 1991.
10. Leakey 1984:143.
11. Stalker 1977; Wilson, Harvey, and Forbis 1983; R. Brown et al. 1983.
12. Good introductions to the various radiometric dating methods are a series of papers in *Evolutionary Anthropology*, including Dieno, Renne, and Fisher 1998; Feathers 1996; Grun 1993; Kappleman 1993; Schwarz 1992; Taylor 1995; and Wagner 1996.
13. Summarized in Gonzalez, Huddart, and Bennett 2006; see also Irwin-Williams 1981; Steens-McIntyre, Fryxell, and Malde 1981.
14. Renne et al. 2005.
15. Hardaker 2007.
16. Gonzalez et al. 2006.

17. Renne et al. 2005; Duller 2006; Schwenninger et al. 2006; see also the "Mexican foot-prints" webpage, http://www.mexicanfootprints.co.uk/default.htm.

18. Lynch 1991a. The same argument is made by Martin (2005:136), who observes that Australia has lower rainfall, and hence supported fewer people creating fewer sites, and yet Aussies found them despite not having the massive CRM programs we have here in North America. But flip this around: lower rainfall means better surface exposures, and though we do an awful lot of CRM work (Martin calls it by the old-fashioned name "salvage archaeology") how much of it has exposed Clovis age sites? Hardly any. Although there has been, as Jelinek notes, an "enormous amount of earth" in the United States moved by construction, providing "a random sampling of the topography and environments of the continent" (Jelinek 1992:346), it's doubtful that sample is truly random. It is true it hasn't yielded any pre-Clovis sites. But then it has not turned up very many Clovis sites either. The issue is not the amount of earth moved, but the amount moved that exposes surfaces of the proper age. We need an age-representative sample, not just a random one; then the absence of evidence becomes evidence of absence.

19. O'Connell and Allen 2004. After re-dating, Jinmium turned out to be only 20,000 years old. For a broader summary of Australia's dating controversy and discussions of peopling of that continent—including the debate over the role of the first Australians in Pleistocene extinctions there—see Hiscock (2008).

20. Martin 1987:11.

21. Hassan's analysis of population growth rates appears in his *Demographic Archaeology* (1981). A similar point is made by Bettinger and Young (2004).

22. Irving 1985; Stanford, Bonnichsen, and Moran 1981.

23. See, for example, Lyman 1984; Myers, Voorhies, and Corner 1980.

24. A colorful, if not wholly one-sided précis of the Meadowcroft controversy can be found in Adovasio and Page (2002), which is accompanied by a thorough bibliography of the Meadowcroft team's work on the site. Publications by the site's critics are not listed.

25. Adovasio and Page 2002:186.

26. C.V. Haynes 1991b.

27. Adovasio and Page 2002:224.

28. Adovasio and Page 2002:186.

29. Guidon and Arnaud 1991:170

30. Guidon and Arnaud 1991:177.

31. Meltzer, Adovasio, and Dillehay 1994.

32. The Pedra Furada debate took place in the pages of *Antiquity*. See Meltzer, Adovasio, and Dillehay 1994; the responses by Guidon and Pessis 1996 and Parenti, Fontugue, and Guerin 1996, with additional thoughtful comments by Borrero 1995; and an almost-tongue-in-cheek commentary by Dennell and Hurcombe 1995 on how to test whether nature flakes stone (they also raise the Shakespeare-typing monkey analogy).

33. Recent radiocarbon and TL analyses on material from Pedra Furada appear in Santos et al. 2003 and Valladas et al. 2003, respectively.

34. The Monte Verde evidence is laid out in the two volumes (Dillehay 1989, 1997). Lynch's critique and reply appeared in *American Antiquity* (Lynch 1990, 1991b) along with Dillehay and Collins's (1991) rebuttal.

35. Dillehay 1989:xx.

36. Dincauze 1991:116.

37. Lynch 1990:26.

38. Dillehay, personal communication, 2007.

39. Dillehay and Collins 1991; Lynch 1991b.

40. Saunders 1998:145.

41. The Monte Verde site visit was reported in the October 1997 issue of *National Geographic*, which sent a writer and photographer along with us.

42. Wilford 1997.

43. A point raised by Haynes, but one he himself would later demonstrate to be a nonissue; see Taylor et al. 1999.

44. Adovasio and Page 2002:224.

45. D. Thomas 2000:166.

46. See On Monte Verde: Fiedel's Confusions and Misrepresentions (www.uky.edu/ Projects/MonteVerde/). In the course of their response, Dillehay's team discovered that many of the errors Fiedel cited had actually been corrected by them in their page proofs, but the Smithsonian Institution Press (which published *Monte Verde*) failed to make the corrections before printing the book. The net result of all the brouhaha? Grayson put it this way: "The only substantive thing Fiedel accomplished was forcing the Smithsonian to issue a volume correcting a bunch of typos."

47. The assertion of Paul Martin—who wasn't a part of the site visit—that "not all in the group agreed on the claims of antiquity" is merely his wishful thinking.

48. Meltzer et al. 1997.

49. C. Haynes 1999.

50. Dillehay et al. 2008; D. Sandweiss, quoted in Balter 2008.

51. J. Johnson et al. 2002; Rick, Erlandson, and Vellanoweth 2001. The precise age of the Arlington Springs specimen is not fully settled, since this individual may have fed on marine resources, which can skew a radiocarbon age. That possibility can be taken into account by careful study of the skeleton's isotopic composition; however, that's not yet been done.

52. Davis 2006.

53. Fedje et al. 2004. See also Davis 2006; Mandryk et al. 2001.

54. Dillehay 2000:128–31.

55. Gilbert et al. 2008; Jenkins 2007.

56. Pitulko et al. 2004.

57. Martin (2005:136–37) essentially argues thus: If humans were here before 11,500 years ago, they would have killed off the ground sloths. Ground sloths do not become extinct until ca. 11,000 BP. Therefore, humans were not here before 11,000 BP. But whether humans were here and what caused the extinction of the ground sloths are independent issues.

58. Butzer 1991.

59. Collins 2002. See also Davis 2006; Dillehay and Meltzer 1991; Fedje et al. 2004; Punke and Davis 2006.

CHAPTER 5: NON-ARCHAEOLOGICAL ANSWERS
TO ARCHAEOLOGICAL QUESTIONS

1. Goddard 1927.
2. Goddard 1927:263.
3. Holmes 1919:55; Hrdlička 1925:485.
4. Sapir 1916.
5. Greenberg, Turner, and Zegura 1986.
6. Details on the number and status of American Indian languages come from Kaufman and Golla 2000:47–48; Nichols 1990, 2002; and Goddard 1996.
7. Goddard 2004.
8. For a good summary of language change, see Kaufman and Golla 2000; Golla 2000; and Ringe 2000 in Renfrew. For recent efforts to push back the time depth in which one can still see shared ancestry (now at 8,000–10,000 years), see Dunn et al. 2005 and Gray 2005. The change has striking parallels to the evolution of biological species: they split into new languages, mutate, and go extinct.
9. Nettle (1999) argues that as small colonizing groups fill new niches, linguistic diversity increases sharply, and as population density increases and isolated groups rejoin, linguistic diversity decreases.
10. Ruhlen 1994.
11. J. W. Powell 1891:26.
12. Sapir 1921.
13. Greenberg 1987. Greenberg provides an overview of this history of linguistic classification, as does Campbell 1997. For reasons that will shortly become clear, they differ in their take on that history. It might be added that Greenberg didn't stop with the Americas: whereas most linguists grouped the world's approximately 6,000 languages into about 300 families, he collapsed them into about 18 families (Pennisi 2004; Ruhlen 1994).
14. Which is why some of our words are obviously from German (foot = *fuss,* as opposed to the French *pied*), while others are clearly French in derivation (human = *humain,* as opposed to the German *menschlich*). To complicate matters, English and French were also related in the more distant past: both are Indo-European derivatives, and both have elements of Latin, which was the primary language of formal discourse and thus also parachuted directly into English without first being filtered through French.
15. Ringe 2000.
16. Numbers from Golla 2000, Kaufman and Golla 2000; Algic distribution from map in Goddard 1996; also Foster 1997:97; Goddard 1996:4; Campbell 1997:152.
17. J. Hill 2004:39.
18. J. Hill 2004:39.
19. See Greenberg 1987 for a description of his method.

20. That connection was recently drawn tightly by Edward Vajda, comparing tonal forms, words, and sound shifts between Ket (which is in the Yeniseian language family), with several Na-Dene languages (including Tlingit and Eyak) (Holden 2008).
21. Greenberg 1987:55.
22. Quoted in Lewin 1988.
23. As Lyle Campbell announced in his comments that immediately followed the paper by Greenberg et al. 1986.
24. Ringe, quoted in Pennisi 2004:1321.
25. J. Hill 2004. For searing criticisms of Greenberg's *Language in the Americas,* see Campbell 1988, 1997; Matisoff 1990; for a more recent summary, see J. Hill 2004. Greenberg responded to his critics in Greenberg 1990.
26. Greenberg 1987:ix.
27. Campbell 1997. The lack of a tally for just how often *n/m* and *m/t* occur or fail to occur in both American and non-American languages, makes it impossible to judge whether there are statistically significant differences in the frequency of the *n/m* pronoun here as opposed to elsewhere. Greenberg insists the *n/m* pattern is real, citing as evidence on this point the fact the *n/m* discussion takes up "five full pages" of *Language of the Americas* (1987). That's a rather unconventional measure of the strength of evidence. In the book, however, he reports "over one hundred" *n/m* occurrences. If, as it happens, *n/m* and *m/t* are both abundant, and not the result of common ancestry, it raises interesting questions: Why are there only a limited number of pronoun forms? Could these reflect sound patterns basic to all human language?
28. Campbell 1997:254.
29. Greenberg's defense has been carried forward by his one-time student Merritt Ruhlen, who has taken the tack that the critics have it all wrong: Greenberg wasn't trying to present a finished product, but rather was merely classifying like languages as a first step, from which traditional historical linguists could then do sound correspondences, reconstructions, and the like. And one could do so even with meager evidence—as in the case of Shebayo, a language that vanished in the 1600s for which only fifteen words are known. In effect, Ruhlen argued, Greenberg was doing American linguists a favor by pointing to where detailed connections might be found (Ruhlen 1994:130–32). The critics still weren't buying: multilateral comparison, Ringe snarled, is "not even a reasonable heuristic for interesting similarities deserving further study; it ought to be abandoned" (Ringe 2000:155).
30. Turner 1985:36.
31. Turner, Manabe, and Hawkey 2000. Ages approaching 30,000 years ago are provided in Chen, Hedges, and Yuan 1989; favoring a younger age are Wu and Wang (1985).
32. Turner 1986. There may be less here than meets the eye, since in many cases the independent divergence ages were based on the presumed association of the teeth with archaeological complexes that were dated—but which are not necessarily associated with a dental group—or in another case, on the breaching of Beringia, which is assumed to have separated northeast Asians and ancestral Aleutians. Why a seafaring people would have been stopped cold in their tracks by rising post-glacial seas

is unclear; for that matter, the ages used for the breaching (i.e., 14,000–12,000) are incorrect (Turner 1985:37–40).

33. Greenberg, Turner, and Zegura 1986.
34. Ruhlen 1994:166.
35. Greenberg, Turner, and Zegura 1986.
36. Turner 1985.
37. Turner 1985:43.
38. Hanihara and Ishida 2005.
39. Paabo 2003.
40. Cavalli-Sforza, Menozzi, and Piazza 1994.
41. Greenberg, Turner, and Zegura 1986:487.
42. Wang et al. 2007.
43. Turner 2005:171.
44. Or so we thought. There has been some recent work that hints at the possibility that bits of the father's mtDNA might infiltrate the mtDNA of the child. That claim has been hotly disputed, and it is still not clear just how much of a problem, if any, exists. It seems reasonable to operate as if there is no recombination of maternal and paternal genes in the mtDNA molecule until proven otherwise.
45. For those keeping count, the human replication error rate is between 10^{-9} and 10^{-10} base pairs per nucleotide per generation (Zegura, personal communication, 2007).
46. This example is not entirely arbitrary: the mutation at np 16327 (C → T) happens to be the defining mutation for haplogroup **C**. The mutation at np 16325 (C → T) is the defining mutation for the subset of haplotypes C1b, C1c, and C1d within haplogroup C, while the mutation at np 16356 defines haplotype C1a. As it happens, C1a is the haplotype within that haplogroup that remained in Asia; C1b–C1d represent New World founding haplotypes (Tamm et al. 2007).
47. Eshleman, Malhi, and Smith 2003; Merriwether 2002; Schurr 2004a:237.
48. Torroni et al. 1993:581–82; see also Forster et al. 1996:938; Merriwether, Rothhammer, and Ferrell 1995:425. Kemp et al. 2007 recently added that founding lineages in theory should be detected in ancient DNA, and the older the sample, the more likely the lineage is a founder.
49. Seielstad and colleagues sparked an intense debate in the late 1990s when their survey of African and European populations showed mtDNA markers were more widespread, while NRY variants tended to be more restricted geographically. Their conclusion: women tend to marry out (Seielstad, Minch, and Cavalli-Sforza 1998). They assumed this was a universal pattern, but others were quick to criticize: differences in mutation rates between mtDNA and NRY and in the sample sizes they analyzed meant that Seielstad's team were comparing "apples and pears." Michael Hammer and colleagues subsequently better aligned the mtDNA and NRY data and samples, and found no evidence of higher (or lower) migration rates for females on a global scale (Wilder et al. 2004). Locally, of course, there could be significant differences in movements of males and females, depending on cultural practices: for example, in farming societies where land was inherited father to son, and the oldest son (who often stood first in line to inherit) stayed put while younger brothers left to find new lands, joined new groups, and married out.

50. It's the *divergence method* that does pair-wise comparison of all haplotypes within groups, and multiplies the distance by the mutation rate to estimate the time required to produce this much diversity. The *coalescent method* estimates the amount of time necessary to track two groups that are genetically separate today back to their most recent common ancestor; in effect, it can be viewed as genetic drift run backward. Ages derived by the divergence method require firmer control on sequence diversity than coalescent ages, while coalescent ages will not provide as clear a picture of the process of diversification over time (Schurr 2004a:200–201).

51. Mutation rates of mtDNA and NRY are calibrated using cases where the moment in time when two populations split from one another is dated independently. Ironically, one of the splits originally used to calculate the mtDNA mutation rate was based on the genetic distance between Asians and American Indians, which was assumed to have occurred 12,000 years ago. It yielded a mutation rate of 2% –4% per million years (Cann et al. 1987; Schurr 2004a:229–30).

52. Schurr 2004a:229–30, 2004b:559; see also Ho and Larson 2006.

53. In part for this reason, more recent studies have focused on sequencing complete mtDNAs (Tamm et al. 2007). Full details on the mitochondrial molecule are available on the web at MITOMAP: A Human Mitochondrial Genome Database *(www.mitomap .org)*, which includes a compendium of polymorphisms and mutations of the human mitochondrial DNA.

54. Wallace's pathbreaking work appeared in D. Wallace et al. 1985. A bottleneck is also apparent in autosomal DNA (Wang et al. 2007).

55. Torroni collaborated with Wallace, and together they published a number of the central papers on American Indian mtDNA during this time. Torroni (2000) summarizes the evolution of their results and thinking during the first dozen years of their work. The first four letters (haplogroups) of the mtDNA alphabet were based on this work (Tamm et al. 2007).

56. Fix 2005:432.

57. Torroni 2000.

58. At one point, Torroni and colleagues suspected there might have been four migrations to account for migration of Eskimo-Aleut.

59. Merriwether, Rothhammer, and Ferrell 1995; also Eshleman, Malhi, and Smith 2003; Mulligan et al. 2004.

60. For a helpful discussion of "The Curious Case of Haplogroup X," see Eshleman, Malhi, and Smith 2003:10.

61. Haplogroup **X** was recognized as a founding lineage by several laboratories in the latter 1990s. It was M. Brown et al. (1998) who speculated it might have been brought to America by Europeans in Pleistocene times, a possibility subsequently discounted by Smith et al. (1999) and others (e.g., Mulligan et al. 2004). The presence of haplogroup X in Siberia was documented by Derenko et al. 2001, but more recent work indicates that the Altai–American Indian haplogroup X connection is not as close as once thought (Zegura, personal communication, 2007).

62. To be more precise, the five Native American haplogroups are A2, B2, C1, D1, and X2a (Tamm et al. 2007; Fagundes et al. 2008).

63. See, for example, Karafet et al. 1997, 1999.

64. Bolnick et al. 2004; Hunley and Long 2005; Zegura et al. 2004.

65. Hunley and Long 2005:1316.

66. Studies of autosomal DNA likewise document a DNA mismatch with Greenberg's language families (K. Schroeder et al. 2007) and, in more general terms, a low correlation between genetics and linguistics (Wang et al. 2007).

67. Distribution data summarized from Merriwether 2002; Schurr 2004b; Smith et al. 2006.

68. Schurr 2004b:555–56. Smith et al. 2006:248

69. Torroni et al. 1993.

70. Tamm et al. 2007; for slightly different age estimates of the haplogroups, see Achilli et al. 2008. On Paisley Cave DNA, see Gilbert et al. 2008.

71. Eshleman, Malhi, and Smith 2003; Mulligan et al. 2004; Schurr 2004a; Tamm et al. 2007.

72. Merriwether et al. 1996; Schurr 2004a.

73. Malhi et al. 2002:905–6; Merriwether et al. 1996:205, 210; Schurr 2004a.

74. A good introduction to NRY work is provided in Hammer and Zegura (1996), from which this discussion is largely drawn. Over 95% of the Y chromosome is non-recombining; overall, the Y chromosome possesses about forty separate functional genes (S. Zegura, personal communication, 2007), essentially giving it fewer targets for natural selection.

75. Nomenclature was standardized by the Y Chromosome Consortium in 2002.

76. Zegura et al. 2004:Table 1.

77. NRY data from Lell et al. 2002; Zegura et al. 2004:170. The quote is from Zegura et al. 2004.

78. Seielstad et al. 2003:704; Zegura et al. 2004:1720.

79. Hey 2005; Kitchen, Miyamoto, and Mulligan 2008; see also Fagundes et al. 2008; Moore 2001; Wang et al. 2007.

80. Merriwether, Rothhammer, and Ferrell 1995.

81. Malhi et al. 2007:642. Also A. Merriwether 2002:299.

82. For a variety of reasons, however, American Indians have been reluctant to participate in the Geonographic Project (Harmon 2006).

83. That potential, however, may be truly tiny: Nick Patterson calculates that of the 65,536 ancestors one has sixteen generations back, approximately 99% of them might have contributed nothing to your genome (Patterson, personal communication, 2007). Of course, that still leaves about 1% who did, and their histories can be quite varied genetically, and not necessarily concordant with the two persons who were one's mtDNA and NRY ancestors.

84. Malhi et al. 2007:642. Kemp et al. 2007.

85. To date, none of the samples of earliest occurrences of ancient DNA have had sequences deposited in GenBank, the National Institutes of Health genetic sequence database, an annotated collection of all publicly available DNA sequences (Noreen Tuross, personal communication, 2007).

86. Mulligan 2006.

87. Malhi et al. 2002:906; Schurr 2004a; On Windover, see Doran 2002. Given the scarcity of *a*DNA samples, it would be meaningless to compare frequencies of *a*DNA haplogroups with modern frequencies (Mulligan 2006).

88. Kemp et al. 2007; K. Schroeder et al. 2007.

89. Tamm et al. 2007. See also Fagundes et al. 2008.

90. Kitchen, Miyamoto, and Mulligan 2008; Tamm et al. 2007. The standstill inference might be correct, Noreen Tuross observes, but before concluding it is, we must ensure that the sample on which it is based is large, representative, and comprised of nonrelated individuals (otherwise we run the risk of underestimating genetic diversity); that there was no lineage loss (which would also reduce diversity); that the haplotypes are robust (repeatable) and not rare random mutations; and that the same signal is seen in NRY and autosomal data (Tuross, personal communication, 2007).

91. Hooton 1933.

92. Powell and Neves 1999:168.

93. Marks 2002:235. For the opposing view, see A. Nelson 2006:287 and especially Sparks and Jantz (2002), who argue that Boas misinterpreted his immigrant data—largely for want of adequate quantitative methods—and that the differences between the crania of immigrants and those of their children were insignificant, save in a small number of cases. Rightly so. But then Sparks and Jantz leap to the conclusion that this result "supports what morphologists and morphometricians have known for a long time: most of the variation is genetic variation" (Sparks and Jantz 2002:14637). It does no such thing. Showing that change fails to occur over the few years of Boas's study is not tantamount to proving that craniofacial change cannot happen as a consequence of selection, genetic drift, and the like on a scale of centuries or millennia. Boas may have erred on the side of mutability, but Sparks and Jantz err on the side of stability. See also Gravlee, Bernard, and Leonard 2003.

94. G. Steele and Powell 2002:101; also Powell and Neves 1999:163.

95. Powell and Neves 1999:156, 161, 178; Turner 2002:126; Williams and Armelagos 2007. Zegura observes that the global F_{ST} for crania is approximately 0.09 or 9%, which means that about 10% of cranial variation occurs between groups, while 90% occurs within groups (F_{ST} measures variation within a species, ranging from F_{ST} = 100%, indicating variability occurs entirely between groups, to F_{ST} = 0%, which suggests that all variability occurs within the group) (Zegura, personal communication, 2007).

96. Owsley and Jantz 2006.

97. Roseman and Weaver 2004.

98. Powell and Neves 1999:161.

99. Brace et al. 2001.

100. Neves and Hubbe 2005; González-José et al. 2003.

101. Jantz and Owsley 2001:149–51

102. G. Steele and Powell 2002:98–101, 112; Powell and Rose 1999; Powell and Neves 1999:162.

103. Jantz and Owsley 2006:291; G. Steele and Powell 2002:100.
104. The Kennewick story's been told in several places: Chatters (2001) offers a firsthand account; others, which include interviews and comments from others involved, are Downey (2000) and D. Thomas (2000).
105. Egan 1996.
106. Downey 2000:11.
107. Quoted in Preston 1997.
108. Chatters 2001:69, 72–73.
109. Downey 2000:42.
110. *Bonnichsen et al. v. United States* 1996.
111. Powell 2005:11.
112. Lahr 1995; Powell and Neves 1999; G. Steele and Powell 2002.
113. Marks 2002:235. Powell and Rose (1999) make the same point in their analysis of the Kennewick remains.
114. Smith et al. 2006; Jantz and Owsley 2003.
115. Van Vark, Kuizenga, and Williams 2003.
116. Jantz and Owsley 2003:187.

CHAPTER 6: AMERICAN ORIGINS: THE SEARCH FOR CONSENSUS

1. My discussion here comes from Nicholas Thomas's superb, anthropologically informed *Cook: The Extraordinary Voyages of Captain James Cook* (2003).
2. Eshleman, Malhi, and Smith 2003; Merriwether 2002; K. Schroeder et al. 2007; G. Steele and Powell 2002; Straus, Meltzer, and Goebel 2005; Turner 2002; Wang et al. 2007.
3. Eiseley 1942:417.
4. Bliss 1940a, 1940b; Brand 1940; Hibben 1941.
5. Preston 1995; Crane 1955.
6. Bradley, quoted in Preston 1995.
7. Tony Hillerman's *Dance Hall of the Dead* (1973).
8. Stanford and Bradley (2002) early on enthusiastically hyped haplogroup X as proof of a genetic link between America and Europe. After haplogroup X was spotted in groups living in Asia, and geneticists showed the haplotype distinctions between American and European X, Bradley and Stanford (2004) chose to quietly ignore the genetic evidence. Funny thing, that.
9. The discussion that follows is based on Straus, Meltzer, and Goebel (2005), responding to Bradley and Stanford 2004 (also Stanford and Bradley 2002).
10. Stafford et al. 2003:81.
11. Schledermann 2000; Wallace 2000.
12. Bradley and Stanford observe that the distance from Siberia to Clovis is greater than the ocean span separating Europe to America. Yes, it's a shorter distance, but it's also a silly comparison. Fall down 16,000 years ago on the grassy plains of Beringia and you can get back up, dust yourself off, and keep walking. Fall out of a raft 16,000 years ago into the frigid Pleistocene North Atlantic, and most of the possible story endings are fatal.

13. Quoted in D. Thomas 2000:170.
14. King and Slobodin 1996.
15. Goebel 2004; Straus, Meltzer, and Goebel 2005.
16. King and Slobodin 1996.
17. Pitulko et al. 2004.
18. Hoffecker, Powers, and Goebel 1993:52.
19. Jeffrey C. Long, personal communication, 2007. See K. Schroeder et al. 2007; see also Wang et al. 2007.
20. Zegura et al. 2004; Mulligan et al. 2004:302.
21. Nichols 1990; Nichols quoted in Petit 1998.
22. Bever 2001:128; Hamilton and Goebel 1999:156; Slobodin 2001:39–40; Clark 2001:68; Dumond 2001.
23. So who were the bearers of the Paleoarctic assemblage? As it happens, this archaeological distribution overlaps nicely with the historic range of northern Na-Dene speakers. But then it also overlaps with the distribution of some Eskimo and the eastern Aleut peoples. Although such linguistic-to-archaeological correspondences are intriguing, we don't know which language was spoken by makers of those artifacts, only when the makers arrived where, and by what route.
24. Fedje et al. 2004.
25. Mandryk et al. 2001; see also Clague, Mathewes, and Ager 2004.
26. Fladmark, Driver, and Alexander 1988; Fedje et al. 1995.
27. California may have harbored as many as 80 mutually unintelligible tongues, and the northwest coast another 45, for a total of 125 out of the 330 estimated to have been spoken in North America (Shipley 1978; Thompson and Kinkade 1990).
28. Nettle 1998; Rogers, Rogers, and Martin (1992) consider the possible role of population density leading to great language diversity, but dismiss it saying, "China has a vastly larger population than Siberia but is not any more diverse linguistically" (1992:289). Even if this undocumented assertion is true, the point backfires, since China was unquestionably occupied far earlier and longer than Siberia; by their own reasoning, it ought to have much *greater* linguistic diversity.
29. Eshleman, Malhi, and Smith 2003:12. Recently detected patterns in autosomal DNA indicate greater genetic diversity in coastal regions of South America, relative to eastern South American populations, which suggest a west-to-east movement of people, and hence are also compatible with a population starting out on the west coast of the continent (Wang et al. 2007). But that does not mean they entered the continent down the coast; that's a separate issue.
30. Anderson and Gillam 2000.
31. Surovell 2003.
32. Fix 2005.
33. Fix (2005:433) rightly stresses this point.
34. Greenberg, Turner, and Zegura 1986:478.
35. Meltzer 1989.
36. Hrdlička 1926:9.
37. Eshleman, Malhi, and Smith 2003:8–9. See also Fagundes et al. 2008:586.

38. Eshleman, Malhi, and Smith 2003:8.

39. Including autosomal marker D9S1120, mtDNA haplotype A2, and NRY lineage Q-M3 (see Karafet et al. 1997; Schroeder et al. 2007; Tamm et al. 2007). The idea of back-migrations is not the "novel" scenario geneticists paint it to be (e.g., Tamm et al. 2007); the possibility of two-way traffic across the Bering Sea or Beringia was a hypothesis raised decades ago (e.g., Hrdlička 1926; also Meltzer 1989).

40. Schurr 2004a. See also Eshleman, Malhi, and Smith 2003:8; Fix 2005:431; Merriwether, Rothhammer, and Ferrell 1995.

41. K. Schroeder et al. 2007. See also Fagundes et al. 2008.

42. New radiocarbon evidence suggests the migration may have started even later, perhaps as late as AD 1200 (Friesen and Arnold 2008).

43. C. Haynes 1971:10.

44. Meltzer 1995:29; Storck 1991.

45. Tamm et al. 2007; Mulligan, personal communication, 2008.

46. Kitchen, Miyamoto, and Mulligan 2008; Wang et al. 2007.

47. Quoted in Gibbons 1993.

48. Tamm et al. 2007.

49. Fagundes et al. 2008:589; Hey 2005.

50. Fix 2002:3–4; Meltzer 1995:30. Such changes likely render comparisons of the frequency of haplogroups between ancient and modern DNA meaningless.

51. Recall that teeth are less susceptible to evolutionary change than crania (Chapter 5). The problems with the dental evidence are primarily ones of analysis and results, not inherent ambiguity.

52. Thomas 2000.

53. Woodbury 1984:56. Not surprisingly, the greatest dialect diversity occurred at the extremes of the range (Alaska and Greenland), while the middle (Canada) had relatively low dialect diversity, this a by-product of the greater population density and interaction with non-Eskimo peoples.

54. Eskimo groups are, for example, primarily mtDNA haplogroups A and D, but as Saillard et al. found, "there is an Eskimo-specific mtDNA subgroup characterized by nucleotide position 16265G within mtDNA group A2. This subgroup is found in all Eskimo groups analyzed so far and is estimated to have originated <3,000 years ago" (Saillard et al. 2000; see also Forster et al. 1996; Shields and Jones 1998; Szathmary 1984, 1996).

55. McGhee 1984:376; see also Dumond 1984; McGhee 1984:370, 373; Odess, Loring, and Fitzhugh 2000; Park 1993. For a recent discussion of the timing of the Thule expansion and the possible climatic trigger for it, see Friesen and Arnold 2008, and Chapter 7.

56. Kirch 2000:211–16.

57. In general, it's thought there was no significant genetic exchange between Eskimo populations and other groups, save perhaps on the extremes of their range (Szathmary 1996). Helgason et al. (2006) have shown that some genetic exchange is likely to have occurred. See also Fix 2005:431.

58. Tuross, personal communication, 2007.

1. Making it rather odd that some colleagues complain about models for colonization of the Americas that lack ethnographic examples (G. Haynes 2002). Of course they do. That's hardly a fault of the model—just a reflection of reality.

2. Anthony 1990; Irwin 1992:211–12; Keegan and Diamond 1987:69; Kirch 1997:64–65.

3. Expansion beyond what's needed to settle an expanding population seems true of Oceania as well, as noted by Kirch 1997:64; Irwin 1992:212–15

4. Gamble 1994:94.

5. Kirch 1997:65.

6. Irwin 1992:214; also Gamble 1994:182, 241–42, 245.

7. For example, Barton et al. 2004:149.

8. McGhee 2005.

9. McGhee 2005:44–49. Musk oxen fell victim to one of evolution's nasty tricks: successfully adapting to one set of circumstances is no guarantee of success when the circumstances change. An organism has to keep evolving in order to survive. The *Red Queen Hypothesis* it's called, because, as the Queen told Alice in Lewis Carroll's *Through the Looking Glass*, "It takes all the running you can do, to keep in the same place."

10. McGhee 1984:370–73. After advocating for many years the idea the Thule expansion was driven by the movement of whales and Thule whale hunters, McGhee recently (2005:121) admitted he "no longer believed" the thesis. He changed his mind based on his belief that humans "do not behave in such a deterministic manner," but then went on to suggest there was a singular motive, just not whales: the discovery of iron in the eastern Arctic. He also suggested that sea-ice conditions never fully opened passage for the bowhead whales, a suggestion not supported by recent glacial geological work in the Arctic (Arthur Dyke, personal communication, 2007). Friesen and Arnold (2008) discuss the implications of the new radiocarbon ages for the timing and cause of the Thule migration and make the point that if the new radiocarbon dates for the onset of the Thule expansion (AD 1200 as opposed to AD 1000) prove secure, it would remove the onset of the Medieval Warm Period as the trigger for Thule expansion but still preserve a role for an expanded bowhead whale range in the cause of that expansion. See also Odess, Loring, and Fitzhugh 2000:198–200.

11. Kelly and Todd 1988; also Kelly 1996. Kelly and Todd expressly disavowed the exploitation of Pleistocene megafauna, but their model essentially demands it—and most others accept the notion that these were the prey in question (see Meltzer 2004 for a fuller discussion).

12. Van Dyne et al. 1980: 285–98; also Frison 1991:141; Johnson et al. 1992; Winterhalder et al. 1988.

13. Frison 2004:226.

14. Silberbauer 1981:292–93.

15. Much of the discussion that follows comes from Meltzer 2002 and 2004, and references therein.

16. Yesner 2001:320–23.

17. Andre, Karst, and Turner 2006.

18. Bird and O'Connell 2006; also Lupo 2007.

19. Bird and O'Connell 2006:150. Grass seed, by contrast, is routinely a very low-ranked resource because it takes so much time and energy to gather and process the seeds for consumption; this is why it is in the diet only in times of stress when other, higher-ranked resources are unavailable.

20. Resource depression, as more formally defined, is "the situation in which the activities of a predator lead to reduced capture rates of prey by that predator" (Charnov, Orians, and Hyatt 1976). As Donald Grayson explains, "These reduced capture rates may result from 'behavioral depression,' in which prey adopt behaviors—for instance, increased vigilance or altered periods of activity—that decrease the likelihood that they will be preyed upon. They may also result from 'microhabitat depression,' in which prey decrease their vulnerability by moving out of geographic reach of their predators, or from 'exploitation depression,' in which numbers of prey individuals decrease because harvest rates exceed both reproductive rates and rates of in-migration. From the point of view of the predator, all have the same impact—reduced prey availability" (Grayson 2001:5).

21. Van Dyne et al. 1980:286, 320.

22. Martin 2005:138; G. Haynes (2002:28) offers a similar caricature of how archaeologists—but not he—have portrayed the first Americans. They think, he says, the first Americans were like the "pictures of stone age people migrating to Antarctica [sic]—step by step bands barely move forward in the face of horribly alien landscape. The people stumble and fall on the ice. They bleed from the bottoms of their feet; they have to eat the dogs. After a couple thousand years someone invents skis." Right.

23. Stahle et al. 1998. Drought aside, the English colonists were put in that vulnerable position by virtue of having crossed a vast distance in short period of time, but also because they assumed Virginia was England, only warmer, ignoring the great differences in the ecosystems. That's the downside of prior knowledge. Because your knowledge as a colonist on a new landscape is limited, there can be a strong temptation to assume things about the new place based on experiences in prior, seemingly similar environments: *false analogy*, McGovern (1994) calls it. When doing so, colonizers run the risk of applying preconceived and perhaps faulty models to a landscape, and in so doing, fail to consider the actual patterns on that landscape and may make bad predictions about anticipated changes, which can have disastrous consequences. This might prove particularly disadvantageous to colonists on a new landscape, for reliance on generic knowledge will assuredly be greatest during the early phases of movement into a region, and it is at those stages in the colonization process that snap judgments about a landscape or habitat are often made. Making a snap judgment with imperfect knowledge can be fatal (Orians and Heerwagen 1992:562–63).

24. Golledge 2003.

25. See R. Nelson (1969:99–104) on being lost in the Arctic; Silberbauer (1981) on desert foragers.

26. Aporta and Higgs 2005:731–32.

27. Binford 1983:110, 115; Kelly 1995:Table 4-1, 2003.

28. Aporta and Higgs (2005) document the loss of traditional wayfinding skills because of the introduction of satellite technology—and the consequences when instruments or batteries die.

29. Golledge 2003.

30. R. Nelson 1969, 1986; also Kelly 2003.

31. Golledge 2003.

32. Bolton 1949; Weber 1992. Coronado designated a soldier to count his every step in order to track the distance they had travelled.

33. Bolton 1949:298–99.

34. Epstein 1999; McGovern 1994:149; Stahle et al. 1998.

35. Wyckoff 1993; Anderson and Faught 2000.

36. Moermann, personal communication, 2000. Also Berlin and Berlin 1996:53. Moermann estimates the number of medicinal plants among the North American flora at approximately 2,700 species.

37. Dillehay (1991) examines the role of disease in the colonization of America.

38. The medically proper conditions are known as hyperammonemia and hyperaminoacidemia, marked by excess ammonia in the urea and amino acids in the blood (respectively), which trigger nausea, diarrhea, and death. Hunter-gatherers, though unaware of the physiological basis of rabbit starvation, were nonetheless acutely aware of its effects; as Speth and Spielmann (1983) have shown, they avoided excess protein in their diets and sought animals high in body fat or carbohydrates from plant foods. Virtually every western explorer, from Lewis and Clark's Corps of Discovery onward, recorded experiencing rabbit starvation at one time or another.

39. MacNutt 1912, letter of November 1493.

40. Snyder and Leonard 2006.

41. Lewis and Clark had heard tales of bear and were anxious to see one, mostly, it appears, to see if a bear could possibly be that fearsome. Their first encounter was with a 300-pound cub, which impressed Lewis but hardly intimidated him. American Indians might fear this animal, he said, since they were equipped with bows and arrows, but "in the hands of skilled riflemen they are by no means as formidable or dangerous as they have been represented." He changed his tune over the next two weeks following four more encounters with full-grown grizzlies. On May 5, 1805, Clark killed the largest one they'd yet seen, an animal 8.5 feet long, weighing 500–600 pounds, and with 4.5-inch claws. Not only was it, as Lewis described, "a most tremendous looking animal," it proved "extremely hard to kill notwithstanding he had five balls through his lungs and five others in various parts[:] he swam more than half the distance across the river to a sandbar & it was at least twenty minutes before he died; [he] made the most tremendous roaring from the moment he was shot." "I find," Lewis wrote in his diary on May 6, 1805, "that the curiossity of our party is pretty well satisfyed with rispect to this animal" (sic) (Moulton 2003:122–25).

42. Kelly and Todd 1988:234.

43. Bird and O'Connell 2006:155.

44. Kaplan and Hill 1992:186; Kelly 1995:96–98; Stephens and Krebs 1986:75.

45. Silberbauer 1981:247–48.

46. On gender differences in wayfinding among moderns, see Lawton 1996:141; on division of hunting labor, see Borgerhoff Mulder 1992:363; Chilton 2004; Foley 1988:218–20; Kelly 1995:269; and Silberbauer 1981:93; on costly signaling and the role of men's hunting, see Bird and O'Connell 2006:154, 159, 163–65; Hawkes and Bliege Bird 2002.

47. Silverman and Eals 1992:534–35.

48. R. Nelson 1969:374.

49. Silberbauer 1981:249.

50. Beaton 1991:220–21.

51. Binford 1983:34; Kelly 1995:150–51.

52. Ellis 2008:310; also Harpending 1999:517.

53. Moore 2001; see also Keegan and Diamond 1987:78; S. Black 1978:67; McArthur, Saunders, and Tweedie 1976:322–23.

54. Beaton 1991.

55. Beaton 1991:223; see also Martin 1973; Szathmary 1993.

56. Borgerhoff Mulder 1992:341–42; MacArthur and Wilson 1967:78, 80, 88; Meltzer 1995, 1999; Whallon 1989:434–35.

57. Anderson 1995; Meltzer 2002; Lourandos 1997:28.

58. Beaton 1991:224.

59. Beaton 1991:224.

60. Bird and O'Connell 2006:155–56.

61. Kirch 1988:113–14, 1997:254.

62. One of my archaeological colleagues has proclaimed, with great gravitas and a just hint of melancholy, that the "absence of permanent records doomed . . . to anonymity" the Paleoindian achievements in landscape learning. Hardly. After all, ours is a business routinely conducted without permanent records, and though many of the details are surely long gone, we can nonetheless see archaeologically some of their broad strategies to learning the new landscape, and certainly witness the results.

CHAPTER 8: CLOVIS ADAPTATIONS AND PLEISTOCENE EXTINCTIONS

1. Eiseley 1975:95. Eiseley was by then embittered with Howard, whose presence at the University of Pennsylvania had attracted Eiseley east in the first place. Howard had been uninterested in a find Eiseley made, ostensibly because it wasn't old enough: "We've got to go deeper, much deeper," Howard had exhorted.

2. Howard 1932.

3. E. Howard to H. Jayne, July 28, 1933. General Correspondence, Academy of Natural Sciences Archives, Philadelphia.

4. E. Howard to H. Jayne, August 3, 1933. Curatorial Correspondence, University Museum Archives, University of Pennsylvania.

5. Howard 1933.

6. Johnston 1933.

7. Bever 2001; Reanier 1995 on Alaska; Gryba 2001 on Clovis in the ice-free corridor.

8. Sellet 2004.

9. Ahler and Geib 2000:803.

10. Bever and Meltzer 2007.

11. Lyman, O'Brian, and Hayes 1998.

12. Adovasio, Hyland, and Soffer 2001:213.

13. Agenbroad and Hesse 2004.

14. Frison 2004:58.

15. Frison and Bradley 1999.

16. Wilke, Flenniken, and Ozbun 1991:254.

17. Gramly 1993.

18. Wilke, Flenniken, and Ozbun 1991.

19. A half dozen or so caches of Dalton age have been reported from eastern North America, and these are discussed in Chapter 9.

20. Collins, Lohse, and Shoberg 2007. Unfortunately, the deGraffenreid cache was collected in the 1930s or 1940s, and its exact find-spot is unknown. It may have come from the Gault site, or not. The stone is virtually the same as that at Gault, but then such stone outcrops elsewhere in this part of central Texas.

21. Gardner 1974.

22. Seeman 1994:237.

23. D. Anderson and Faught 2000; Bever and Meltzer 2007. The "Paleoindian database of the Americas" with accompanying distribution maps can be found on the web at http://pidba.utk.edu/.

24. On staging areas, see D. Anderson 1995; Dincauze 1993.

25. Lepper 1983; Meltzer and Bever 1995; Shott 2002.

26. Waters and Stafford 2007.

27. G. Haynes 2002.

28. Martin and Steadman 1999:34.

29. Martin 1973 remains the classic statement, but see his *Twilight of the Mammoths* (2005).

30. J. Berger, Swenson, and Persson 2001.

31. In an ingenious twist on this theme, Gary Belovsky (1988) argues animal extinctions are more likely to occur when hunting-gatherers are able to exploit rich plant resources: more-abundant, gathered foods can support larger human populations, which in turn will be more likely to overexploit local animal prey. Case in point: the extinction of gazelle by Neolithic farmers in the Middle East. Of course, a Clovis diet dominated by plant foods is hardly what overkill envisions.

32. Martin 1984:363.

33. Grayson 2001.

34. The lack of predators on remote oceanic islands results from the fact that few can cross long stretches of open ocean, and also need large islands and correspondingly large prey populations to support them. Animals that could reach these islands arrived by flying or floating (in the air or water), which largely restricted the classes of animals that can make the trip (birds, mostly). And once there, they no longer needed to "invest energy in the embryogenesis and maintenance of functional wings" (MacArthur and Wilson 1967:158). Simply put, over evolutionary time, they lost their ability to fly.

35. Fiedel and Haynes 2004.
36. Grayson and Meltzer 2002.
37. The ossicles adhere to the skin, not the skeleton, of the animal, and were recovered at the Kimmswick site (Missouri). Russell Graham (personal communication) argues they came from hides of animals brought to the sites. The ossicles testify to humans skinning Paramylodon, but where the deed was done has not yet been discovered.
38. Grayson and Meltzer 2003.
39. Wheat 1972.
40. Martin 2005:155.
41. My son's room inspired this observation.
42. Martin 2005:133; compare to his comments on absence of evidence for overkill (2005:157).
43. The exchange plays out in Grayson and Meltzer 2003; Fiedel and Haynes 2004; Grayson and Meltzer 2004; and G. Haynes 2007. We did not bother to respond to G. Haynes 2007.
44. Taphonomy is the study of the processes that affect organic remains after the death of the individual as, for example, when the bones of the carcass are scavenged, scattered, weathered, become buried, and so on. Neotaphonomy, in Haynes's usage, refers to the study of fresh elephant kills, bison kills, and the like.
45. Frison 2004 discusses these at length.
46. Jones et al. 2008.
47. Martin 2005:52.
48. Meltzer and Mead 1985:147–48
49. Shapiro et al. 2004.
50. Guthrie 2003, 2006.
51. Grayson 1977; Martin and Steadman 1999; Jackson and Weng 1999.
52. Grayson 2007; also Barnosky et al. 2004.
53. We don't know the details of the reproductive strategies of these now-extinct animals; after all, they are extinct. But it is not unreasonable to surmise, given the size of these animals and some basic facts of biology, that they must have had long gestation periods (like elephants), had only single births (twins are rare among large mammals), and had reproductive spans. Elephants typically do not reach sexual maturity until 10–12 years of age, and will only give birth every 4–9 years. Animals with this kind of reproduction pattern are known as K strategists, and are particularly vulnerable when their young die before reaching reproductive maturity, since it took so long to breed another. On large mammal susceptibility to extinction, see Cardillo et al. 2005; C. Johnson 2002.
54. Grayson 2007; also Barnosky et al. 2004.
55. After several failed attempts, the Clovis site was finally sold in 1979 to Eastern New Mexico University, and the gravel operations ceased. By then, most of the original pond had been mined out. Still, intact deposits cling to the side walls of the gravel pit, and even today, those produce valuable data on Clovis and later occupations. The site, now open to the public, is on the National Register of Historic Places.
56. Huckell and Judge 2006:151.
57. C. Haynes 1991; C. Haynes et al. 1999; Holliday 2000.

58. Huckell and Judge 2006.
59. Haury, Sayles, and Wasley 1959.
60. C. Haynes and Huckell 2007.
61. Colby is one of the very few mammoth Clovis sites in which the point was found before the mammoth bones, in this case more than a decade earlier (Frison 2004:44).
62. C. Haynes et al. 1998.
63. Leonhardy 1966.
64. Holliday et al. 1994.
65. Western New York's Hiscock site has yielded mastodon bones and a fluted point, but the point proved to be closer to some elk bones than to mastodon bones. In fact, the site's excavator, Richard Laub (2002), wonders if the association was of a different kind: that the mastodon were dead by the time humans arrived, and their remains were quarried by people to fashion bone tools.
66. Cannon and Meltzer 2004.
67. O'Connell, personal communication; Silberbauer 1981:292–93.
68. Frison 1991:146–47, Frison 2004:54–58; Roosevelt 1910:299. Roosevelt was hunting for display specimens for the Smithsonian Institution's National Museum of Natural History; times were different then for ex-presidents, and for museums.
69. Bird and O'Connell 2006:160.
70. Hawkes and Bliege Bird 2002:59–61; Bird and O'Connell 2006:164–65.
71. G. Haynes (2002) thinks this statement a caricature and "nonsense," and that "perhaps professional archaeologists in the modern world are frightened to death of hunting proboscideans . . . but real people foraging in the world today do not think that way." O'Connell and Silberbauer, and Roosevelt, of course, show that's not necessarily so.
72. Roosevelt 1910:289. Roosevelt was not unaware of the analogy of Africa to the Paleolithic, though naturally he conceived of it in highly simplified and typically Victorian fashion: his opening chapter describing his travels through Africa was entitled "Through the Pleistocene." The trash-talk quote is from Hawkes and Bliege Bird 2002:64.
73. Ellis 1997:47–50.
74. Harris lines were spotted in the femur, which are normally caused by seasonal nutritional deficiencies or serious illness. She was between 17 and 21 years of age at the time she died. Green et al. 1998:448–49.
75. Crook and Harris 1957:17.
76. G. Haynes 2002:179.
77. Gramly 1982, 2005, and personal communication, 2007.
78. West 1983.
79. Which is not to say nut collecting leaves no record under any circumstances: in Late Prehistoric times in California and the Great Basin, when resource depression had set in and there was intensive exploitation of foods such as acorn and piñon, substantial archaeological records built up. But then these were harder times, those were critical resources, and villages were often established on the spot.
80. G. Haynes 2002:236.

81. G. Haynes 2002:270. See Hollenbach 2007 for an interesting and nuanced discussion of the role of plant gathering among Paleoindians.

CHAPTER 9: SETTLING IN: LATE PALEOINDIANS AND THE WANING ICE AGE

1. Meltzer 2006b; Seebach 2006. One of the most famous of the Dust Bowl photographs, so emblematic that the U.S. Postal Service used it as one of the images to symbolize the decade, was Arthur Rothstein's *Fleeing a Dust Storm*, which showed a father and his two young sons (the Coble family) running for cover in the panhandle of Oklahoma, ground zero of the Dust Bowl. As it happens, that corrosive storm and others like it soon exposed the Nall Paleoindian site just a few miles away from where that photograph was taken.

2. Shetrone 1936:4.

3. Kroeber 1939:3, 205.

4. Quimby 1954:317.

5. Aschmann 1955:377; Quimby 1955:379.

6. This can get confusing, since the time periods are named after point types, which in turn are named after the sites in which they were first found. Thus, there is a Folsom site, a Folsom point, a Folsom period, and a Folsom culture. Confusion is compounded at sites occupied at different times: the Agate Basin site has Folsom, Agate Basin, and Hell Gap points and periods represented (among others).

7. Beck and Jones 1997:189–97.

8. D. Anderson 2002:44.

9. Meltzer 2006a; Stanford 1999.

10. Amick 1996; G. Jones et al. 2003. See also Stanford 1999:303.

11. This emphasis is also strongly influenced by my own research interests and fieldwork over the last two decades on the plains and Rocky Mountains, which has included investigations at the sites of Bonfire Rockshelter, Folsom, Lindenmeier, Midland, and Mountaineer (among others). For those looking for comparably detailed coverage of other areas, my apologies.

12. Lott 2002:69.

13. As Lott explains it, bisons' natural predators are wolves, and wolves hunt mostly by sight: the best way to avoid them is to hide, no easy task on a treeless landscape where the grass is ankle high, so "the only thing big enough to hide behind is another bison—or better still, a bunch of bison" (Lott 2002:33).

14. Lott suggests the long, hooked shape and forward orientation of *Bison antiquus* horns indicates it may have met its Pleistocene predators by standing its ground and fighting. But then two new predators—humans and gray wolves—appeared on the scene, one that could kill at a distance and the other a pack hunter that could overcome a lone bison. Massive horns were not as valuable as being able to make a fast getaway. Lott 2002:65–66; also Frison 2004:62–64

15. Frison 2004; Lott 2002:102.

16. LaBelle 2005; also Andrews, LaBelle, and Seebach 2008; M. E. Hill 2007.

17. Byerly et al. 2005. Depending on the route a herd was stampeded, the cliff edge at Bonfire might not have been visible until the animals were within 24 meters; traveling at top speed, they could cover that distance in less than 2 seconds.
18. Meltzer 2006a; Wheat 1972.
19. The material in this sidebar is drawn from Meltzer 2006a.
20. Don Yeager, personal communication, 2007; Frison 1991:167–70.
21. Haury, Sayles, and Wasley 1959:15. See also Ellis's study of ethnographic weaponry, which shows that point size or form varied little with the size of the game animals being hunted (Ellis 1997:65).
22. Wilmsen and Roberts 1978:171, 176. In this instance, it's suspected the point was thrust into the animal, and had it been standing at the time, the point would have come in from behind and below.
23. Frison 2004:107 discusses the requirements of a projectile point.
24. Frison 2004:110.
25. Ellis 1997:61; but see Churchill 2002:186 for a different result.
26. Data on velocities and the quote from Garcilaso de la Vega come from Hutchings 1997:136.
27. Frison 1991:211–12; Frison 2004:213.
28. Bamforth 2002.
29. M. E. Hill 2007.
30. M. G. Hill 2001.
31. Reeves 1990:170.
32. Wheat 1972:114. Olsen-Chubbuck had 110 females, of which 63 were adult, the remainder immature and calves; and 80 males, of which 46 were adult, the remainder immature bulls and calves.
33. Frison 2004:113–14.
34. Lawrence C. Todd, personal communication, 2005.
35. Jodry 1999. Similar activities took place at Shifting Sands, though not on as large a scale (Hofman, Amick, and Rose 1990).
36. Sellet 2004; M. G. Hill 2001.
37. Stiger 2006.
38. Frison 2004.
39. LaBelle 2005. See also M. E. Hill 2007.
40. Bamforth 2007; E. Johnson 1987.
41. Lewis Binford, personal communication, 2005.
42. Jason LaBelle, personal communication, 2006.
43. Wilmsen and Roberts 1978:45–48.
44. Wilmsen and Roberts 1978:114.
45. Shapiro et al. 2004.
46. The discussion of Mesa Complex comes from Kunz, Bever, and Adkins 2003; Bever 2001; Mann et al. 2001.
47. Hoffecker 2005.
48. Anderson 2002; Dansie and Jerrems 2006:72; Dunbar and Vojnovski 2007.
49. Eren 2008.

50. Newby et al. 2005. Their "guarantee" comes on page 150.
51. Newby et al. 2005:151.
52. Newby et al. 2005:145, 150, 152.
53. Ellis and Deller 1990:62, 1997.
54. Ellis and Deller 1997:17. On postglacial fish restocking, see Curry 2007.
55. Goodyear 1982; Walthall 1998.
56. Goodyear 1999:378; Yerkes and Gaertner, in Morse 1997.
57. Walthall 1998:232; Grover 1994:126. At Dust Cave, a variety of birds dominate the recovered animal bones, and that includes a cache of humeri (upper wing bones) of Canadian geese, presumably stored there for later toolmaking (Walker 2007:105–6). On the plant side, hickory nuts were a mainstay of the diet the fall that the site was occupied (Hollenbach 2007).
58. Walthall and Koldehoff 1998.
59. Kelly and Todd 1988.
60. Collins 1991; Sherwood et al. 2004.
61. Walthall 1998.
62. Beck and Jones 1997; much of my discussion is taken from this important source, but see also Aikens 2006.
63. Dansie and Jerrems 2006:68.
64. On crescents, see Beck and Jones 1997:206–7 (from which the Clewlow quote comes) and Amick 1999.
65. Harrison and Killen 1978.
66. Stanford 1978. Whether it was a single or multiple occupation is unclear; a detailed report on Jones-Miller has never been published.
67. Stafford et al. 2003.
68. Morse 1997.
69. Sutherland 2000.
70. Skraeling was a pejorative Norse term for a weakling, which the Norse applied without distinction to all Native Americans they encountered, which likely included both Thule (Paleo-Eskimo) and Algonquian speakers, and perhaps the Beothuk Indians (Schledermann 2000:192).
71. Schledermann (2000); Wallace 2000.
71. Wallace 2000.
73. Odess, Loring, and Fitzhugh 2000:203.

CHAPTER 10: WHEN PAST AND PRESENT COLLIDE

1. C. Darwin to S. Darwin, April 23, 1835, in Burkhardt and Smith 1985:445–48. Darwin saw in the sandstone of the Uspallata Range petrified trees at 7,000 feet in elevation; for thousands of feet above the sandstone were sedimentary beds bearing fossil shells and atop them a lava that had been spewed into deep ocean. The trees had once "waved their branches on the shores of the Atlantic, when that ocean (now driven back 700 miles) came to the foot of the Andes," while the marine sediments were testimony that the once dry land had been "let down into the depths of the ocean," by

at least as many thousands of feet as the top of the marine deposits (Darwin 1839a, notes of March 30, 1835).

2. Darwin 1839a, notes of March 25, 1835.

3. Barrazo and Lazzari 2004.

4. Medawar points out that Darwin also suffered from the treatments administered by his physicians, including low levels of chronic arsenic and mercury poisoning (1982:146).

5. The earlier Norse contact did not involve any disease transmission, since infectious pathogens like the plague and smallpox did not reach Norway, Iceland, or Greenland until the thirteenth and fourteenth centuries, long after the Norse had abandoned their Vinland enterprise (Crosby 1986:52).

6. In 1967, the World Health Organization resolved to eradicate smallpox from the world's diseases. Through an ambitious program of vaccination and treatment, that goal was met, eradication being possible because smallpox has only one host: humans. The last "natural" case occurred in Somalia in 1977, and the last laboratory death occurred in 1978, when a photographer in a British University (Birmingham) lab was accidentally infected. In October 1979, the world was declared free of smallpox. All remaining stocks of smallpox virus were transferred to two laboratories: at the Centers for Disease Control (CDC) in Atlanta, and at the Research Institute for Viral Preparations in Moscow. The virus stocks have been there in hermetically sealed containers since 1983 (at least the hope is that the Moscow samples remained protected after the breakup of the former Soviet Union). The samples were slated for destruction over a dozen years ago, but for a variety of reasons never were. Now, with fears of global bioterror, their "execution" has been stayed indefinitely, so as to better cope with the virus (and understand its extraordinarily ability to evade the natural human defense mechanisms) in the event it re-emerges—especially if in a variant form.

7. Le Page du Pratz 1758; Crosby 1986:200.

8. Jefferson 1788.

9. Jefferson, Instructions to Captain Lewis, June 20, 1803, in Jackson 1962. Cowpox, from which Jenner derived his vaccine, has multiple animal hosts (unlike smallpox), including humans, cat, mice, cattle, and elephants. It's far less deadly than smallpox, though a mild bout with cowpox would confer immunity from smallpox. Fenn (2001) details the use of *variolation* (the administration of cowpox) in colonial America.

10. Gentleman of Elvas 1686, Chapter XIV: "How the governor left the province of Patofa and came upon an uninhabited region, where he and all his men experienced great vicissitudes and extreme need."

11. Crosby 1986:201; Li et al. 2007.

12. Fenn 2001.

13. Which is not to say that reliable population numbers are unavailable. The Spanish colonial censuses, for example, are reasonably accurate.

14. Dobyns 1966, 1983.

15. Ubelaker 2006b:696–97.

16. The United States 2000 census puts American Indian numbers at 2.5 million, and the Canada census reports 1.4 million (Snipp 2006:710–11)

17. F. Black 1992.
18. Levine and LeBauve 1997; Schroeder 1979.
19. C. Mann 2005.
20. Ruddiman 2005:121, 132–40. Ruddiman makes the point that the increase in atmospheric CO_2 as a result of deforestation would have been a long, gradual process tied to the slow Neolithic rhythm of the spread of agriculture; in contrast, an epidemic of mass mortality, coupled with the rapidity with which forests re-colonize open ground, would lead—as it does in the isotope record—to a much faster rate of decrease in atmospheric CO_2 (2005:140). Also, Nevle and Bird 2008.
21. McNeill (1976) has an insightful discussion of the process by which pathogens and their hosts co-evolve, and is the basis for what follows.
22. Li et al. 2007.
23. Li et al. 2007.
24. Crosby 1986:208.
25. Le Page du Pratz 1758.
26. Fenn 2001:221.
27. Crosby 1986:286.
28. Stewart 1973:19–20.
29. F. Black 1992. Of course, there are also within American Indian populations—especially those in South America—novel alleles of the major histocompatibility complex (MHC), which are generally attributed to disease-driven selection, likely a response on the part of founding populations to newly encountered pathogens such as acute malarial infection, *Trypanosoma cruzi,* and *Leishmania braziliensis* (see Belich et al. 1992; Watkins et al. 1992).
30. Diamond 1999:168–74.
31. McNeill 1976.
32. Boone 1994:18.
33. Crosby 1986:197, 215–16. There has long been debate over the origins of syphilis. The first recorded epidemic only occurred in 1495, raising the possibility that the bacterium that causes the disease, *Treponema pallidum pallidum,* was not homegrown in the Old World, but was a new and virulent strain that developed in Europe from a non-venereal type of yaws brought back by Christopher Columbus's men. Recent molecular sequencing attempted to trace the genetic history of the syphilis-causing bacterium, and concluded it was indeed a recent strain, and its closest relative was non-venereal New World yaws (Harper et al. 2008). However, that study was based on a limited and degraded DNA sample of New World yaws, and its conclusions remain in need of further testing (Mulligan, Norris, and Lukehart 2008). Dillehay (1991) explores the role and effect of disease in the peopling of the Southern Hemisphere.
34. Crosby 1986:283; Blanton 2003; Stahle et al. 1998.
35. McNutt 1912, Peter Martyr, letter of November 1493.
36. Crosby 1986:151, 155–56.
37. Elliot 2006:103–4
38. Crosby 1986.

39. Swanton 1911:100. Gentleman of Elvas, 1686, Chapter XXIX: "Of the message sent by the governor to Quigaltam and of the answer given by the latter, and of what happened during this time."

40. Brain 1971; Galloway and Jackson 2004; Swanton 1911:107.

41. Swanton 1911:105.

42. Galloway and Jackson 2004:603; also Swanton 1911:185–86.

43. Brain 1971:221–22. The Natchez kin system would not have the opportunity to run its course. After a long series of skirmishes with the French, compounded by complex internecine warfare with both allies and enemies (the sides shifting so fluidly it must have been hard sometimes to keep track), in 1731 some 400 Natchez surrendered and were sent to the Caribbean as slaves. Others scattered throughout the Mississippi Valley, joining with the Chickasaw and Cherokee. Although the Natchez attempted to maintain their own language, it became extinct in the 1930s (Campbell 1997; Galloway and Jackson 2004).

44. Dobyns 1991:551.

45. Goddard 2004.

46. Ronda 1984:52.

47. Abler 1999; Eccles 1988; Secoy 1953:30–32; Swagerty 1988:356; Cronon and White 1988:425.

48. Dunnell 1991:570–71.

49. Native American Graves Protection and Repatriation Act, 25 U.S.C. 3002(a)1–2.

50. Native American Graves Protection and Repatriation Act, 25 U.S.C. 3005(a)4.

51. Thomas 2000:232.

52. Echo-Hawk 1997; also Archambault 2006; Thomas 2000:251–53.

53. B. Babbitt to L. Caldera, September 21, 2000. Available on the National Park Service Archaeology Program "Kennewick Man" Web site, www.nps.gov/archeology/kennewick/babb_letter.htm.

54. Jelderks, Opinion and Order, Civil No. 96-1481-JE, August 30, 2002.

55. USC § 3001(9). (Emphasis mine.)

56. Gould, Opinion for the United States Court of Appeal for the Ninth Circuit, No. 02-35996. D.C. No. CV-96-01481-JE, February 4, 2004. (Emphasis in the original.)

57. In 2005 Senator John McCain offered an amendment (S.536) to change the definition of American Indian in NAGPRA by adding the words, "or was" after "is," (i.e., "of, or relating to, a tribe, people, or culture that is or was indigenous to the United States"). The amendment failed.

58. One is reminded of the rhetoric from the height of the Great Paleolithic War, when the proponents of American Paleolithic railed against government scientists, saying that "of all the arrogant things in the world official science is perhaps the most arrogant, and of all the obstructive things official science is perhaps the most obstructive" (Youmans 1893:841).

59. Schneider and Bonnichsen 2006:301, 307. It's hard to swallow the conspiracy claim, especially since two of the plaintiffs are government officials at the Smithsonian Institution (we've met the enemy, and he is us!), and two of the scientists brought in by the Department of the Interior (a different government branch) in 1998–99 to study the Kennewick remains were PhDs trained by two of the plaintiffs.

60. Deloria 1995:41, Deloria quoted in Thomas 2000; Marks 2002:237.

61. Thomas 2000:xxvi.

62. Thomas 2000:275.

63. Deloria 1995:41; Schneider and Bonnichsen stress the importance of that conversation, but then simultaneously proclaim that those who believe the "assault on science can be defused through conciliation and compromise . . . are mistaken" (2006:307).

64. Gould 1989:283, 289.

65. On Cortés's legal maneuvering, see Elliot 2006.

66. The Spaniards soon enough found out why there was Native resistance: when they entered the Aztec capital, they were stunned and sickened by its blood-stained pyramids and scaffolds, covered by some 136,000 human skulls, gruesome evidence of a thriving industry in human sacrifice and testimony to how the Empire controlled the other peoples in the region (Boone 1994).

67. Boone 1994:142.

68. Boone 1994:14.

69. Elliot 2006:5.

70. A similar scenario played out in the collapse of the Inca Empire a few decades later, and though the circumstances differed in important respects, the outcome was the same. Less militarily decisive, though important all the same, disease influenced the outcome of the American Revolution: British troops under General Cornwallis were hard hit by malaria. "Revolutionary mosquitoes," J. R. McNeill calls them (cited in Mann 2007). See also Fenn 2001.

71. Crosby 1986:200.

REFERENCES

Abbott, C. C. (1889). Evidences of the antiquity of man in eastern North America. *Proceedings American Association for the Advancement of Science* 37:293–315.

Abbott, C. C. (1892a). Paleolithic man in America. *Science* 20:270–71.

Abbott, C. C. (1892b). Paleolithic man: a last word. *Science* 20:344–45.

Abler, T. (1999). Beavers and muskets: Iroquois military fortunes in the face of European colonization. In *War in the tribal zone*, edited by R. Ferguson and N. Whitehead, pp. 151–74. School of American Research, Santa Fe, NM.

Achilli, A., U. Perego, C. Bravi, M. Coble, Q-P. Kong, S. Woodward, A. Salas, A. Torroni, and H-J. Bandelt (2008). The phylogeny of the four pan-American mtDNA haplogroups: implications for evolutionary and disease studies. *PLoS ONE* 3:e1764.

Adovasio, J. and J. Page (2002). *The first Americans: in pursuit of archaeology's greatest mystery*. Random House, New York.

Adovasio, J., D. Hyland, and O. Soffer (2001). Perishable technology and early human populations in the New World. In *On being first: cultural innovation and environmental consequences of first peoplings*. Proceedings of the 31st Annual Chacmool Conference, edited by J. Gillespie, S. Tupakka, and C. de Mille, pp. 201–21. Archaeological Association of the University of Calgary, Calgary, AB, Canada.

Agassiz, E. (1887). *Louis Agassiz: his life and correspondence*. Houghton Mifflin, Boston.

Agenbroad, L. and I. Hesse (2004). Megafauna, Paleoindians, petroglyphs, and pictographs of the Colorado Plateau. In *The settlement of the American continents: a multidisciplinary approach to human biogeography*, edited by C. M. Barton, G. A. Clark, D. R. Yesner, and G. A. Pearson, pp. 189–95. University of Arizona Press, Tucson.

Ahler, S. A. and P. R. Geib (2000). Why flute? Folsom point design and adaptation. *Journal of Archaeological Science* 27:799–820.

Aikens, C. M. (2006). Paleo-indian: west. In *Handbook of North American Indians, Vol. 3, environment, origins, and population,* edited by D. Ubelaker, pp. 194–207. Smithsonian Institution Press, Washington, D.C.

Alley, R. (2000a). *The two-mile time machine: ice cores, abrupt climate change, and our future.* Princeton University Press, Princeton, NJ.

Alley, R. (2000b). Ice-core evidence of abrupt climate changes. *Proceedings of the National Academy of Sciences* 97:1331–34.

Alley, R. (2007). Wally was right: predictive ability of the North Atlantic "Conveyor Belt" hypothesis for abrupt climate change. *Annual Review of Earth and Planetary Sciences* 35:241–72.

Alley, R., P. Clark, P. Huybrechts, and I. Joughin (2005). Ice-sheet and sea level changes. *Science* 310:456–60.

Amick, D. (1996). Regional patterns of Folsom mobility and land use in the American Southwest. *World Archaeology* 27:411–26.

Amick, D. (1999). Using lithic artifacts to explain past behavior. In *Models for the millennium: current research in Great basin anthropology,* edited by C. Beck, pp. 161–70. University of Utah Press, Salt Lake City.

Anderson, D. (1995). Paleoindian interaction networks in the eastern woodlands. In *Native American interaction: multiscalar analyses and interpretations in the eastern woodlands,* edited by M. Nassaney and K. Sassaman, pp. 1–26. University of Tennessee Press, Knoxville.

Anderson, D. (2002). Southeast context. In *Earliest Americans theme study for the eastern United States,* compiled and edited by E. M. Seibert, pp. 44–80. United States Department of Interior, National Park Service, Washington, D.C.

Anderson, D. and M. Faught (2000). Palaeoindian artifact distribution: evidence and implications. *Antiquity* 74:507–13.

Anderson, D. and J. C. Gillam (2000). Paleoindian colonization of the Americas: implications from an examination of physiography, demography, and artifact distribution. *American Antiquity* 65:43–66.

Anderson, D. and K. Sassaman, eds. (1996). *The Paleoindian and Early Archaic Southeast.* University of Alabama Press, Tuscaloosa.

Anderson, E. (1984). Who's who in the Pleistocene: a mammalian bestiary. In *Quaternary extinctions: a prehistoric revolution,* edited by P. S. Martin and R. G. Klein, pp. 40–89. University of Arizona Press, Tucson.

Anderson, P., M. Edwards, and L. Brubaker (2004). Results and paleoclimatic implications of 35 years of paleoecological research in Alaska. In *The Quaternary period in the United States,* edited by A. Gillespie, S. C. Porter, and B. Atwater, pp. 427–40. Elsevier Science, New York.

Andre, A., A. Karst, and N. Turner (2006). Arctic and subarctic plants. In *Handbook of North American Indians, Vol. 3, environment, origins, and population,* edited by D. Ubelaker, pp. 222–35. Smithsonian Institution Press, Washington, D.C.

Andrews, B., J. LaBelle, and J. Seebach (2008). Spatial variability in the Folsom archaeological record: a multi-scalar approach. *American Antiquity* 73:464–490.

Anthony, D. W. (1990). Migration in archaeology: the baby and the bathwater. *American Anthropologist* 92:895–914.

Anyon, R., T. J. Ferguson, L. Jackson, L. Lane, and P. Vicenti (1997). Native American oral tradition and archaeology: issues of structure, relevance, and respect. In *Native Americans and archaeologists: stepping stones to common ground,* edited by N. Swidler, K. Dongoske, R. Anyon, and A. Downer, pp. 77–87. AltaMira Press, Walnut Creek, CA.

Aporta, C. and E. Higgs (2005). Global Positioning Systems, Inuit wayfinding, and the need for a new account of technology. *Current Anthropology* 46:729–53.

Archambault, J. (2006). Native views of origins. *Handbook of North American Indians, Vol. 3, environment, origins, and population,* edited by D. Ubelaker, pp. 4–15. Smithsonian Institution Press, Washington, D.C.

Aschmann, H. (1955). Comment on Quimby's cultural and natural areas before Kroeber. *American Antiquity* 20:377–78.

Baker, V. R. and D. Nummedal (1978). *The Channeled Scabland.* NASA, Washington, D.C.

Balter, M. (2008). Ancient algae suggest sea route for first Americans. *Science* 320:729.

Bamforth, D. (2002). High-tech foragers? Folsom and later Paleoindian technology on the Great Plains. *Journal of World Prehistory* 16:55–98.

Bamforth, D. (2007). *The Allen site: A Paleoindian camp in southwestern Nebraska.* University of New Mexico Press, Albuquerque.

Barnosky, A. D., P. L. Koch, R. S. Feranec, S. L. Wing, and A. B. Shabel (2004). Assessing the causes of late Pleistocene extinctions on the continents. *Science* 306:70–75.

Barrazo, R. and C. R. Lazzari (2004). The response of the blood-sucking bug *Triatoma infestans* to carbon dioxide and other host odours. *Chemical Senses* 29:319–29.

Barton, M., G. A. Clark, D. R. Yesner, and G. A. Pearson, eds. (2004). *The settlement of the American continents: a multidisciplinary approach to human biogeography.* University of Arizona Press, Tucson.

Barton, M., S. Schmich, and S. James (2004). The ecology of human colonization in pristine landscapes. In *The settlement of the American continents: a multidisciplinary approach to human biogeography,* edited by C. M. Barton, G. A. Clark, D. R. Yesner, and G. A. Pearson, pp. 138–61. University of Arizona Press, Tucson.

Bassett, S., G. Milne, J. Mitrovica, and P. Clark (2005). Ice sheet and solid earth influences on far-field sea-level histories. *Science* 309:925–28.

Beaton, J. (1991). Colonizing continents: some problems from Australia and the Americas. In *The first Americans: search and research,* edited by T. D. Dillehay and D. J. Meltzer, pp. 209–30. CRC Press, Baton Rouge, LA.

Beck, C. and G. T. Jones (1997). The terminal Pleistocene/Early Holocene archaeology of the Great Basin. *Journal of World Prehistory* 11:161–236.

Belich M. P., J. A. Madrigal, W. Hildebrand, J. Zemmour, R. Williams, R. Luz, M. L. Petzl-Erler, and P. Parham (1992). Unusual HLA-B alleles in two tribes of Brazilian Indians. *Nature* 357:326–29.

Belovsky, G. (1988). An optimal foraging-based model of hunter-gatherer population dynamics. *Journal of Anthropological Archaeology* 7:329–72.

Benson, L. (2004). Western lakes. In *The Quaternary period in the United States,* edited by A. Gillespie, S. C. Porter, and B. Atwater, pp. 185–204. Elsevier Science, New York.

Berger, A. and M. Loutre (1991). Insolation values for the climate of the last 10 million years. *Quaternary Science Reviews* 10:297–317.

Berger, J., J. Swenson, and I. Persson (2001). Recolonizing carnivores and naïve prey: conservation lessons from Pleistocene extinctions. *Science* 291:1036–39.

Berger, R. (1975). Advances and results in radiocarbon dating: early man in America. *World Archaeology* 7:174–84.

Berger, R. and R. Protsch (1989). UCLA radiocarbon dates XI. *Radiocarbon* 31:55–67.

Berlin, E. and B. Berlin (1996). *Medical ethnobiology of the Highland Maya of Chiapas, Mexico: the gastrointestinal diseases*. Princeton University Press, Princeton, NJ.

Bettinger, R. and D. Young (2004). Hunter-gatherer population expansion in north Asia and the New World. In *Entering America: Northeast Asia and Beringia before the Last Glacial Maximum*, edited by D. Madsen, pp. 239–51. University of Utah Press, Salt Lake City.

Bever, M. (2001). An overview of Alaskan late Pleistocene archaeology: historical themes and current perspectives. *Journal of World Prehistory* 15:125–91.

Bever, M. and M. Kunz, eds. (2001). Between two worlds: late Pleistocene cultural and technological diversity in Beringia. *Arctic Anthropology* 38.

Bever, M. and D. J. Meltzer (2007). Investigating variation in Clovis Paleoindian lifeways: the third revised edition of the Texas Clovis Fluted Point Survey. *Bulletin of the Texas Archeological Society* 78:65–99.

Binford, L. R. (1983). Long term land use patterns: some implications for archaeology. In *Lulu linear punctated: essays in honor of George Irving Quimby*, edited by R. C. Dunnell and D. K. Grayson. *University of Michigan Anthropological Papers* 72:27–53.

Bird, D. and J. F. O'Connell (2006). Behavioral ecology and archaeology. *Journal of Archaeological Research* 14:143–88.

Black, F. L. (1992). Why did they die? *Science* 258:1739–40.

Black, S. (1978). Polynesian outliers: a study in the survival of small populations. In *Simulation studies in archaeology*, edited by I. Hodder, pp. 63–76. Cambridge University Press, Cambridge.

Blanton, D. (2003). The weather is fine, wish you were here, because I'm the last one alive: learning the environment in the English New World colonies. In *Colonization of unfamiliar landscapes: the archaeology of adaptation*, edited by M. Rockman and J. Steele, pp. 190–200. Routledge, London.

Bliss, W. (1940a) A chronological problem presented by Sandia Cave, New Mexico. *American Antiquity* 5:200–201.

Bliss, W. (1940b) Sandia Cave. *American Antiquity* 6:77–78.

Bolnick, D. A., B. A. Shook, L. Campbell, and I. Goddard (2004). Problematic use of Greenberg's linguistic classification of the Americas in studies of Native American genetic variation. *American Journal of Human Genetics* 75:519–22.

Bolton, H. E. (1949). *Coronado, knight of pueblos and plains*. Whittlesey House, New York.

Bonnichsen, R., B. Lepper, D. Stanford, and M. Waters (2006). *Paleoamerican origins: beyond Clovis*. Center for the Study of the First Americans, College Station, TX.

Bonnichsen, R. and K. Turnmire, eds. (1991). *Clovis: origins and adaptations*. Center for the Study of the First Americans, Corvallis, OR.

Bonnichsen, R. and K. Turnmire, eds. (1999). *Ice-Age people of North America: environments, origins, and adaptations.* Center for the Study of the First Americans, Corvallis, OR.

Bonnichsen et al. v. United States 1996 Civil No. 96-1481JE, U.S. District Court, District of Oregon.

Boone, E. (1994). *The Aztec world.* Smithsonian Books, Washington, D.C.

Booth, D., K.G. Troost, J.J. Clague, and R. Wiatt (2004). The Cordilleran ice sheet. In *The Quaternary period in the United States,* edited by A. Gillespie, S.C. Porter, and B. Atwater, pp. 17–43. Elsevier Science, New York.

Borgerhoff Mulder, M. (1992). Reproductive decisions. In *Evolutionary ecology and human behavior,* edited by E. Smith and B. Winterhalder, pp. 339–74. Aldine de Gruyter, New York.

Borrero, L. (1995). Human and natural agency: some comments on Pedra Furada. *Antiquity* 69:602–3.

Bowler, P. (1976). *Fossils and progress: paleontology and the idea of progressive evolution in the nineteenth century.* Science History Publications, New York.

Brace, C.L., A. Nelson, N. Seguchi, H. Oe, L. Sering, P. Qifeng, L. Yongyi, and D. Tumen (2001). Old World sources of the first New World human inhabitants: a comparative craniofacial view. *Proceedings of the National Academy of Sciences* 98:10017–22.

Bradley, B. and D. Stanford (2004). The North Atlantic ice-edge corridor: a possible Palaeolithic route to the New World. *World Archaeology* 36:459–78.

Bradley, R. and J. England (2008). The Younger Dryas and the sea of ancient ice. *Quaternary Research* 70:1–10.

Brain, J.P. (1971). The Natchez paradox. *Ethnology* 10:215–22.

Brand, D. (1940). Regarding Sandia Cave. *American Antiquity* 5:339.

Breternitz, D., A. Swedlund, and D. Anderson (1971). An early burial from Gordon Creek, Colorado. *American Antiquity* 36:170–82.

Brigham-Grette, J., A. Lozhkin, P.M. Anderson, and O.Y. Glushkova (2004). Paleoenvironmental conditions in western Beringia before and during the Last Glacial Maximum. In *Entering America: Northeast Asia and Beringia before the Last Glacial Maximum,* edited by D. Madsen, pp. 29–61. University of Utah Press, Salt Lake City.

Broecker, W.S. (1995). Chaotic climate. *Scientific American* (November): 62–68.

Broecker, W.S. (2003). Does the trigger for abrupt climate change reside in the ocean or in the atmosphere? *Science* 300:1519–22.

Broecker, W.S. (2006). Was the Younger Dryas triggered by a flood? *Science* 312:1146–48.

Broecker, W.S. and G. Denton (1990). What drives glacial cycles? *Scientific American* (January):49–56.

Broecker, W. and R. Kunzig (2008). *Fixing climate: what past climate changes reveal about the current threat—and how to counter it.* Hill and Wang, New York.

Brown, J.H. (1971). Mammals on mountaintops: nonequilibrium insular biogeography. *American Naturalist* 105:467–78.

Brown, M., S. Hoseini, A. Torroni, H. Bandeldt, J. Allen, T. Schurr, R. Scozzari, F. Cruciani, and D. Wallace (1998). mtDNA haplogroup X: an ancient link between Europe/western Asia and North America? *American Journal of Human Genetics* 63:1852–70.

Brown, R.M., H. Andrews, G. Ball, N. Burn, Y. Imahori, and J. Milton (1983). Accelerator ^{14}C dating of the Taber Child. *Canadian Journal of Archaeology* 7:233–37.

Bryan, A., ed. (1978). *Early man in America*. Archaeological Researches International, Edmonton, AB, Canada.

Bryan, A., ed. (1986). *New evidence for the Pleistocene peopling of the Americas*. Center for the Study of Early Man, University of Maine, Orono.

Buckland, W. (1823). *Reliquiæ diluvianæ; or observations on the organic remains contained in caves, fissures, and diluvial gravel, and on other geological phenomena, attesting the action of an universal deluge*. J. Murray, London.

Burkhardt, F., J. Browne, D. Porter, and M. Richmond (1994). *The correspondence of Charles Darwin, Vol. 9, 1861*. Cambridge University Press, Cambridge.

Burkhardt, F. and S. Smith (1985). *The correspondence of Charles Darwin, Vol. 1, 1821–36*. Cambridge University Press, Cambridge.

Burkhardt, F. and S. Smith (1986). *The correspondence of Charles Darwin, Vol. 2, 1837–43*. Cambridge University Press, Cambridge.

Burkhardt, F. and S. Smith (1991). *The correspondence of Charles Darwin, Vol. 7, 1858–59*. Cambridge University Press, Cambridge.

Butzer, K. (1991). An Old World perspective on potential mid-Wisconsinan settlement of the Americas. In *The first Americans: search and research*, edited by T. Dillehay and D. J. Meltzer, pp. 137–56. CRC Press, Boca Raton, FL.

Byerly, R. M., J. R. Cooper, D. J. Meltzer, M. E. Hill, and J. M. LaBelle (2005). On Bonfire shelter (Texas) as a Paleoindian bison jump: an assessment using GIS and zooarchaeology. *American Antiquity* 70:595–629.

Campbell, L. (1988). Review of *Language in the America*. *Language* 64:591–615.

Campbell, L. (1997). *American Indian languages: the historical linguistics of native America*. Oxford University Press, Oxford.

Cann, R., M. Stoneking, and A. Wilson (1987). Mitochondrial DNA and human evolution. *Nature* 325:31–36.

Cannon, M. D. and D. J. Meltzer (2004). Early Paleoindian foraging: examining the faunal evidence for large mammal specialization and regional variability in prey choice. *Quaternary Science Reviews* 23 (18/19):1955–87.

Cardillo, M., G. M. Mace, K. Jones, J. Bielby, O. R. Bininda-Emonds, W. Sechrest, C. D. Orme, and A. Purvis (2005). Multiple causes of high extinction risk in large mammal species. *Science* 309:1239–41.

Carlson, A., P. U. Clark, B. A. Haley, G. P. Klinkhammer, K. Simmons, E. Brook and K. Meissner (2007). Geochemical proxies of North American freshwater routing during the Younger Dryas cold event. *Proceedings of the National Academy of Sciences* 104:6556–61.

Carlson, R. and L. Bona, eds. (1996). *Early human occupation in British Columbia*. University of British Columbia Press, Vancouver.

Cavalli-Sforza, L., P. Menozzi, and A. Piazza (1994). *The history and geography of human genes*. Princeton University Press, Princeton, NJ.

Chamberlin, T. C. (1892). Geology and archaeology mistaught. *Dial* 13:303–6.

Charnov, R., G. Orians, and K. Hyatt (1976). The ecological implications of resource depression. *American Naturalist* 110:247–59.

Chatters, J. C. (2000). The recovery and first analysis of an Early Holocene human skeleton from Kennewick, Washington. *American Antiquity* 65:291–316.

Chatters, J.C. (2001). *Ancient encounters. Kennewick man and the first Americans*. Simon & Schuster, New York.

Chen, T., R.E.M. Hedges, and Z. Yuan (1989). Accelerator radiocarbon dating for the Upper Cave of Zhoukoudian. *Acta Anthropologica Sinica* 8:216–21.

Chilton, E. (2004). Beyond "Big": gender, age, and subsistence diversity in Paleoindian studies. In *The settlement of the American continents: a multidisciplinary approach to human biogeography*, edited by C.M. Barton, G.A. Clark, D.R. Yesner, and G.A. Pearson, pp. 162–72. University of Arizona Press, Tucson.

Churchill, S. (2002). Of assegais and bayonets: reconstructing prehistoric spear use. *Evolutionary Anthropology* 11:185–86.

Clague, J. (2006). Open letter by INQUA executive committee. *Quaternary International* 154/155:158–59.

Clague, J.J., R.W. Mathewes, and T.A. Ager (2004). Environments of northwestern North America before the Last Glacial Maximum. In *Entering America: Northeast Asia and Beringia before the Last Glacial Maximum*, edited by D. Madsen, pp. 63–94. University of Utah Press, Salt Lake City.

Clark, D. (2001). Microblade culture systematics in the far interior Northwest. *Arctic Anthropology* 38:64–80.

Clark, P.U., R. Alley, and D. Pollard (1999). Northern hemisphere ice-sheet influences on global climate change. *Science* 286:1104–11.

Clark, P.U., A.M. McCabe, A.C. Mix, and A.J. Weaver (2004). Rapid rise of sea level 19,000 years ago and its global implications. *Science* 304:1141–44.

Clark, P.U., N. Pisias, T. Stocker, and A. Weaver (2002). The role of thermohaline circulation in abrupt climate change. *Nature* 415:863–69.

Clausen, C., J. Brooks, and A. Wesolowsky (1975). The early man site at Warm Mineral Springs, Florida. *Journal of Field Archaeology* 2:191–213.

Collins, M.B. (1991). Rockshelters and the early archaeological record in the Americas. In *The first Americans: search and research*, edited by T. Dillehay and D.J. Meltzer, pp. 157–82. CRC Press, Boca Raton, FL.

Collins, M.B. (2002). The Gault site, Texas, and Clovis research. *Athena Review* 3:31–41.

Collins, M.B., J. Lohse, and M. Shoberg (2007). The deGraffenreid collection: a Clovis bifaces cache from the Gault site, central Texas. *Bulletin of the Texas Archeological Society* 78:101–23.

Cook, H.J. (1925). Definite evidence of human artifacts in the American Pleistocene. *Science* 62:459–60.

Cook, H.J. (1927). New trails of ancient men. *Scientific American* 138:114–17.

Crane, H.R. (1955). Antiquity of the Sandia Culture: carbon-14 measurements. *Science* 122:689–90.

Cronon, W. and R. White (1988). Ecological change and Indian-White relations. *Handbook of North American Indians, Vol. 4, history of Indian-White relations*, edited by W.E. Washburn, pp. 417–29. Smithsonian Institution Press, Washington, D.C.

Crook, W. and R. Harris (1957). Hearths and artifacts of early man near Lewisville, Texas, and associated faunal material. *Bulletin of the Texas Archeological Society* 28:7–97.

Crosby, A. (1986). *Ecological imperialism: the biological expansion of Europe, 900–1900*. Cambridge University Press, Cambridge.

Curry, R. A. (2007). Late glacial impacts on dispersal and colonization of Atlantic Canada and main by freshwater fishes. *Quaternary Research* 67:225–33.

Cybulski, J., D. Howes, J. Haggarty, and M. Eldridge (1981). An early human skeleton from south central British Columbia: dating and bioarchaeological inference. *Canadian Journal of Archaeology* 5:59–60.

Dalton, R. (2007). Archaeology: blast in the past? *Nature* 447:256–7.

Dansie, A. and W. Jerrems (2006). More bits and pieces: a new look at Lahontan chronology and human occupation. In *Paleoamerican origins: beyond Clovis,* edited by R. Bonnichsen, B. Lepper, D. Stanford, and M. Waters, pp. 51–79. Center for the Study of the First Americans, College Station, TX.

Darwin, C. (1839a) *Journal of the researches into the natural history and geology of the countries visited during the voyage of H. M. S. Beagle round the world, under the command of Capt. Fitz Roy, R. N., from 1832 to 1836.* H. Colburn, London.

Darwin, C. (1839b) Observations on the Parallel Roads of Glen Roy, and of other parts of Lochaber in Scotland, with an attempt to prove they are of marine origin. In *Philosophical Transactions of the Royal Society of London,* Pt. 1:39–81.

Darwin, C. (1859). *On the origin of species.* J. Murray, London.

Darwin, C. (1871). *The descent of man, and selection in relation to sex.* J. Murray, London.

Davis, L. G. (2006). Geoarchaeological insights from Indian Sands, a late Pleistocene site on the southern northwest coast, USA. *Geoarchaeology* 21:351–61.

Dawkins, W. B. (1863). Wookey Hole hyaena-den. *Somersetshire Archaeological and Natural History Society, Proceedings* XI:197–219.

Deloria, V. (1995). *Red earth, white lies: Native Americans and the myth of scientific fact.* Scribner, New York.

Dennell, R. and L. Hurcombe (1995). Comment on Pedra Furada. *Antiquity* 69:604.

Denton, G. H., R. B. Alley, G. Comer, and W. Broecker (2005). The role of seasonality in abrupt climate change. *Quaternary Science Reviews* 24:1159–82.

Derenko, M. V., T. Grzybowski, B. A. Malyarchuk, J. Czarny, D. Mishcicka-Shliwka, and I. A. Zakharov (2001). The presence of mitochondrial haplogroup X in Altaians from South Siberia. *American Journal of Human Genetics* 69:237–41.

Diamond, J. (1999). *Guns, germs, and steel: the fates of human societies.* W. W. Norton, New York.

Dieno, A., P. Renne, and C. Swisher (1998). $^{40}Ar/^{39}Ar$ dating in paleoanthropology and archaeology. *Evolutionary Anthropology* 6:63–75.

Dillehay, T. D. (1989). *Monte Verde: a late Pleistocene settlement in Chile, Vol. 1, Palaeoenvironment and site context.* Smithsonian Institution Press, Washington, D.C.

Dillehay, T. D. (1991). Disease ecology and initial human migration. In *The first Americans: search and research,* edited by T. Dillehay and D. Meltzer, pp. 231–64. CRC Press, Boca Raton, FL.

Dillehay, T. D. (1997). *Monte Verde: a late Pleistocene settlement in Chile, Vol. 2, The archaeological context and interpretation.* Smithsonian Institution Press, Washington, D.C.

Dillehay, T. D. (2000). *The settlement of the Americas: a new prehistory.* Basic Books, New York.

Dillehay, T. D. and M. Collins (1991). Monte Verde, Chile: a comment on Lynch. *American Antiquity* 56:333–41.

Dillehay, T. D. and D. J. Meltzer (1991). Finale: processes and prospects. In *The first Americans: search and research*, edited by T. Dillehay and D. Meltzer, pp. 287–94. CRC Press, Boca Raton, FL.

Dillehay, T. D., C. Ramírez, M. Pino, M. B. Collins, J. Rossen, and J. Pino-Navarro (2008). Monte Verde: seaweed, food, medicine, and the peopling of South America. *Science* 320:784–86.

Dincauze, D. (1991). Review of "Monte Verde: a late Pleistocene settlement in Chile, Volume 1: Palaeoenvironment and site context," by Tom D. Dillehay. *Journal of Field Archaeology* 18:116–19

Dincauze, D. (1993). Fluted points in the eastern forests. In *From Kostenki to Clovis: Upper Paleolithic—Paleo-indian adaptations,* edited by O. Soffer and N. Praslov, pp. 279–92. Plenum Press, New York.

Dixon, E. J. (1999). *Bones, boats & bison: Archaeology and the first colonization of western North America.* University of New Mexico Press, Albuquerque.

Dobyns, H. (1966). Estimating aboriginal American population: an appraisal of techniques with a new hemispheric estimate. *Current Anthropology* 7:395–416.

Dobyns, H. (1983). *Their number become thinned: Native American population dynamics in eastern North America.* University of Tennessee Press, Knoxville.

Dobyns, H. (1991). New native world: links between demographic and cultural changes. In *Columbian consequences, Vol. 3, the Spanish borderlands in Pan-American perspective,* edited by D. Thomas, pp. 541–59. Smithsonian Institution Press, Washington, D.C.

Doran, G., ed. (2002). *Windover. Multidisciplinary investigations of an Early Archaic Florida cemetery.* University Press of Florida.

Downey, R. (2000). *Riddle of the bones: politics, science, race, and the story of Kennewick man.* Copernicus Books.

Duller, G. A. (2006). Comment on "Human footprints in central Mexico older than 40,000 years," by S. Gonzalez, D. Huddart, M. R. Bennett, and A. Gonzalez-Huesca. *Quaternary Science Reviews* 25:3074–76.

Dumond, D. (1984). Prehistory: summary. In *Handbook of North American Indians, Vol. 5, Arctic,* edited by D. Damas, pp. 72–79. Smithsonian Institution Press, Washington, D.C.

Dumond, D. (2001). The archaeology of eastern Beringia: some contrasts and connections. *Arctic Anthropology* 38:196–205.

Dunbar, J. and P. Vojnovski (2007). Early Floridians and late megamammals: some technological and dietary evidence from four North Florida Paleoindian sites. In *Foragers of the terminal Pleistocene in North America*, R. Walker and B. Driskell, eds., pp. 166–202. University of Nebraska Press, Lincoln.

Dunn, M., A. Terrill, G. Reesink, R. A. Foley, and S. C. Levinson (2005). Structural phylogenetics and the reconstruction of ancient language history. *Science* 309:2072–75.

Dunnell, R. (1991). Methodological impacts of catastrophic depopulation on American archaeology and ethnology. In *Columbian consequences, Vol. 3, the Spanish borderlands in Pan-American perspective,* edited by D. Thomas, pp. 561–80. Smithsonian Institution Press, Washington, D.C.

Dyke, A. (2004). An outline of North American deglaciation with emphasis on central and northern Canada. In *Quaternary Glaciations–Extent and Chronology, Part II*, edited by J. Ehlers and P. L. Gibbard. *Developments in Quaternary Science, Vol. 2b*, 373–424, Elsevier, Amsterdam.

Eccles, W. J. (1988). The fur trade in the colonial Northeast. *Handbook of North American Indians, Vol. 4, history of Indian-White relations*, edited by D. Damas, pp. 324–34. Smithsonian Institution Press, Washington, D.C.

Echo-Hawk, R. (1997). Forging a new ancient history for Native America. In *Native Americans and archaeologists: stepping stones to common ground*, edited by N. Swidler, K. Dongoske, R. Anyon, and A. Downer, pp. 88–102. AltaMira Press, Walnut Creek, CA.

Egan, T. (1996). Tribe stops study of bones that challenge history. *New York Times*, September 30, 1996.

Eiseley, L. (1942). Review of "Evidences of early occupation in Sandia Cave, New Mexico, and other sites in the Sandia Monzano region," by Frank C. Hibben. *American Antiquity* 7:415–17.

Eiseley, L. C. (1975). *All the strange hours: the excavation of a life.* Scribner, New York.

Elliot, J. H. (2006). *Empires of the Atlantic world: Britain and Spain in America 1492–1830.* Yale University Press, New Haven, CT.

Ellis, C. J. (1997). Factors influencing the use of stone projectile tips: an ethnographic perspective. In *Projectile technology*, edited by H. Knecht, pp. 37–74. Plenum Press, New York.

Ellis, C. J. (2008). The fluted point tradition and the Arctic Small Tool tradition: what's the connection? *Journal of Anthropological Archaeology* 27:298–314.

Ellis, C. J. and D. B. Deller (1990). Paleo-indians. In *The archaeology of southern Ontario to A. D. 1650*, edited by C. Ellis and N. Ferris. *Occasional Publications of the London Chapter, Ontario Archaeological Society* No. 5:37–64.

Ellis, C. J. and D. B. Deller (1997). Variability in the archaeological record of northeastern early Paleoindians: a view from southern Ontario. *Archaeology of Eastern North America* 25:1–30.

Ellison, C., M. Chapman, and I. Hall (2006). Surface and deep ocean interactions during the cold climate event 8200 years ago. *Science* 312:1929–32.

Epstein, P. R. (1999). Climate and health. *Science* 285:347–48.

Eren, M. (2008). Paleoindian stability during the Younger Dryas in the North American lower Great Lakes. In *Transitions in Prehistory: papers in honor of Ofer Bar-Yosef*, edited by J. Shea and D. Lieberman. American School of Prehistoric Research Press, Cambridge, MA.

Eshleman, J. A., R. S. Malhi, and D. G. Smith (2003). Mitochondrial DNA studies of Native Americans: conceptions and misconceptions of the population prehistory of the Americas. *Evolutionary Anthropology* 12:7–18.

Evans, J. (1943). *Time and chance; the story of Arthur Evans and his forebears.* London, Longmans, Green.

Fagundes, N., R. Kanitz, R. Eckert, A. Valls, M. Bogo, F. Salzano, D. Smith, W. Silva, M. Zago, A. Ribeiro-dos-Santos, S. Santos, M. Petzl-Erler, and S. Bonatto (2008). Mitochondrial population genomics supports a single pre-Clovis origin with a coastal route for the peopling of the Americas. *American Journal of Human Genetics* 82:583–92.

Fall, P. L. (1997). Timberline fluctuations and late Quaternary paleoclimates in the southern Rocky Mountains, Colorado. *Geological Society of America Bulletin* 109:1306–10.

FAUNMAP Working Group (1996). Spatial response of mammals to Late Quaternary environmental fluctuations. *Science* 272:1601–6.

Feathers, J. (1996). Luminescence dating and modern human origins. *Evolutionary Anthropology* 5:25–36.

Fedje, D. and H. Josenhans (2000). Drowned forests and archaeology on the continental shelf of British Columbia, Canada. *Geology* 28:99–102.

Fedje, D., Q. Mackie, E. J. Dixon, and T. H. Heaton (2004). Late Wisconsin environments and archaeological visibility on the northern Northwest Coast. In *Entering America: Northeast Asia and Beringia before the Last Glacial Maximum,* edited by D. Madsen, pp. 97–138. University of Utah Press, Salt Lake City.

Fedje, D. and R. Matthewes (2005). *Haida Gwaii: human history and environment from the time of loon to the time of iron people.* University of British Columbia Press, Vancouver.

Fedje, D., J. M. White, M. C. Wilson, D. E. Nelson, J. S. Vogel, and J. R. Southon (1995). Vermilion Lakes site: adaptations and environments in the Canadian Rockies during the latest Pleistocene and Early Holocene. *American Antiquity* 60:81–108.

Fenn, E. (2001). *Pox Americana: the great smallpox epidemic of 1775–82.* Hill & Wang, New York.

Ferguson, T. J., J. Watkins, and G. Pullar (1997). Native Americans and archaeologists: commentary and personal perspectives. In *Native Americans and archaeologists: stepping stones to common ground,* edited by N. Swidler, K. Dongoske, R. Anyon, and A. Downer, pp. 237–52. AltaMira Press, Walnut Creek, CA.

Ferring, C. R. (2001). *The archaeology and paleoecology of the Aubrey Clovis site (41DN479) Denton County, Texas.* Center for Environmental Archaeology, University of North Texas, Denton.

Fiedel, S. and G. Haynes (2004). A premature burial: comments on Grayson and Meltzer's "Requiem for overkill." *Journal of Archaeological Science* 31:121–31.

Firestone, R. B. and W. Topping (2001). Terrestrial evidence of a nuclear catastrophe in Paleoindian times. *Mammoth Trumpet* 16.2:9–16.

Firestone, R. B., A. West, J. P. Kennett, L. Becker, T. E. Bunch, Z. S. Revay, P. H. Schultz, et al. (2007). Evidence for an extraterrestrial impact 12,900 years ago that contributed to the megafaunal extinctions and the Younger Dryas cooling. *Proceedings of the National Academy of Sciences* 104:16016–21.

Firestone, R. B., A. West, and S. Warwick-Smith (2006). *The cycle of cosmic catastrophes: how a Stone-Age comet changed the course of world culture.* Bear, Rochester, VT.

Fisher, T., C. Yansa, T. Lowell, K. Lepper, I. Hajdas, and A. Ashworth (2008). The chronology, climate, and confusion of the Moorhead Phase of glacial Lake Agassiz: new results from the Ojata Beach, North Dakota, USA. *Quaternary Science Reviews* 27:1124–35.

Fix, A. G. (2002). Colonization models and initial genetic diversity in the Americas. *Human Biology* 74:1–10.

Fix, A. G. (2005). Rapid deployment of the five founding Amerind mtDNA haplogroups via coastal and riverine colonization. *American Journal of Physical Anthropology* 128:430–36.

Fladmark, K., J. Driver, and D. Alexander (1988). The Paleoindian component at Charlie Lake Cave (HbRf39), British Columbia. *American Antiquity* 53:371–84.

Foley, R. (1988). Hominids, humans and hunter-gatherers: an evolutionary perspective. In *Hunters and gatherers: history, evolution, and social change*, edited by T. Ingold, D. Riches, and J. Woodburn, pp. 207–21. Berg, Oxford, UK.

Forster, P., R. Harding, A. Torroni, and H. Bandelt (1996). Origin and evolution of Native American mtDNA variation: a reappraisal. *American Journal of Human Genetics* 59:935–45.

Foster, M. (1996). Language and the culture history of America. *Handbook of North American Indians, Vol. 17, Languages*, edited by I. Goddard, pp. 64–110. Smithsonian Institution Press, Washington, D.C.

Friesen, T. and C. Arnold (2008). The timing of the Thule migration: new dates from the western Canadian Arctic. *American Antiquity* 73:527–38.

Frison, G. C. (1991). *Prehistoric hunters of the high plains*, 2nd ed. Academic Press, New York.

Frison, G. C. (2004). *Survival by hunting: prehistoric human predators and animal prey*. University of California Press, Berkeley.

Frison, G. C. and B. Bradley (1999). *The Fenn cache: Clovis weapons and tools*. One Horse Land and Cattle Company, Santa Fe.

Galloway, P., and J. Jackson (2004). Natchez and neighboring groups. *Handbook of North American Indians, Vol. 14, Southeast*, edited by R. Fogelson, pp. 598–615. Smithsonian Institution Press, Washington, D.C.

Gamble, C. (1994). *Timewalkers: the prehistory of global colonization*. Harvard University Press, Cambridge, MA.

Gardner, W. M. (1974). *The Flint Run Paleo-indian complex: a preliminary report*. Occasional Publication No. 1, Archaeology Laboratory, Catholic University.

Gentleman of Elvas (1686). *A relation of the invasion and conquest of Florida by the Spaniards, under the command of Fernando de Soto, written in Portuguese by a gentleman of the town of Elvas; now Englished*. J. Lawrence, London.

Gibbons A. (1993). Geneticists trace the DNA trail of the first Americans. *Science* 259:312–13.

Gilbert, T., D. Jenkins, A. Götherstrom, N. Naveran, J. Sanchez, M. Hofreiter, P. Thomsen, J. Binladen, T. Higham, R. Yohe, R. Parr, L. Cummings, and E. Willerslev (2008). DNA from pre-Clovis human coprolites in Oregon, North America. *Science* 320:786–89.

Gilder, R. (1911). Scientific "inaccuracies" in reports against probability of geological antiquity of remains of Nebraska loess man considered by its discoverer. *Records of the Past* 10:157–69.

Gillespie, A., S. Porter, and B. Atwater, eds. (2004). *The Quaternary period in the United States*. Elsevier Science, New York.

Goddard, I., ed. (1996). *Handbook of North American Indians, Vol. 17, Languages*. Smithsonian Institution Press, Washington, D.C.

Goddard, I. (2004). Endangered knowledge: what we can learn from Native American languages. *AnthroNotes* 25.2:1–8.

Goddard, P. E. (1927). Facts and theories concerning Pleistocene man in America. *American Anthropologist* 29:262–66.

Goebel, T. (2004). The search for a Clovis progenitor in sub-Arctic Siberia. In *Entering America: Northeast Asia and Beringia before the Last Glacial Maximum*, edited by D. Madsen, pp. 311–56. University of Utah Press, Salt Lake City.

Golla, V. (2000). Language families of North America. In *America past, America present: genes and languages in the Americas and beyond*, edited by C. Renfrew, pp. 59–72. McDonald Institute for Archaeological Research, Cambridge, UK.

Golledge, R. (2003). Human wayfinding and cognitive maps. In *Colonization of unfamiliar landscapes: the archaeology of adaptation*, edited by M. Rockman and J. Steele, pp. 25–43. Routledge, London.

Gonzáles-José, R., A. Gonzales-Martin, M. Hernandez, H. Pucciarelli, M. Sardi, A. Rosales, and S. van der Molen (2003). Craniometric evidence for Paleoamerican survival in Baja California. *Nature* 425:62–65.

Gonzalez, S., D. Huddart, and M. Bennett (2006). Valsequillo Pleistocene archaeology and dating: ongoing controversy in central Mexico. *World Archaeology* 38:611–27.

Gonzalez, S., D. Huddart, M. Bennett, and A. Gonzalez-Huesca (2006). Human footprints in central Mexico older than 40,000 years. *Quaternary Science Reviews* 25:201–22.

Goodyear, A. C. (1982). The chronological position of the Dalton Horizon in the southeastern United States. *American Antiquity* 47:382–95.

Goodyear, A. C. (1999). The Early Holocene occupation of the southeastern United States: a geoarchaeological summary. In *Ice-Age people of North America: environments, origins, and adaptations*, edited by R. Bonnichsen and K. Turnmire, pp. 432–81. Center for the Study of the First Americans, Corvallis, OR.

Gould, R. (2004). Opinion for the United States Court of Appeal, February 4, 2004. *Bonnichsen et al. v. United States et al.*, 357 F.3d 962 (9th Cir. 2004).

Gould, S. J. (1989). *Wonderful life: the Burgess Shale and the nature of history.* W. W. Norton, New York.

Gramly, R. M. (1982). *The Vail site: a Paleo-indian encampment in Maine.* Buffalo Museum of Science, Buffalo, NY.

Gramly, R. M. (1993). *The Richey Clovis cache: earliest Americans along the Columbia River.* Persimmon Press, Buffalo, NY.

Gramly, R. M. (2005). Recent archaeological fieldwork at the Vail Clovis site. *Amateur Archaeologist* 11:19–38.

Gravlee, C., H. R. Bernard, and W. Leonard (2003). Boas's Changes in bodily form: the immigrant study, cranial plasticity, and Boas's physical anthropology. *American Anthropologist* 105:326–32.

Gray, R. (2005). Pushing the time barrier in the quest for language roots. *Science* 309:2007–8.

Grayson, D. K. (1977). Pleistocene avifaunas and the overkill hypothesis. *Science* 193:691–93.

Grayson, D. K. (1983). *The establishment of human antiquity.* Academic Press, New York.

Grayson, D. K. (1993). *The desert's past: a natural history of the Great Basin.* Smithsonian Institution Press, Washington DC.

Grayson, D. K. (2001). The archaeological record of human impacts on animal populations. *Journal of World Prehistory* 15:1–68.

Grayson, D. K. (2006a). Brief histories of some Great Basin mammals: extinctions, extirpations, and abundance histories. *Quaternary Science Reviews* 25:2964–91.

Grayson, D. K. (2006b). Late Pleistocene faunal extinctions. *Handbook of North American Indians, Vol. 3, environment, origins, and population*, edited by D. Ubelaker, pp. 208–18. Smithsonian Institution Press, Washington, D.C.

Grayson, D. K. (2007). Deciphering North American Pleistocene extinctions. *Journal of Anthropological Research* 63:185–213.

Grayson, D. K. and D. J. Meltzer (2002). Clovis hunting and large mammal extinction: a critical review of the evidence. *Journal of World Prehistory* 16:313–59.

Grayson, D. K. and D. J. Meltzer (2003). Requiem for North American overkill. *Journal of Archaeological Science* 30:585–93.

Grayson, D. K. and D. J. Meltzer (2004). North American overkill continued? *Journal of Archaeological Science* 31:133–36.

Green, T., B. Cochran, T. Fenton, J. Woods, G. Titmus, L. Tieszen, M. Davis, and S. Miller. (1998). The Buhl burial: a Paleoindian woman from southern Idaho. *American Antiquity* 63:437–56.

Greenberg, J. H. (1987). *Language in the Americas*. Stanford University Press, Stanford.

Greenberg, J. H. (1990). The American Indian language controversy. *Review of Archaeology* 11:514.

Greenberg, J. H., C. Turner, and S. Zegura (1986). The settlement of the Americas: a comparison of the linguistic, dental, and genetic evidence. *Current Anthropology* 27:477–97.

Griffin, J. B., D. J. Meltzer, B. Smith, and W. C. Sturtevant (1988). A mammoth fraud in science. *American Antiquity* 53:578–82.

Grimm, E. and G. Jacobson (2004). Late Quaternary vegetation history of the eastern United States. In *The Quaternary period in the United States*, edited by A. Gillespie, S. C. Porter, and B. Atwater, pp. 381–402. Elsevier Science, New York.

Grover, J. (1994). Faunal remains from Dust Cave. *Journal of Alabama Archaeology* 40:116–34.

Grun, R. (1993). Electron spin resonance dating in paleoanthropology. *Evolutionary Anthropology* 2:172–81.

Gryba, E. (2001). Evidence of the fluted point tradition in western Canada. In *On being first: cultural innovation and environmental consequences of first peoplings*. Proceedings of the 31st Annual Chacmool Conference, edited by J. Gillespie, S. Tupakka, and C. de Mille, pp. 251–84. Archaeological Association of the University of Calgary, Calgary, AL, Canada.

Guidon, N. (2002). Pedra Furada, Brazil: Paleoindians, paintings, and paradoxes. *Athena Review* 3:42–52.

Guidon, N. and B. Arnaud (1991). The chronology of the New World: two faces of one reality. *World Archaeology* 23:167–78.

Guidon, N. and A-M. Pessis (1996). Falsehood or untruth? *Antiquity* 70:408–21.

Guthrie, R. (1990). *Frozen fauna of the mammoth steppe: the story of Blue Babe*. University of Chicago Press, Chicago.

Guthrie, R. (2003). Rapid body size decline in Alaskan Pleistocene horses before extinction. *Nature* 426:169–71.

Guthrie, R. (2006). New carbon dates link climatic change with human colonization and Pleistocene extinctions. *Nature* 441:207–9.

Haber, F. (1959). *The age of the world: Moses to Darwin*. Johns Hopkins Press, Baltimore.

Hajdas, I., D. Lowe, and R. Newnham (2006). High-resolution radiocarbon chronologies and synchronization of records. *PAGES News* 14:17–18.

Hamilton, T. and T. Goebel (1999). Late Pleistocene peopling of Alaska. In *Ice-Age people of North America: environments, origins, and adaptations*, edited by R. Bonnichsen and K. Turnmire, pp. 156–99. Center for the Study of the First Americans, Corvallis, OR.

Hammer, M. F. and S. Zegura (1996). The role of the Y chromosome in human evolutionary studies. *Evolutionary Anthropology* 5:116–34.

Hanihara, T. and H. Ishida (2005). Metric dental variation of major human populations. *American Journal of Physical Anthropology* 128:287–98.

Hardaker, C. (2007). *The first American: the suppressed story of the people who discovered the New World*. New Page Books, Franklin Lakes, NJ.

Harmon, A. (2006). DNA gatherers hit snag: tribes don't trust them. *New York Times*, December 10, 2006.

Harpending, H. (1999). Comment on "reproductive interests and forager mobility" by D. MacDonald and B. Hewlett. *Current Anthropology* 40:517.

Harper, K., P. Ocampo, B. Steiner, R. George, M. Silverman, S. Bolotoni, A. Pillay, N. Saunders, and G. Armelagos (2008). On the origin of the Treponematoses: a phylogenetic approach. *PLoS Neglected Tropical Diseases* 2(1): e148 doi:10.1371/journal.pntd.0000148.

Harrison, B. and K. Killen (1978). *Lake Theo: a stratified, early man bison butchering and camp site, Briscoe County, Texas*. Special Archaeological Report 1. Panhandle-Plains Historical Museum, Canyon, Texas.

Hassan, F. (1981). *Demographic archaeology*. Academic Press, New York.

Haury, E. W., Sayles, E., and Wasley, W. (1959). The Lehner mammoth site, southeastern Arizona. *American Antiquity* 25:2–30.

Haven, S. (1856). The archaeology of the United States. *Smithsonian Contributions to Knowledge* 8:1–168.

Hawkes, K. and R. Bliege Bird (2002). Showing off, handicap signaling, and the evolution of men's work. *Evolutionary Anthropology* 11:58–67.

Hay, O. P. (1918). Doctor Aleš Hrdlička and the Vero man. *Science* 47:459–62.

Haynes, C. V. (1964). Fluted projectile points: their age and dispersion. *Science* 145:1408–13.

Haynes, C. V. (1967). Quaternary geology of the Tule Springs area Clark County, Nevada. In *Pleistocene studies in southern Nevada*, edited by H. M. Wormington and D. Ellis. Anthropological Papers 13:16–104. Nevada State Museum, Las Vegas.

Haynes, C. V. (1969). The earliest Americans. *Science* 166:709–715.

Haynes, C. V. (1971) .Time, environment and early man. *Arctic Anthropology* 8:3–14.

Haynes, C. V. (1991a). Geoarchaeological and paleohydrological evidence for a Clovis-age drought in North America. *Quaternary Research* 35:438–50.

Haynes, C. V. (1991b). More on Meadowcroft radiocarbon chronology. *The Review of Archaeology* 12:8–14.

Haynes, C. V. (1999). Monte Verde and the pre-Clovis situation in America. *Discovering Archaeology* (November-December):17–19.

Haynes, C. V. and B. Huckell (2007). Murray Springs: a Clovis site with multiple activity areas in the San Pedro Valley, Arizona. *Anthropological Papers of the University of Arizona*, Number 71. Tucson, AZ.

Haynes, C. V., M. McFaul, R. Brunswig, and K. Hopkins (1998). Lersey-Kuner terrace investigations at the Dent and Bernhardt sites, Colorado. *Geoarchaeology* 13:201–18.

Haynes, C. V., S. Stanford, M. Jodry, J. Dickenson, J. Montgomery, P. Shelley, J. Rovner, and G. Agogino (1999). A Clovis well at the type site 11,500 B.C.: the oldest prehistoric well in America. *Geoarchaeology* 14:455–70.

Haynes, G. (2002). *The early Settlement of North America: the Clovis era*. Cambridge University Press, Cambridge.

Haynes, G. (2007). A review of some attacks on the overkill hypothesis, with special attention to misrepresentation and doubletalk. *Quaternary International* 169–70:84–94.

Hays, J. D., J. Imbrie, and N. J. Shackleton (1976). Variations in the earth's orbit: pacemaker of the Ice Ages. *Science* 194:1121–32.

Helgason, A., G. Pálsson, H. Pedersen, E. Angulalik, E. Gunnarsdóttir, B. Yngvadóttir, and K. Stefánsson. (2006). mtDNA variation in Inuit populations of Greenland and Canada: migration history and population structure. *American Journal of Physical Anthropology* 130:123–34.

Hey, J. (2005). On the number of New World founders: a population genetic portrait of the peopling of the Americas. *PLoS Biology* 3:0001–11.

Hibben, F. (1941). Sandia Cave. *American Antiquity* 6:266.

Hill, J. (2004). Evaluating historical linguistic evidence for ancient human communities in the Americas. In *The settlement of the American continents: a multidisciplinary approach to human biogeography*, edited by C. M. Barton, G. A. Clark, D. R. Yesner, and G. A. Pearson, pp. 39–48. University of Arizona Press, Tucson.

Hill, M. E. (2007). A moveable feast: variation in faunal resource use among central and western North American Paleoindian sites. *American Antiquity* 72:417–38.

Hill, M. G. (2001). Paleoindian diet and subsistence behavior on the northwestern Great Plains of North America. PhD dissertation, Department of Anthropology, University of Wisconsin–Madison.

Hillerman, T. (1973). *Dance hall of the dead*. Harper and Row, New York.

Hiscock, P. (2008). *Archaeology of ancient Australia*. Routledge, London.

Hitt, J. (2005). Mighty white of you: racial preferences color America's oldest skulls and bones. *Harper's Magazine*, July 2005, pp. 39–55.

Ho, S. and G. Larson (2006). Molecular clocks: "when times are a-changin." *TRENDS in Genetics* 22:79–83.

Hoffecker, J. F. (2005). Incredible journey: plains bison hunters in the Arctic. *Review of Archaeology* 26:18–23.

Hoffecker, J. F. and S. Elias (2003). Environment and archeology in Beringia. *Evolutionary Anthropology* 12:34–49.

Hoffecker, J. F. and S. Elias (2007). *Human Ecology of Beringia*. Columbia University Press, New York.

Hoffecker, J. F., W. R. Powers, and T. Goebel (1993). The colonization of Beringia and the peopling of the New World. *Science* 259:46–53.

Hofman, J. L., D. Amick, and R. Rose (1990). Shifting sands: a Folsom-Midland assemblage from a campsite in western Texas. *Plains Anthropologist* 35:221–53.

Holden, C. (2008). Verbs across the Bering Strait. *Science* 319:1595.

Hollenbach, K. D. (2007). Gathering in the Late Paleoindian period: archaeobotanical remains from Dust Cave, Alabama. In *Foragers of the terminal Pleistocene in North America*, edited by R. Walker and B. Driskell, pp. 132–47. University of Nebraska Press, Lincoln.

Holliday, V. T. (2000). Folsom drought and episodic drying on the southern high plains from 10,900–10,200 ¹⁴C yr B. P. *Quaternary Research* 53:1–12.

Holliday, V. T., C. V. Haynes, J. L. Hofman, and D. J. Meltzer (1994). Geoarchaeology and geochronology of the Miami (Clovis) site, southern high plains of Texas. *Quaternary Research* 41:234–44.

Holmes, W. H. (1890). A quarry workshop of the flaked-stone implement makers in the District of Columbia. *American Anthropologist* 3:1–26.

Holmes, W. H. (1893a). Are there traces of man in the Trenton gravels? *Journal of Geology* 1:15–37.

Holmes, W. H. (1893b). Traces of glacial man in Ohio. *Journal of Geology* 1:147–63.

Holmes, W. H. (1919). Handbook of aboriginal American antiquities. Part I. *Bureau of American Ethnology Bulletin 60*.

Hooton, E. (1930). *The Indians of Pecos Pueblo: a study of their skeletal remains*. Yale University Press, New Haven, CT.

Hopkins, D., J. Matthews, C. Schweger, and S. Young, eds. (1982). *Paleoecology of Beringia*. Academic Press, New York.

Howard, E. B. (1932). Arrowheads found with New Mexican fossils. *Science* 76:12–13.

Howard, E. B. (1933). Association of artifacts with mammoth and bison in eastern New Mexico. *Science* 78:524.

Hrdlička, A. (1907). Skeletal remains suggesting or attributed to early man in North America. *Bureau of American Ethnology Bulletin 33*.

Hrdlička, A. (1925). The origin and antiquity of the American Indian. *Smithsonian Institution Annual Report for 1923*:481–494.

Hrdlička, A. (1926). The race and antiquity of the American Indian. *Scientific American* 135:7–9.

Huckell, B. and W. J. Judge (2006). Paleo-indian: plains and Southwest. *Handbook of North American Indians, Vol. 3, environment, origins, and population*, edited by D. Ubelaker, pp. 148–70. Smithsonian Institution Press, Washington, D.C.

Huckell, B. and D. Kilby, eds. (2004). *Readings in Late Pleistocene and Early Holocene Paleoindians: selections from* American Antiquity. Society for American Archaeology, Washington, D.C.

Hughen, K., J. Southon, S. Lehman, and J. Overpeck. (2000). Synchronous radiocarbon and climate shifts during the last deglaciation. *Science* 290:1951–54.

Hull, D. (1988). *Science as a process: an evolutionary account of the social and conceptual development of science*. University of Chicago Press, Chicago.

Hunley, K. and J. Long (2005). Gene flow across linguistic boundaries in Native North American populations. *Proceedings of the National Academy of Sciences* 102:1312–17.

Hutchings, W. K. (1997). The Paleoindian fluted point: dart or spear armature? PhD dissertation, Department of Archaeology, Simon Fraser University, Vancouver.

Imbrie, J. and K. Imbrie (1976). *Ice Ages: solving the mystery*. Enslow, Berkeley Heights, New Jersey.

Irving, W. N. (1985). Context and chronology of early man in the Americas. *Annual Review of Anthropology* 14:529–55.

Irwin, G. (1992). *The prehistoric exploration and colonisation of the Pacific*. Cambridge University Press, Cambridge.

Irwin-Williams, C. (1981). Commentary on "Geologic evidence for age of deposits at Hueyatlaco archaeological site, Valsequillo, Mexico." *Quaternary Research* 16:258.

Ivester, A., D. Godfrey-Smith, M. J. Brooks, and B. E. Taylor (2003). Concentric sand rims document the evolution of a Carolina Bay in the middle coastal plain of South Carolina. Paper presented at the Geological Society of America Annual Meeting, Seattle, WA.

Jablonski, N., ed. (2002). *The first Americans: the Pleistocene colonization of the New World*. Memoirs of the California Academy of Sciences 27, University of California Press, Berkeley.

Jablonski, D., ed. (1962). *Letters of the Lewis and Clark expedition*. University of Illinois Press, Urbana.

Jackson, S. and C. Weng (1999). Late Quaternary extinction of a tree species in eastern North America. *Proceedings of the National Academy of Sciences* 96:13847–52.

Jantz, R. L. and D. Owsley (2001). Variation among early North American crania. *American Journal of Physical Anthropology* 114:146–55.

Jantz, R. L. and D. Owsley (2003). Reply to Van Vark et al.: is European Upper Paleolithic cranial morphology a useful analogy for early Americans? *American Journal of Physical Anthropology* 121:185–88.

Jantz, R. L. and D. Owsley (2006). Circumpacific populations and the peopling of the New World: evidence from cranial morphometrics. In *Paleoamerican origins: beyond Clovis*, edited by R. Bonnichsen, B. Lepper, D. Stanford, and M. Waters, pp. 267–75. Center for the Study of the First Americans, College Station, TX.

Jefferson, T. (1788). *Notes on the state of Virginia*. Prichard and Hall, Philadelphia.

Jelderks, J. (2002). *Opinion and Order*, August 30, 2002. *Bonnichsen et al. v. United States et al.*, 217 F. Supp. 2d 1116 (Dist. Or. 2002).

Jelinek, A. J. (1992). Perspectives from the Old World on the habitation of the New. *American Antiquity* 57:345–47.

Jenkins, D. (2007). Distribution and dating of cultural and paleontological remains at the Paisley 5 Mile Point Caves in the northern Great Basin: an early assessment. In *Paleoindian or Paleoarchaic? Great basin human ecology at the Pleistocene/Holocene transition*, edited by K. Graf and D. Schmitt, pp. 57–81. University of Utah Press, Salt Lake City.

Jenks, A. (1937). Minnesota's Browns Valley man and associated burial artifacts. *American Anthropological Association Memoir* 49. Menasha, Wisconsin.

Jodry, M. (1999). Folsom technological and socioeconomic strategies: views from Stewart's Cattle Guard and the Upper Rio Grande Basin, Colorado. PhD Dissertation, Department of Anthropology, American University, Washington, D.C.

Johnson, A. R., J. Wiens, B. Milne, and T. Crist (1992). Animal movements and population dynamics in heterogeneous landscapes. *Landscape Ecology* 7:63–75.

Johnson, C. N. (2002). Determinants of loss of mammal species during the late Quaternary "megafauna" extinctions: life history and ecology, but not body size. *Proceedings of the Royal Society of London*, Series B 269:2221–27.

Johnson, E. (1987). *Lubbock Lake: late Quaternary studies on the southern high plains*. Texas A&M Press, College Station.

Johnson, J., T. Stafford, H. Aije, and D. Morris (2002). Arlington Springs revisited. In *Proceedings of the Fifth California Islands Symposium*, edited by D. Browne, L. Mitchell, and H. Chaney, pp. 541–45. Santa Barbara Museum of Natural History, Santa Barbara, CA.

Johnston, W. A. (1933). Quaternary geology of North America in relation to the migration of man. In *The American Aborigines: their origin and antiquity*, edited by D. Jenness, pp. 9–45. University of Toronto Press, Toronto.

Jones, G. T., C. Beck, E. Jones, and R. Hughes (2003). Lithic source use and Paleoarchaic foraging territories in the Great Basin. *American Antiquity* 68:5–38.

Jones, T. and K. Klar (2007). *California prehistory: colonization, culture, and complexity*. AltaMira Press, Lanham, MD.

Jones, T., J. Porcasi, J. Erlandson, H. Dallas, T. Wake, and R. Schwader (2008). The protracted Holocene extinction of California's flightless sea duck *(Chendytes lawi)* and its implications for the Pleistocene overkill hypothesis. *Proceedings of the National Academy of Sciences* 105:4105–08.

Jouzel, J. (1999). Calibrating the isotopic paleothermometer. *Science* 286:910–11.

Kaplan, H. and K. Hill (1992). The evolutionary ecology of food acquisition. In *Evolutionary ecology and human behavior*, edited by E. Smith and B. Winterhalder, pp. 167–201. Aldine de Gruyter, New York.

Kappleman, J. (1993). The attraction of paleomagnetism. *Evolutionary Anthropology* 2:89–99.

Karafet, T., S. Zegura, O. Posukh, L. Ospiva, A. Bergen, J. Long, D. Goldman, et al. (1999). Ancestral Asian source(s) of New World Y-chromosome founder haplotypes. *American Journal of Human Genetics* 64:817–31.

Karafet, T., S. Zegura, J. Vuturo-Brady, O. Posukh, L. Ospiva, V. Wiebe, F. Romero, et al. (1997). Y chromosome markers and trans-Bering Strait dispersals. *American Journal of Physical Anthropology* 102:301–14.

Karner, D. and R. Muller (2000). A causality problem for Milankovitch. *Science* 288:2143–44.

Kaufman, T. and V. Golla (2000). Language groupings in the New World: their reliability and usability in cross-disciplinary studies. In *America past, America present: genes and languages in the Americas and beyond*, edited by C. Renfrew, pp. 47–57. McDonald Institute for Archaeological Research, Cambridge, UK.

Keegan, W. and J. Diamond (1987). Colonization of islands by humans: a biogeographic perspective. *Advances in Archaeological Method and Theory* 10:49–92.

Keigwin, L., J. P. Donnelly, M. S. Cook, N. W. Driscoll, and J. Brigham-Grette (2006). Rapid sea-level rise and Holocene climate in the Chukchi Sea. *Geology* 34:861–64.

Kelly, R. L. (1995). *The foraging spectrum: diversity in hunter-gatherer lifeways*. Smithsonian Institution Press, Washington, D.C.

Kelly, R. L. (1996). Ethnographic analogy and migration to the western hemisphere. In *Prehistoric Mongoloid dispersals*, edited by T. Akazawa and E. Szathmary, pp. 228–40. Oxford University Press, Oxford.

Kelly, R. L. (2003). Maybe we do know when people first came to North America; and what does it mean if we do? *Quaternary International* 109/110:133–145.

Kelly, R. L. and L. Todd (1988). Coming into the country: early Paleoindian hunting and mobility. *American Antiquity* 53:231-244.

Kemp, B., R. Malhi, J. McDonough, D. Bolnick, J. Eshleman, O. Rickards, C. Martinez-Labraga, et al. (2007). Genetic analysis of Early Holocene skeletal remains from Alaska and implications for the settlement of the Americas. *American Journal of Physical Anthropology* 132:605–21.

Kerr, R. (2006). Atlantic mud shows how melting ice triggered an ancient chill. *Science* 312:1860.

Kerr, R. (2007). Mammoth-killer impact gets mixed reception from earth scientists. *Science* 316:1264–65.

Kerr, R. (2008). Experts find no evidence for a Mammoth-killer impact. *Science* 319:1331–32.

Kidder, A. V. (1936). Speculations on New World prehistory. In *Essays in anthropology*, edited by R. Lowie, pp. 143–51. University of California Press, Berkeley.

King, M. L. and S. Slobodin (1996). A fluted point from the Uptar site, northeastern Siberia. *Science* 273:634–36.

Kirch, P. V. (1988). Long-distance exchange and island colonization: the Lapita case. *Norwegian Archaeological Review* 21:103–17.

Kirch, P. V. (1997). *The Lapita peoples: ancestors of the Oceanic World*. Blackwell, New York.

Kirch, P. V. (2000). *On the road of the winds: an archaeological history of the Pacific Islands*. University of California Press, Berkeley.

Kitchen, A., M. Miyamoto, and C. Mulligan (2008). A three-stage colonization model for the peopling of the Americas. *PLoS ONE* 3:e1596.

Kittler, R., M. Kayser, and M. Stoneking (2003). Molecular evolution of *Pediculus humanus* and the origin of clothing. *Current Biology* 13:1414–17.

Kohfeld, K. E., C. Le Quere, S. P. Harrison, and R. F. Anderson (2005). Role of marine biology in glacial-interglacial CO_2 cycles. *Science* 308:74–78.

Krajick, K. (2002). Melting glaciers release ancient relics. *Science* 296:454–56.

Krieger, A. D. (1953). New World culture history: Anglo-America. In *Anthropology today: an encyclopedic inventory*, prepared under the chairmanship of A. L. Kroeber, pp. 238–64. University of Chicago Press, Chicago.

Krieger, A. D. (1964). Early man in the New World. In *Prehistoric man in the New World*, edited by J. Jennings and E. Norbeck, pp. 23–81. University of Chicago Press, Chicago.

Kring, D. (2007). The Chicxulub impact event and its environment consequences at the Cretaceous-Tertiary boundary. *Palaeogeography, Palaeoclimatology, Palaeoecology* 255:4–21.

Kroeber, A. L. (1939). *Cultural and natural areas of native North America*. University of California Press, Berkeley.

Kroeber, A. L. (1962). The Rancho La Brea skull. *American Antiquity* 27:416–17.

Kunz, M. L., Bever, M., and C. Adkins (2003). *The Mesa Site: Paleoindians above the Arctic Circle*. BLM-Alaska Open File Report 86, Anchorage, AK.

Kurten, B. and E. Anderson (1980). *Pleistocene mammals of North America*. Columbia University Press, New York.

LaBelle, J.M. (2005). Hunter-gatherer foraging variability during the Early Holocene of the Central Plains of North America. PhD dissertation, Department of Anthropology, Southern Methodist University, Dallas.

Labeyrie, L., J. Cole, K. D. Alverson, and T. Stocker (2003). The history of climate dynamics in the Late Quaternary. In *Paleoclimate, global change, and the future*, edited by K. D. Alverson, R. S. Bradley, and T. F. Pederson, pp. 33–61. Springer, Berlin.

Lahr, M. (1995). Patterns of modern human diversification: implications for Amerindian origins. *Yearbook of Physical Anthropology* 38:163–98.

Lambeck, K., Y. Yokoyama, and T. Purcell (2002). Into and out of the Last Glacial Maximum: sea level change during Oxygen Isotope Stages 3 and 2. *Quaternary Science Reviews* 21:343–60.

Laub, R. (2002). The Paleoindian presence in the northeast: the view from the Hiscock site. In *Ice Age peoples of Pennsylvania*, edited by K. Carr and J. Adovasio, pp. 105–21. Pennsylvania Historical & Museum Commission, Harrisburg, PA.

Lavallée, D. (2000). *The first South Americans: the peopling of a continent from the earliest evidence to high culture*. University of Utah Press, Salt Lake City.

Lawlor, T. E. (1998). Biogeography of Great Basin mammals: paradigm lost? *Journal of Mammalogy* 79:1111–30.

Lawton, C. (1996). Strategies for indoor wayfinding: the role of orientation. *Journal of Environmental Psychology* 16:137–45.

Leakey, L., R. Simpson, and T. Clements (1969). Man in America: the Calico Mountains excavations. *1970 Britannica yearbook of science and the future*, pp. 65–79.

Leakey, M. (1984). *Disclosing the past: an autobiography*. Doubleday, New York.

Lell, J., R. Sukernik, Y. Starikovskaya, B. Su, L. Jin, T. Schurr, P. Underhill, and D. Wallace (2002). The dual origin and Siberian affinities of native American Y chromosomes. *American Journal of Human Genetics* 70:192–206.

Leonhardy, F. (1966). *Domebo: a Paleo-indian mammoth kill in the prairie-plains*. Contributions of the Museum of the Great Plains, Number 1. Lawton, Oklahoma.

Le Page du Pratz, A. S. (1758). *Histoire de la Louisiane: contenant la découverte de ce vaste pays; sa description géographique; un voyage dans les terres; l'histoire naturelle, les moeurs coûtumes & religion des naturels, avec leurs origines; deux voyages dans le nord du nouveau Mexique, dont un jusqu'à la Mer du Sud; ornée de deux cartes et de 40 planches en taille douce*. De Bure, Paris.

Lepper, B. T. (1983). Fluted point distributional patterns in the eastern United States: a contemporary phenomena. *Midcontinental Journal of Archaeology* 8:269–85.

Levine, F. and A. LeBauve (1997). Examining the complexity of historic population decline: a case study of Pecos Pueblo, New Mexico. *Ethnohistory* 44:75–112.

Li, Y., D. Carroll, S. Gardner, M. Walsh, E. Vitalis, and I. Damon (2007). On the origin of smallpox: correlating variola phylogenics with historical smallpox records. *Proceedings of the National Academy of Sciences* 104:15787–92.

Liu, J. P., J. Millman, S. Gao, and P. Cheng (2004). Holocene development of the Yellow River's subaqueous delta, North Yellow Sea. *Marine Geology* 209:45–67.

Lott, D. (2002). *American bison: a natural history*. University of California Press, Berkeley.

Lourandos, H. (1997). *Continent of hunter-gatherers: new perspectives in Australian prehistory*. Cambridge University Press, Cambridge.

Lowell, T., T. Fisher, G. Comer, I. Hajdas, N. Waterson, K. Glover, H. Loope, J. Schaeffer, V. Rinterknecht, W. Broecker, G. Denton, and J. Teller (2005). Testing the Lake Agassiz meltwater trigger for the Younger Dryas. *EOS* 86:365–73.

Lubbock, J. (1865). *Prehistoric times*. Williams and Norgate, London.

Lupo, K. (2007). Evolutionary foraging models in zooarchaeological analysis: recent applications and future challenges. *Journal of Archaeological Research* 15:143–89.

Lyell, C. (1860). On the occurrence of works of human art in postpliocene deposits. *British Association for the Advancement of Science Report* 1859:93–95.

Lyell, C. (1863). *The geological evidences of the antiquity of man with remarks on theories of the origin of species by variation.* John Murray, London.

Lyman, R. L. (1984). Broken bones, bone expediency tools, and bone pseudotools: lessons from the blast zone around Mount St. Helens, Washington. *American Antiquity* 49:315–33.

Lyman, R. L., M. O'Brien, and V. Hayes (1998). A mechanical and functional study of bone rods from the Richey-Roberts Clovis cache, Washington, U. S. A. *Journal of Archaeological Science* 25:887–906.

Lynch, T. F. (1990). Glacial age man in South America? a critical review. *American Antiquity* 55:12–36.

Lynch, T. F. (1991a). The peopling of the Americas—a discussion. In *The first Americans: search and research,* edited by T. Dillehay and D. J. Meltzer, pp. 267–74. CRC Press, Boca Raton, FL.

Lynch, T. F. (1991b). Lack of evidence for glacial-age settlement of South America: reply to Dillehay and Collins and to Gruhn and Bryan. *American Antiquity* 56:348–55.

MacArthur, R. and E. O. Wilson (1967). *The theory of island biogeography.* Monographs in Population Biology 1. Princeton University Press, Princeton.

MacDonald, G. M., D. W. Beilman, K. V. Kremenetski, Y. Sheng, L. C. Smith, and A. A. Velichko (2006). Rapid early development of circumarctic peatlands and atmospheric CH_4 and CO_2 variations. *Science* 314:285–88.

Macdougall, D. (2004). *Frozen earth: the once and future story of Ice Ages.* University of California Press, Berkeley.

MacNeish, R. S. (1976). Early man in the New World. *American Scientist* 63:316–27.

MacNutt, F. A. (1912). *De orbe novo, the eight decades of Peter Martyr d'Anghera.* Translated from the Latin with notes and introduction. G. P. Putnam's Sons, New York.

MacPhee, R. D. and H. Sues, eds. (1999). *Extinctions in near time: causes, contexts, and consequences.* Kluwer Academic/Plenum, New York.

Madsen, D., ed. (2004). *Entering America: Northeast Asia and Beringia before the Last Glacial Maximum,* University of Utah Press, Salt Lake City.

Malhi, R., J. Eshleman, J. Greenberg, D. Weiss, B. Shook, F. Kaestle, J. Lorenz, B. Kemp, J. Johnson, and D. Smith (2002). The structure and diversity within New World mitochondrial DNA haplogroups: implications for the prehistory of North America. *American Journal of Human Genetics* 70:905–19.

Malhi, R., B. Kemp, J. Eshleman, J. Cybulski, D. Smith, S. Cousins, and H. Harry (2007). Mitochondrial haplogroup M discovered in prehistoric Americans. *Journal of Archaeological Science* 34:642–48.

Mandryk, C., H. Josenhans, D. Fedje, and R. Mathewes (2001). Late Quaternary paleoenvironments of northwestern North America: implications for inland versus coastal migration routes. *Quaternary Science Reviews* 20:301–14.

Mann, C. (2005). *1491: new revelations of the Americas before Columbus*. Knopf, New York.

Mann, C. (2007). America found and lost. *National Geographic* May 2007, pp. 32–53.

Mann, D. and T. Hamilton (1995). Late Pleistocene and Holocene environments of the North Pacific Coast. *Quaternary Science Reviews* 14:449–71.

Mann, D. and D. Peteet (1994). Extent and timing of the Last Glacial Maximum in southwestern Alaska. *Quaternary Research* 42:136–48.

Mann, D., R. Reanier, D. Peteet, M. Kunz, and M. Johnson (2001). Environmental change and Arctic Paleoindians. *Arctic Anthropology* 38:119–38.

Marks, J. (2002). *What it means to be 98% chimpanzee: apes, people, and their genes*. University of California Press, Berkeley.

Marshall, S., D. Pollard, S. Hostetler, and P. U. Clark (2004). Coupling ice-sheet and climate models for simulation of former ice sheets. In *The Quaternary period in the United States*, edited by A. Gillespie, S. C. Porter, and B. Atwater, pp. 105–26. Elsevier Science, New York.

Martin, P. S. (1973). The discovery of America. *Science* 179:969–74

Martin, P. S. (1984). Prehistoric overkill: The global model. In *Quaternary Extinctions: A Prehistoric Revolution*, edited by P. S. Martin and R. Klein, pp. 354–403. University of Arizona Press, Tucson.

Martin, P. S. (1987). Clovisia the beautiful. *Natural History* 96:10–13.

Martin, P. S. (2005). *Twilight of the mammoths*. University of California Press, Berkeley.

Martin, P. S. and R. G. Klein, eds. (1984). *Quaternary extinctions: a prehistoric revolution*. University of Arizona Press, Tucson.

Martin, P. S. and D. Steadman (1999). Prehistoric extinctions on islands and continents. In *Extinctions in near time: causes, contexts, and consequences*, edited by R. MacPhee, pp. 17–55. Plenum, New York.

Matisoff, J. (1990). On megalocomparison. *Language* 66:106–20.

Mayor, A. (2005). *Fossil legends of the first Americans*. Princeton University Press, Princeton.

McArthur, N., I. Saunders, and R. Tweedie (1976). Small population isolates: a micro-simulation study. *Journal of the Polynesian Society* 85:307–26.

McGee, W J (1893). Man and the glacial period. *American Anthropologist* 6:85–95.

McGhee, R. (1984). Thule prehistory of Canada. *Handbook of North American Indians, Vol. 5, Arctic*, edited by D. Damas, pp. 369–76. Smithsonian Institution Press, Washington, D.C.

McGhee, R. (2004). Between racism and romanticism, scientism and spiritualism: the dilemmas of New World archaeology. In *Archaeology on the edge: new perspectives from the northern plains*, edited by B. Kooyman and J. Kelley, pp. 13–22. University of Calgary Press, Calgary.

McGhee, R. (2005). *The last imaginary place: a human history of the Arctic world*. University of Chicago Press, Chicago.

McGovern, T. (1994). Management for extinction in Norse Greenland. In *Historical ecology*, edited by C. Crumley, pp. 127–54. SAR Press, Santa Fe, NM.

McManus J., R. Francois, J.-M. Gherardi, L. D. Keigwin, and S. Brown-Leger (2004). Collapse and rapid resumption of Atlantic meridional circulation linked to deglacial climate changes. *Nature* 428:833–37.

McNeill, W. (1976). *Plagues and people*. Anchor/Doubleday, Garden City, NJ.

Medawar, P. (1982). *Pluto's republic*. Oxford University Press, Oxford.

Meltzer, D. J. (1983). The antiquity of man and the development of American archaeology. *Advances in Archaeological Method and Theory* 6:1–51.

Meltzer, D. J. (1989). Why don't we know when the first people came to North America? *American Antiquity* 54:471–90.

Meltzer, D. J. (1990). In search of a mammoth fraud. *New Scientist* 127(1725):51–55.

Meltzer, D. J. (1993a). *Search for the first Americans*. Smithsonian Books, Washington, D.C.

Meltzer, D. J. (1993b). The Pleistocene peopling of the Americas. *Evolutionary Anthropology* 1:157–69.

Meltzer, D. J. (1995). Clocking the first Americans. *Annual Review of Anthropology* 24:21–45.

Meltzer, D. J. (1999). Human responses to Middle Holocene (altithermal) climates on the North American Great Plains. *Quaternary Research* 52:404–16.

Meltzer, D. J. (2002). What do you do when no one's been there before? Thoughts on the exploration and colonization of new lands. In *The First Americans: the Pleistocene colonization of the New World*, edited by N. Jablonski. Memoirs of the California Academy of Sciences 27:25–56.

Meltzer, D. J. (2003). Lessons in landscape learning. In *Colonization of unfamiliar landscapes: the archaeology of adaptation*, edited by M. Rockman and J. Steele, pp. 222–41. Routledge, London.

Meltzer, D. J. (2004). Modeling the initial colonization of the Americas: issues of scale, demography, and landscape learning. In *The settlement of the American continents: a multidisciplinary approach to human biogeography*, edited by C. M. Barton, G. A. Clark, D. R. Yesner, and G. A. Pearson, pp. 123–37. University of Arizona Press, Tucson.

Meltzer, D. J. (2005). The seventy-year itch: controversies over human antiquity and their resolution. *Journal of Anthropological Research* 61.4:433–68.

Meltzer, D. J. (2006a). *Folsom: new archaeological investigations of a classic Paleoindian bison kill*. University of California Press, Berkeley.

Meltzer, D. J. (2006b). History of research on the Paleo-Indian. *Handbook of North American Indians, Vol. 3, environment, origins, and population*, edited by D. Ubelaker, pp. 110–28. Smithsonian Institution Press, Washington, D.C.

Meltzer, D. J., J. M. Adovasio, and T. D. Dillehay (1994). On a Pleistocene human occupation at Pedra Furada, Brazil. *Antiquity* 68:695–14.

Meltzer, D. J. and M. Bever (1995). Paleoindians of Texas: an update on the Texas Clovis fluted point survey. *Bulletin of the Texas Archeological Society* 66:17–51.

Meltzer, D. J., D. K. Grayson, G. Ardila, A. Barker, D. Dincauze, C. V. Haynes, F. Mena, L. Núñez, and D. J. Stanford (1997). On the Pleistocene antiquity of Monte Verde, southern Chile. *American Antiquity* 62:659–63.

Meltzer, D. J. and J. I. Mead (1985). Dating late Pleistocene extinctions: theoretical issues, analytical bias and substantive results. In *Environments and extinctions: man in late glacial*

North America, edited by J. I. Mead and D. J. Meltzer, pp. 145–74. Center for the Study of Early Man, University of Maine, Orono.

Meltzer, D. J. and W. C. Sturtevant (1983). The Holly Oak shell game: an historic archaeological fraud. In *Lulu Linear Punctated: essays in honor of G. Irving Quimby*, edited by R. C. Dunnell and D. K. Grayson. *Anthropological Papers* No. 72:325–52. Museum of Anthropology, University of Michigan.

Merriwether, D. A. (2002). A mitochondrial perspective on the peopling of the New World. In *The first Americans: the Pleistocene colonization of the New World*, edited by N. Jablonski. Memoirs of the California Academy of Sciences 27:295–310.

Merriwether, D. A., W. Hall, A. Vahlne, and R. Ferrell (1996). mtDNA variation indicates Mongolia may have been the source for the founding population for the New World. *American Journal of Human Genetics* 59:204–12.

Merriwether, D. A., F. Rothhammer, and R. Ferrell (1995). Distribution of the four founding lineage haplotypes in Native Americans suggests a single wave of migration to the New World. *American Journal of Physical Anthropology* 98:411–30.

Mithen, S. (2003). *After the ice: a global human history 20,000–5000 BC*. Weidenfield & Nicholson, London.

Moore, J. (2001). Evaluating five models of human colonization. *American Anthropologist* 103:395–408.

Morlan, R. E. (1988). Pre-Clovis people: early discoveries of America? *Ethnological Monographs* 12:31–43.

Morrell, V. (1995). *Ancestral passions*. Simon and Schuster, New York.

Morse, D. (1997). *Sloan: a Paleoindian Dalton cemetery in Arkansas*. Smithsonian Institution Press, Washington, D.C.

Moulton, G. (2003). *The Lewis and Clark journals: an American epic of discovery*. University of Nebraska Press, Lincoln.

Mulligan, C. (2006). Anthropological applications of ancient DNA: problems and prospects. *American Antiquity* 71:365–80.

Mulligan, C., K. Hunley, S. Cole, and J. Long (2004). Population genetics, history, and health patterns in Native Americans. *Annual Review Genomics Human Genetics* 5:295–315.

Mulligan, C., S. Norris, and S. Lukehart (2008). Molecular studies in *Treponema pallidum* evolution: toward clarity? *PLoS Neglected Tropical Diseases* 2(1): e184. doi:10.1371/journal.pntd.0000184.

Muscheler, R., B. Kromer, S. Björk, A. Svensson, M. Friedrich, K. F. Kaiser, and J. Southon (2008). Tree rings and ice cores reveal ^{14}C calibration uncertainties during the Younger Dryas. *Nature Geoscience* 1:263–67.

Myers, T., M. Voorhies, and R. Corner (1980). Spiral fractures and bone pseudotools at paleontological sites. *American Antiquity* 45:483–90.

Native American Graves Protection and Repatriation Act, Public Law 101-601; 25 U. S. C. 3001 et seq. (November 16, 1990).

Nelson, A. R. (2006). Patterns of craniometric variation and geographical distribution in North America: a historical comparison. In *Paleoamerican origins: beyond Clovis*, edited by R. Bonnichsen, B. Lepper, D. Stanford, and M. Waters, pp. 277–88. Center for the Study of the First Americans, College Station, TX.

Nelson, R. (1969). *Hunters of the northern ice.* University of Chicago Press, Chicago.

Nelson, R. (1986). *Hunters of the northern forest.* University of Chicago Press, Chicago.

Nettle, D. (1998). Explaining global patterns of language diversity. *Journal of Anthropological Archaeology* 17:354–74.

Nettle, D. (1999). Linguistic diversity of the Americas can be reconciled with a recent colonization. *Proceedings of the National Academy of Sciences* 96:3325–29.

Neves, W. and M. Hubbe (2005). Cranial morphology of early Americans from Lagoa Santa, Brazil: implications for the settlement of the New World. *Proceedings of the National Academy of Sciences* 102:18309–14.

Nevle, R. and D. Bird (2008). Effects of syn-pandemic fire reduction and reforestation in the tropical Americas on atmospheric CO_2 during European conquest. *Palaeogeography, Palaeoclimatology, Palaeoecology* 264:25–38.

Newby, P., J. Bradley, A. Spiess, B. Shuman, and P. Leduc (2005). A Paleoindian response to Younger Dryas climate change. *Quaternary Science Reviews* 24:141–54.

Nichols, J. (1990). Linguistic diversity and the first settlement of the New World. *Language* 66:475–521.

Nichols, J. (2002). The first American languages. In *The first Americans: the Pleistocene colonization of the New World,* edited by N. Jablonski. Memoirs of the California Academy of Sciences 27:273–93.

Noonan, J.P., G. Coop, S. Kudaravalli, D. Smith, J. Krause, J. Alessi, F. Chen, et al. (2006). Sequencing and analysis of Neanderthal genomic DNA. *Science* 314:1113–18.

O'Connell, J.F. and J. Allen (2004). Dating the colonization of Sahul (Pleistocene Australia–New Guinea): a review of recent research. *Journal of Archaeological Science* 31:835–53.

Odess, D., S. Loring, and W. Fitzhugh (2000). Skraeling: first peoples of Helluland, Markland, and Vinland. In *Vikings: the North Atlantic Saga,* edited by W. Fitzhugh and E. Ward, pp. 193–205. Smithsonian Institution Press, Washington, D.C.

Oldfield, F. and K. D. Alverson (2003). The societal relevance of paleoenvironmental research. In *Paleoclimate, global change, and the future,* edited by K. D. Alverson, R. S. Bradley, and T. F. Pederson, pp. 1–11. Springer, Berlin.

Orians, G. and J. Heerwagen (1992). Evolved responses to landscapes. In *The adapted mind: evolutionary psychology and the generation of culture,* edited by J. Barkow, L. Cosmides, and J. Tooby, pp. 555–79. Oxford University Press, Oxford.

Orr, P. C. (1962). The Arlington Springs site, Santa Rosa Island. *American Antiquity* 27:417–19.

Osborn, H. F. (1922). *Hesperopithecus,* the first anthropoid primate found in America. *American Museum Novitates* 37.

Overpeck, J., C. Whitlock, and B. Huntley (2003). Terrestrial biosphere dynamics in the climate system: past and future. In *Paleoclimate, global change, and the future,* edited by K. D. Alverson, R. S. Bradley, and T. F. Pederson, pp. 81–111. Springer, Berlin.

Owsley, D. and D. Hunt (2001). Clovis and Early Archaic period crania from the Anzick site (24 PA 506), Park County, Montana. *Plains Anthropologist* 46:115–24.

Owsley, D. and R. Jantz (2006). Nearsightedness in Paleoamerican research: historical perspectives and contemporary analysis. In *Paleoamerican origins: beyond Clovis,* edited by

R. Bonnichsen, B. Lepper, D. Stanford, and M. Waters, pp. 289–94. Center for the Study of the First Americans, College Station, TX.

Pääbo, S. (2003). The mosaic that is our genome. *Nature* 421:409–12.

Parenti, F., M. Fontugue, and C. Guerin (1996). Pedra Furada in Brazil and its "presumed" evidence: limitations and potential of the available data. *Antiquity* 70:422–27.

Park, R. (1993). The Dorset-Thule succession in Arctic North America: assessing claims for cultural contact. *American Antiquity* 58:203–34.

Peltier, W. R. (2002). On eustatic sea level history: Last Glacial Maximum to Holocene. *Quaternary Science Reviews* 21:377–96.

Peltier, W. R. (2005). On the hemispheric origins of meltwater pulse 1a. *Quaternary Science Reviews* 24:1655–71.

Pennisi, E. (2004). Speaking in tongues. *Science* 303:1321–23.

Petit, C. (1998). Rediscovering America: the New World may be 20,000 years older than experts thought. *US News & World Report* 125:56–64.

Pettitt, P. (2000). The Paviland radiocarbon dating programme. In *Paviland Cave and the "Red Lady": a definitive report*, edited by S. Aldhouse-Green, pp. 63–71. Western Academic and Specialist Press, Bristol, UK.

Pintar, N. and S. Ishman (2008). Impacts, mega-tsunami, and other extraordinary claims. *GSA Today* 18:37–38.

Pitulko, V., P. Nikolsky, E. Girya, E. Basilyan, V. Tumskoy, S. Koulakov, S. Astakhov, E. Pavlova, and M. Anisimov (2004). The Yana RHS site: humans in the Arctic before the Last Glacial Maximum. *Science* 303:52–56.

Poinar, H. N., C. Schwarz, J. Qi, B. Shapiro, R. D. E. MacPhee, B. Buigues, A. Tikhonov, et al. (2006). Metagenomics to paleogenomics: large-scale sequencing of mammoth DNA. *Science* 311:392–94.

Powell, J. F. (2005). *The first Americans: race, evolution, and the origin of Native Americans.* Cambridge University Press, Cambridge.

Powell, J. F. and W. Neves (1999). Craniofacial morphology of the first Americans: pattern and process in the peopling of the New World. *Yearbook of Physical Anthropology* 42:153–88.

Powell, J. F. and J. Rose (1999). *Report on the osteological assessment of the "Kennewick man" skeleton* (CENWW.97.Kennewick). Available at http://www.nps.gov/archeology/ kennewick/powell_rose.htm.

Powell, J. W. (1891). Indian linguistic families of America north of Mexico. *Annual Report of the Bureau of American Ethnology* 7:3–142.

Preston, D. (1995). The mystery of Sandia Cave. *New Yorker*, June 12, 1995, pp. 66–83.

Pringle, H. (2007). Did a comet wipe out prehistoric Americans? *New Scientist*, 2591:28–33.

Punke, M. L. and L. G. Davis (2006). Problems and prospects in the preservation of Late Pleistocene cultural sites in southern Oregon coastal river valleys: implications for evaluating coastal migration routes. *Geoarchaeology* 21:333–50.

Quimby, G. (1954). Cultural and natural areas before Kroeber. *American Antiquity* 19:317–31.

Quimby, G. (1955). Reply to Aschmann's comment. *American Antiquity* 20:378–79.

Raynaud, D., T. Blunier, Y. Ono, and R. Delmas (2003). The Late Quaternary history of atmospheric trace gases and aerosols: interactions between climate and biogeochemical cycles.

In *Paleoclimate, global change, and the future,* edited by K. D. Alverson, R. S. Bradley, and T. F. Pederson, pp. 13–31. Springer, Berlin.

Reanier, R. E. (1995). The antiquity of Paleoindian materials in northern Alaska. *Arctic Anthropology* 32:31–50.

Reasoner, M. and M. Jodry (2000). Rapid response of alpine timberline vegetation to the Younger Dryas climate oscillation in the Colorado Rocky Mountains, USA. *Geology* 28:51–54.

Redder, A. and J. Fox (1998). Excavation and positioning of the Horn Shelter burial and grave goods. *Central Texas Archaeologist* 11:1–12.

Reeves, B. O. (1990). Communal bison hunters of the northern plains. In *Hunters of the recent past,* edited by L. Davis and B. O. Reeves, pp. 168–94. Unwin Hyman, London.

Renfrew, C., ed. (2000). *America past, America present: genes and languages in the Americas and beyond.* MacDonald Institute for Archaeological Research, Cambridge, UK.

Renne, P., J. Feinberg, M. Waters, J. Arroyo-Cabrales, P. Ochoa-Castillo, M. Perez-Campa, and K. Knight (2005). Age of Mexican ash with alleged "footprints." *Nature* 438, E7–E8.

Rick, T., J. Erlandson, and R. Vellanoweth (2001). Paleocoastal fishing along the Pacific Coast of the Americas: evidence from Daisy Cave, San Miguel Island, California. *American Antiquity* 66:595–614.

Rind, D. (2002). The sun's role in climate variations. *Science* 296:673–77.

Ringe, D. (2000). Some relevant facts about historical linguistics. In *America past, America present: genes and languages in the Americas and beyond,* edited by C. Renfrew, pp. 139–62. McDonald Institute for Archaeological Research, Cambridge, UK.

Roberts, F. H. H. (1940). Developments in the problem of the North American Paleo-Indian. *Smithsonian Miscellaneous Collections* 100:51–116.

Robinson, L. F., J. F. Adkins, L. D. Keigwin, J. Southon, D. P. Fernandez, S-L. Wang, and D. S. Scheirer (2005). Radiocarbon variability in the western North Atlantic during the last deglaciation. *Science* 310:1469–73.

Rockman, M. and J. Steele, eds. (2003). *Colonization of unfamiliar landscapes: the archaeology of adaptation.* Routledge, London.

Rogers, R., L. Rogers, and L. Martin (1992). How the door opened: the peopling of the New World. *Human Biology* 64:281–302.

Ronda, J. (1984). *Lewis and Clark among the Indians.* University of Nebraska Press, Lincoln.

Roosevelt, T. (1910). *African game trails: an account of the African wanderings of an American hunter-naturalist.* Charles Scribner and Sons, New York.

Roseman, C. and T. Weaver (2004). Multivariate apportionment of global human craniometric diversity. *American Journal of Physical Anthropology* 125:257–63.

Ruddiman, W. (2005). *Plows, plagues, and petroleum: how humans took control of climate.* Princeton University Press, Princeton, NJ.

Rudwick, M. (1985). *The great Devonian controversy: the shaping of scientific knowledge among gentlemanly specialists.* University of Chicago Press, Chicago.

Rudwick, M., ed. (1997). *Georges Cuvier, fossil bones, and geological catastrophes: new translations & interpretations of the primary texts.* University of Chicago Press, Chicago.

Ruhlen, M. (1994). *The origin of language: tracing the evolution of the mother tongue.* John Wiley and Sons, New York.

Saillard, J., P. Forster, N. Lynnerup, H. Bandelt, and S. Norby (2000). mtDNA variation among Greenland Eskimos: the edge of the Beringian expansion. *American Journal of Human Genetics* 67:718–26.

Santos, G., M. Bird, F. Parenti, L. Fifield, N. Guidon, and P. Hausladen (2003). A revised chronology of the lowest occupation layer of Pedra Furada rock shelter, Piaui, Brazil: the Pleistocene peopling of the Americas. *Quaternary Science Reviews* 22:2303–10.

Sapir, E. (1916). Time perspective in aboriginal American culture: a study in method. *Geological Survey Canada, Anthropological Series* 13. Ottawa, Canada.

Sapir, E. (1921). A bird's-eye view of American languages north of Mexico. *Science* 54:408.

Saunders, N. (1998). Review of "Monte Verde: a late Pleistocene settlement in Chile, Volume 2: archaeological context and interpretation," by Tom D. Dillehay. *Nature* 392:145–46.

Scarre, C., ed. (2005). *The human past: world prehistory and the development of human society.* Thames & Hudson, London.

Schaefer, J., G. Denton, D. Barrel, S. Ivy-Ochs, P. Kubik, B. Anderon, F. Phillis, T. Lowell, and C. Schluter (2006). Near-synchronous interhemispheric termination of the Last Glacial Maximum in mid-latitudes. *Science* 312:1510–13.

Schenninger, J., S. Gonzalez, D. Huddart, M. Bennett, and A. Gonzales-Huesca (2006). The OSL dating of the Xalnene ash: a reply to comments by Duller on "Human footprints in central Mexico older than 40,000 years." *Quaternary Science Reviews* 25:3077–80.

Schledermann, P. (2000). A.D. 1000: East meets West. In *Vikings: the North Atlantic saga,* edited by W. Fitzhugh and E. Ward, pp. 189–192. Smithsonian Institution Press, Washington, D.C.

Schneider, A. and R. Bonnichsen (2006). Where are we going? Public policy and science. In *Paleoamerican origins: beyond Clovis,* edited by R. Bonnichsen, B. Lepper, D. Stanford, and M. Waters, pp. 297–312. Center for the Study of the First Americans, College Station, TX.

Schrag, D. (2000). Of ice and elephants. *Nature* 404:23–24.

Schroeder, A. (1979). Pecos Pueblo. *Handbook of North American Indians, Vol. 9, Southwest,* edited by A. Ortiz, pp. 430–37. Smithsonian Institution Press, Washington, D.C.

Schroeder, K., T. Schurr, J. Long, N. Rosenberg, M. Crawford, L. Tarskaia, L. Osipova, S. Zhadanov, and D. Smith (2007). A private allele ubiquitous in the Americas. *Biology Letters* 3:218–23.

Schuiling, W. (1972). *Pleistocene man at Calico: a report on the international conference on the Calico Mountains excavations, San Bernardino County, California.* San Bernardino County Museum Association, San Bernardino, CA.

Schurr, T.G. (2004a). Molecular genetic diversity in Siberians and Native Americans suggests an early colonization of the New World. In *Entering America: Northeast Asia and Beringia before the Last Glacial Maximum,* edited by D. Madsen, pp. 187–38. University of Utah Press, Salt Lake City.

Schurr, T.G. (2004b). The peopling of the New World: perspectives from molecular anthropology. *Annual Review of Anthropology* 33:551–83.

Schwarz, H. (1992). Uranium series dating in paleoanthropology. *Evolutionary Anthropology* 1:56–62.

Schwenninger, J-L., S. Gonzalez, D. Huddart, M.R. Bennett, and A. Gonzalez-Huesca (2007). The OSL dating of the Xalnene ash: a reply to comments by G. Duller on "Human footprints in central Mexico older than 40,000 years." *Quaternary Science Reviews* 25:3077–80.

Secoy, F. (1953). Changing military patterns of the Great Plains Indians. *Monographs of the American Ethnological Society* 21. Seattle, Washington.

Seebach, J. (2006). Drought or development? patterns of paleoindian site discovery on the Great Plains of North America. *Plains Anthropologist* 51:71–88.

Seeman, M. (1994). Intercluster lithic patterning at Nobles Pond: a case for "disembedded" procurement among early Paleoindian societies. *American Antiquity* 59:273–88.

Seielstad, M., E. Minch, and L. Cavalli-Sforza (1998). Genetic evidence for a higher female migration rate in humans. *Nature Genetics* 20:278–80.

Seielstad, M., N. Yuldasheva, N. Singh, P. Underhill, P. Oefner, P. Shen, and R. Wells (2003). A novel Y-chromosome variant puts an upper limit on the timing of first entry into the Americas. *American Journal of Human Genetics* 73:700–705.

Sellet, F. (2004). Beyond the point: projectile manufacture and behavioral inference. *Journal of Archaeological Science* 31:1553–66.

Severinghaus, J. and E. Brook (1999). Abrupt climate change at the end of the last glacial period inferred from trapped air in polar ice. *Science* 286:930–34.

Shapiro, B., A.J. Drummond, A. Rambaut, M.C. Wilson, P.E. Matheus, A.V. Sher, O.G. Pybus, et al. (2004). Rise and fall of the Beringian Steppe Bison. *Science* 306:1561–65.

Sherwood, S.C., B. Driskell, A. Randall, and S. Meeks (2004). Chronology and stratigraphy at Dust Cave, Alabama. *American Antiquity* 69.3:533–54.

Shetrone, H. (1936). The Folsom phenomena as seen from Ohio. *Ohio State Archaeological and Historical Quarterly* 45:240–56.

Shields, E. and G. Jones (1998). Dorset and Thule divergence from east central Asian roots. *American Journal of Physical Anthropology* 106:207–18.

Shipley, W. (1978). Native languages of California. *Handbook of North American Indians, Vol. 8, California,* edited by R. Heizer, pp. 80–90. Smithsonian Institution Press, Washington, D.C.

Shott, M.J. (2002). Sample bias in the distribution and abundance of Midwestern fluted bifaces. *Midcontinental Journal of Archaeology* 27:89–123.

Shuman, B., P. Bartlein, and T. Webb (2005). The magnitudes of millennial- and orbital-scale climatic change in eastern North America during the Late Quaternary. *Quaternary Science Reviews* 24:2194–2206.

Shuman, B., T. Webb, P. Bartlein, and J.W. Williams (2002). The anatomy of a climatic oscillation: vegetation change in eastern North America during the Younger Dryas chronozone. *Quaternary Science Reviews* 21:1777–91.

Siegenthaler, U., T. Stocker, E. Monnin, D. Luthi, J. Schwander, B. Stauffer, D. Raynaud, et al. (2005). Stable carbon cycle-climate relationship during the late Pleistocene. *Science* 310:1313–17.

Silberbauer, G. (1981). *Hunter & habitat in the central Kalahari Desert.* Cambridge University Press, Cambridge.

Silverman, I. and M. Eals (1992). Sex differences in spatial abilities: evolutionary theory and data. In *The adapted mind: evolutionary psychology and the generation of culture,* edited by J. Barkow, L. Cosmides, and J. Tooby, pp. 533–49. Oxford University Press, Oxford.

Simms, S. R. (1996). North America: the North American West. In *The Oxford Companion to Archaeology*, edited by B. Fagan, pp. 523–25. Oxford University Press, Oxford.

Skinner, M., S. Skinner, and R. Gooris (1977). Stratigraphy and biostratigraphy of late Cenozoic deposits in central Sioux County, western Nebraska. *American Museum of Natural History Bulletin* 158:271–367.

Slobodin, S. (2001). Western Beringia at the end of the Ice Age. *Arctic Anthropology* 38:31–47.

Smith, D., R. Malhi, J. Eshleman, F. Kaestle, and B. Kemp (2006). Mitochondrial DNA haplogroups of Paleoamericans in North America. In *Paleoamerican origins: beyond Clovis*, edited by R. Bonnichsen, B. Lepper, D. Stanford, and M. Waters, pp. 243–54. Center for the Study of the First Americans, College Station, TX.

Smith, D., R. Malhi, J. Eshleman, J. Lorenz, and F. Kaestle (1999). Distribution of mtDNA haplogroup X among native North Americans. *American Journal of Physical Anthropology* 110:271–84.

Snipp, C. M. (2006). Population size, nadir to 2000. *Handbook of North American Indians, Vol. 3, environment, origins, and population*, edited by D. Ubelaker, pp. 702–11. Smithsonian Institution Press, Washington, D.C.

Snyder, L. M. and J. Leonard (2006). Dog. *Handbook of North American Indians, Vol. 3, environment, origins, and population*, edited by D. Ubelaker, pp. 452–62. Smithsonian Institution Press, Washington, D.C.

Southon, J. and R. E. Taylor (2002). Brief comments on "Terrestrial evidence of a nuclear catastrophe in Paleoindian times" by Richard B. Firestone and William Topping. *Mammoth Trumpet* 17.2:14–17.

Sparks, C. and R. Jantz (2002). A reassessment of human cranial plasticity: Boas revisited. *Proceedings of the National Academy of Sciences* 99:14636–39.

Speth, J. and K. Spielmann (1983). Energy source, protein metabolism, and hunter-gatherer subsistence strategies. *Journal of Anthropological Archaeology* 2:1–31.

Stafford, M. D., G. C. Frison, D. Stanford, and G. Zeimans (2003). Digging for the color of life: Paleoindian red ochre mining at the Powars II site, Platte County, Wyoming, U.S.A. *Geoarchaeology* 18:71–90.

Stahle, D., M. Cleaveland, D. Blanton, M. Therrell, and D. Gray (1998). The Lost Colony and Jamestown droughts. *Science* 280:564–67.

Stalker, A. M. (1977). Indications of Wisconsin and earlier man from the southwest Canadian prairies. In *Amerinds and their paleoenvironments in northeastern North America*, edited by W. Newman and B. Salwen. *New York Academy of Sciences Annals* 288:119–36.

Stanford, D. (1978). The Jones-Miller site: an example of Hell Gap bison procurement strategy. *Plains Anthropologist Memoir* 14:90–97.

Stanford, D. (1999). Paleoindian archaeology and late Pleistocene environments in the plains and southwestern United States. In *Ice-Age people of North America: environments, origins, and adaptations*, edited by R. Bonnichsen and K. Turnmire, pp. 281–339. Center for the Study of the First Americans, Corvallis, OR.

Stanford, D., R. Bonnichsen, and R. Morlan (1981). The Ginsberg experiment: modern and prehistoric evidence of a bone-flaking technology. *Science* 212:438–40.

Stanford, D. and B. Bradley (2002). Ocean trails and prairie paths? thoughts about Clovis origins. In *The first Americans: the Pleistocene colonization of the New World*, edited by N. Jablonski. *Memoirs of the California Academy of Sciences* 27:255–71.

Steele, D. G. (1998). Human biological remains. In *Wilson-Leonard: an 11,000-year archaeological record of hunter-gatherers in central Texas, Vol. 5*, edited by M. B. Collins, pp. 1441–62. Studies in Archaeology 31, Texas Archaeological Research Laboratory, University of Texas, Austin.

Steele, G. and J. Powell (2002). Facing the past: a view of the North American human fossil record. In *The first Americans: the Pleistocene colonization of the New World*, edited by N. Jablonski. Memoirs of the California Academy of Sciences 27:93–122.

Steens-McIntyre, V., R. Fryxell, and H. Malde (1981). Geologic evidence for age of deposits at Hueyatlaco archaeological site, Valsequillo, Mexico. *Quaternary Research* 16:1–17.

Stephens, D. and J. Krebs (1986). *Foraging theory*. Princeton University Press, Princeton, NJ.

Stewart, T. D. (1973). *The people of America*. Scribner, New York.

Stiger, M. (2006). A Folsom structure in the Colorado mountains. *American Antiquity* 71:321–52.

Storck, P. (1991). Imperialists without a state: the cultural dynamics of early Paleoindian colonization as seen from the Great Lakes region. In *Clovis origins and adaptations*, edited by R. Bonnichsen and K. Turnmire, pp. 153–62. Center for the Study of the First Americans, Corvallis, OR.

Strauss, L. G., D. J. Meltzer, and T. Goebel (2005). Ice Age Atlantis? exploring the Solutrean-Clovis "connection." *World Archaeology* 37:506–31.

Surovell, T. (2003). Simulating coastal migration in New World colonization. *Current Anthropology* 44:580–91.

Sutherland, P. (2000). The Norse and Native North Americas. In *Vikings: the North Atlantic saga*, edited by W. Fitzhugh and E. Ward, pp. 238–47. Smithsonian Institution Press, Washington, D.C.

Swagerty, W. R. (1988). Indian trade in the trans-Mississippi west to 1870. *Handbook of North American Indians, Vol. 4, history of Indian-White relations*, edited by W. E. Washburn, pp. 351–74. Smithsonian Institution Press, Washington, D.C.

Swanton, J. (1911). Indian tribes of the lower Mississippi Valley. *Bureau of American Ethnology, Bulletin* 43.

Swedlund, A. and D. Anderson (1999). Gordon Creek woman meets Kennewick man: new interpretations and protocols regarding the peopling of the Americas. *American Antiquity* 64:569–76.

Szathmary, E. J. (1984). Human biology of the Arctic. *Handbook of North American Indians, Vol. 5, Arctic*, edited by D. Damas, pp. 64–71. Smithsonian Institution Press, Washington, D.C.

Szathmary, E. J. (1993). Genetics of aboriginal North Americans. *Evolutionary Anthropology* 1:202–20.

Szathmary, E. J. (1996). Ancient migrations from Asia to North America. In *Prehistoric Mongoloid dispersals*, edited by T. Akazawa and E. Szathmary, pp. 149–64. Oxford University Press, Oxford, UK.

Tamm, E., T. Kivisild, M. Reidla, M. Metspalu, D. Smith, C. Mulligan, C. Bravi, et al. (2007). Beringian standstill and spread of Native American founders. *PLoS One* 9:e829.

Tarasov, L. and W. Peltier (2006). A calibrated deglacial drainage chronology for the North American continent: evidence of an Arctic trigger for the Younger Dryas. *Quaternary Science Reviews* 25:659–88.

Taylor, R. E. (1995). Radiocarbon dating: the continuing revolution. *Evolutionary Anthropology* 4:169–81.

Taylor, R. E., C. V. Haynes, D. Kirner, and J. Southon (1999). Radiocarbon analyses of modern organics at Monte Verde, Chile: no evidence for a local reservoir effect. *American Antiquity* 64:455–60.

Teller, J. T. (2004). Controls, history, outbursts, and impact of large Late-Quaternary proglacial lakes in North America. In *The Quaternary period in the United States,* edited by A. Gillespie, S. C. Porter and B. Atwater, pp. 45–61. Elsevier Science, New York.

Teller, J. T., M. Boyd, Z. Yang, P. S. G. Kor, and A. M. Fard (2005). Alternative routing of Lake Agassiz overflow during the Younger Dryas: new dates, paleotopography, and a re-evaluation. *Quaternary Science Reviews* 24:1890–1905.

Thomas, D. H., ed. (1989). *Columbian consequences, Vol. 1: archaeological and historical perspectives on the Spanish borderlands west.* Smithsonian Institution Press, Washington, D.C.

Thomas, D. H., ed. (1990). *Columbian consequences, Vol. 2: archaeological and historical perspectives on the Spanish borderlands east.* Smithsonian Institution Press, Washington, D.C.

Thomas, D. H., ed. (1991). *Columbian consequences, Vol. 3: the Spanish borderlands in Pan-American perspective.* Smithsonian Institution Press, Washington, D.C.

Thomas, D. H. (2000). *Skull wars: Kennewick man, archeology, and the battle for Native American identity.* New York, Basic Books.

Thomas, D. H. (2006). Finders keepers and deep American prehistory. In *Imperialism, art and restitution,* edited by J. Merryman, pp. 218–53. Cambridge University Press, Cambridge.

Thomas, N. (2003). *Cook: the extraordinary voyages of Captain James Cook.* Walker, New York.

Thompson, L. and M. Kinkade (1990). Languages. *Handbook of North American Indians, Vol. 7, Northwest Coast,* edited by W. Suttles, pp. 30–51. Smithsonian Institution Press, Washington, D.C.

Torroni, A. (2000). Mitochondrial DNA and the origin of Native Americans. In *America past, America present: genes and languages in the Americas and beyond,* edited by C. Renfrew, pp. 77–87. McDonald Institute for Archaeological Research, Cambridge, UK.

Torroni, A., T. Schurr, M. Cabell, M. Brown, J. Neel, M. Larson, D. Smith, C. Vullo, and D. Wallace (1993). Asian affinities and continental radiation of the four founding Native American mtDNA. *American Journal of Human Genetics* 53:563–90.

Toth, N. (1991). The material record. In *The first Americans: search and research,* edited by T. Dillehay and D. J. Meltzer, pp. 53–76. CRC Press, Boca Raton, FL.

Touhy, D. and A. Dansie (1997). Spirit Cave, Wizards Beach and Grimes burial shelter. *Nevada Historical Society Quarterly* 40:24–53.

Turner, C. G. (1985). The dental search for Native American origins. In *Out of Asia: peopling of the Americas and the Pacific,* edited by R. Kirk and E. Szathmary, pp. 31–78. *Journal of Pacific History,* Canberra.

Turner, C. G. (1986). Dentochronological separation estimates for Pacific Rim populations. *Science* 232:1140–42.

Turner, C. G. (2002). Teeth, needles, dogs, and Siberia: bioarchaeological evidence for the colonization of the New World. In *The first Americans: the Pleistocene colonization of the New World,* edited by N. Jablonski. Memoirs of the California Academy of Sciences 27:123–58.

Turner, C. G. (2005). A synoptic history of physical anthropological studies on the peopling of Alaska and the Americas. *Alaska Journal of Anthropology* 3:157–79.

Turner, C. G., Y. Manabe, and D. E. Hawkey (2000). The Zhoukoudian Upper Cave dentition. *Acta Anthropologica Sinica* 19, 253–68.

Ubelaker, D. ed. (2006a). *Handbook of North American Indians, Vol. 3, environment, origins, and population.* Smithsonian Institution Press, Washington, D.C.

Ubelaker, D. (2006b). Population size, Contact to nadir. In *Handbook of North American Indians, Vol. 3, environment, origins, and population,* edited by D. Ubelaker, pp. 694–701. Smithsonian Institution Press, Washington, D.C.

Valladas, H., N. Mercier, M. Michab, J. Joron, J. Reyss, and N. Guidon (2003). TL age-estimates of burnt quartz pebbles from the Toca do Boquierao da Pedra Furada (Piaui, northeastern Brazil). *Quaternary Science Reviews* 22:1257–63.

Van Dyne, G. M., N. Brockington, Z. Szocs, J. Duek, and C. Ribic (1980). Large herbivore system. In *Grasslands, systems analysis, and man,* edited by A. I. Breymeyer and G. M. Van Dyne, pp. 269–537. Cambridge University Press, Cambridge.

Van Riper, A. B. (1993). *Men among the mammoths: Victorian science and the discovery of human prehistory.* University of Chicago Press, Chicago.

Van Vark, G. N., D. Kuizenga, and F. Williams (2003). Kennewick and Luzia: lessons from the European Upper Paleolithic. *American Journal of Physical Anthropology* 121:181–84.

Wagner, G. (1996). Fission-track dating in paleoanthropology. *Evolutionary Anthropology* 5:165–71.

Waitt, R. B. (1985). Case for periodic, colossal jökulhlaups from Pleistocene glacial Lake Missoula. *Geological Society of America Bulletin* 96:1271–1286.

Walker, R. B. (2007). Hunting in the Late Paleoindian period: faunal remains from Dust Cave, Alabama. In *Foragers of the terminal Pleistocene in North America,* edited by R. Walker and B. Driskell, pp. 99–115. University of Nebraska Press, Lincoln.

Wallace, B. (2000). The Viking settlement at L'Anse aux Meadows. In *Vikings: the North Atlantic Saga,* edited by W. Fitzhugh and E. Ward, pp. 208–16. Smithsonian Institution Press, Washington, D.C.

Wallace, D., K. Garrison, and W. C. Knowler (1985). Dramatic founder effects in Amerindian mitochondrial DNAs. *American Journal of Physical Anthropology* 68:149–55.

Walthall, J. (1998). Rockshelters and hunter-gatherer adaptation to the Pleistocene/Holocene transition. *American Antiquity* 63:223–38.

Walthall, J. and B. Koldehoff (1998). Hunter-gatherer interaction and alliance formation: Dalton and the cult of the long blade. *Plains Anthropologist* 43:257–73.

Wang, S., C. Lewis, M. Jakobsson, S. Ramachandran, N. Ray, G. Bedoya, W. Rojas, et al. (2007). Genetic variation and population structure in Native Americans. *PLoS Genetics* 3:2049–67.

Waters, M. R. (1986). Sulphur Springs woman: an early human skeleton from southeastern Arizona. *American Antiquity* 51:361–365.

Waters, M. R., and T. W. Stafford, Jr. (2007). Redefining the age of Clovis: Implications for the peopling of the New World. *Science* 315:1122–26.

Watkins D. I., N. McAdam, X. Liu, C. Strang, E. Milford, C. Levine, T. Garber, et al. (1992). New recombinant HLA-B alleles in a tribe of South American Amerindians indicate rapid evolution of MHC class I loci. *Nature* 357:329–30.

Webb, S. D., R. W. Graham, A. D. Barnosky, C. J. Bell, R. Franz, E. Hadly, E. Lundelius, et al. (2004). Vertebrate paleontology. In *The Quaternary period in the United States*, edited by A. Gillespie, S. C. Porter, and B. Atwater, pp. 519–38. Elsevier Science, New York.

Webb, T., B. Shuman, and J. W. Williams (2004). Climatically forced vegetation dynamics in eastern North America during the late Quaternary Period. In *The Quaternary period in the United States*, edited by A. Gillespie, S. C. Porter, and B. Atwater, pp. 459–78. Elsevier Science, New York.

Weber, D. (1992). *The Spanish frontier in North America.* Yale University Press, New Haven, CT.

West, F. (1983). The antiquity of man in America. In *Late Quaternary environments of the United States, Vol. 1, the Late Pleistocene,* edited by S. Porter, pp. 364–82. University of Minnesota Press, Minneapolis.

West, F., ed. (1996). *American beginnings: the prehistory and palaeoecology of Beringia.* University of Chicago Press, Chicago.

Whallon, R. (1989). Elements of culture change in the Later Palaeolithic. In *The human revolution: behavioral and biological perspectives on the origins of modern humans,* edited by P. Mellars and C. Stringer, pp. 433–54. Princeton University Press, Princeton, NJ.

Wheat, J. B. (1972). The Olsen-Chubbuck site: a Paleo-indian bison kill. *American Antiquity Memoir* 26. Washington, D.C.

Wilder, J., S. Kingan, Z. Mobasher, M. Pilkington, and M. Hammer (2004). Global patterns of human mitochondrial DNA and Y-chromosome structure are not influenced by higher migration rates of females versus males. *Nature Genetics* 36:1122–25.

Wilford, J. (1997). Human presence in Americas is pushed back a millennium. *New York Times,* February 11, 1997.

Wilke, P., J. Flenniken, and T. Ozbun (1991). Clovis technology at the Anzick site. *Journal of California and Great Basin Anthropology* 13:242–72.

Williams, F. and G. Armelagos (2007). Reply to "On the misclassification of human crania," by M. Hubbe and W. Neves. *Current Anthropology* 48:285–288.

Williams, J. W., B. Shuman, and T. Webb (2001). Dissimilarity analyses of Late-Quaternary vegetation and climate in eastern North America. *Ecology* 82:3346–62.

Wilmsen, E. and F. H. H. Roberts (1978). *Lindenmeier, 1934–1974: concluding report on investigations.* Smithsonian Contributions to Anthropology, Number 24, Washington, D.C.

Wilson, M. C., D. W. Harvey, and R. G. Forbis (1983). Geoarchaeological investigations of the age and context of the Stalker (Taber Child) site, DiPa 4, Alberta. *Canadian Journal of Archaeology* 7:179–207.

Winterhalder, B., W. Baillargeon, F. Cappalleto, R. Daniel, and C. Prescott (1988). The population dynamics of hunter-gatherers and their prey. *Journal of Anthropological Archaeology* 7:289–328.

Woodbury, A. (1984). Eskimo and Aleut languages. *Handbook of North American Indians, Vol. 5, Arctic,* edited by D. Damas, pp. 49–63. Smithsonian Institution Press, Washington, D.C.

Wright, G. F. (1890). Report. In *Discovery of a Paleolithic implement at New Comerstown, Ohio,* edited by W. C. Mills and G. F. Wright, pp. 5–14. Western Reserve Historical Society, *Tract* 75.

Wright, G. F. (1892). *Man and the glacial period.* D. Appleton, New York.

Wright, G. F. (1912). *Origin and antiquity of man.* Bibliotheca Sacra, Oberlin.

Wu, X. and L. Wang (1985). Chronology in Chinese palaeoanthropology. In *Palaeoanthropology and Palaeolithic archaeology in the People's Republic of China,* edited by R. Wu and J. W. Olsen, pp. 29–51. New York: Academic Press.

Wyckoff, D. (1993). Gravel sources of knappable Alibates silicified dolomite. *Geoarchaeology* 8:35–58.

Yesner, D. (2001). Human dispersal into interior Alaska: antecedent conditions, mode of colonization, and adaptations. *Quaternary Science Reviews* 20:315–27.

Youmans, W. J. (1893). The insolence of office. *Popular Science Monthly* 42:841–42.

Young, D. (1998). An osteological analysis of the Paleoindian double burial from Horn Shelter, Number 2. *Central Texas Archaeologist* 11:13–105.

Yu, Z. and H. E. Wright, Jr. (2001). Response of interior North America to abrupt climate oscillations in the North Atlantic region during the last deglaciation. *Earth Science Reviews* 52:333–69.

Zegura, S., T. Karafet, L. Zhivotovsky, and M. Hammer (2004). High-resolution SNPs and microsatellite haplotypes point to a single, recent entry of Native American Y Chromosomes into the Americas. *Molecular Biology and Evolution* 21:164–75.

Zimmerman, L. (1997). Remythologizing the relationship between Indians and archaeologists. In *Native Americans and archaeologists: stepping stones to common ground,* edited by N. Swidler, K. Dongoske, R. Anyon, and A. Downer, pp. 44–56. AltaMira Press, Walnut Creek, CA.

INDEX

Ice Age, end of, 18, 48–50
  landscape learning and, 232f
  population loss and, 326
  rapid, and glacial movement, 28
  reproduction, effects on, 267
  thermohaline circulation and, 51–52
  vegetation responses to, 42, 52–53
climatic stress, as possible motivation for
    migration, 212
clothing, 16, 225, 342
  body lice, as indicators of, 19
  as disease carrier, 323
  insulated, 21, 225
  manufacture and repair of, 303, 313
Clovis and Beyond conference (1999), 127–128
Clovis period
  antiquity of, 92, 254–255
  art, 246–247
  diet, 274–279
  diffusion, of technology, 202
  dispersal of, 5, 242, 255, 257, 280, 283
  geological knowledge, 248
  habitation sites, 246, 252–253, 311
  megafaunal hunting by, 268–271, 273–274
  migration, 201–202, 213
  mobility, plant collecting and, 279
  movement, predictability of, 251
  population densities, 254
  role in megafaunal extinction, supposed,
    255–265
  sites across the continent, 241, 248m
  toolkit, 5, 97, 190, 213, 244–246
Clovis projectile points, 91–92, 237, 244f, 269,
    281
  caches, 251, 310
  fluting, 242–243
  and Folsom projectile points compared, 284,
    295
  impact damage, scarcity of, 271
  multiple uses of, 279
  small-caliber, 295
  tallying by state, 253
  used against bison, 270
  variations in, 244
Clovis site (New Mexico), 4–5, 13, 91, 186, 239,
    268–269
  human occupation, chronology of, 254–255

  initial discovery and fieldwork, 239–241
  mammoth remains, 4–5, 268, 271
  sale of to Eastern New Mexico University,
    376
  well-digging, proposed evidence of, 226, 269
coalescent method, haplogroup age estimation
    using, 365
coastal colonization, 130, 196
Cody knife, 297
cognates, language, 141–142
Colbert projectile points, 308
Colby site (Wyoming), 267, 271, 377
Collins, Michael, 120, 122, 134, 251, 311
colonization, 1, 5, 214, 233–238
  big-game hunting, hypothesized role of, 215
  bold vs. cautious colonizing strategies,
    234–236
  challenges of, xii, 210, 219–229
  coastal, 130, 196
  competing demands of, 235–236
  defined, xv
  different ecosystems, failure to appreciate,
    372
  Norse, attempted, 319–320
  process of, 17
  reproduction, effects on, 233
colonizing group, 132, 196, 207
Columbus, Christopher, xiv, 4, 228, 322, 333,
    349, 382
coming of age celebrations, 302
commodities exchanges, late prehistoric period,
    318
common ancestry, 64, 184, 363
  tracing, 157–158
communal hunts and kills, 289, 298, 302
competitive release, 60
Concannon, Joseph and Michael, 79–80
conflict and warfare, effects of European contact
    on, 336
Conrad, Timothy Abbott, 26, 68
conspiracy against early sites in America,
    accusations of, xii, 74, 106, 383
context, importance in identifying artifacts, 14,
    71, 95, 98–99, 109, 117
continental climate, 38
continental shelf, 3, 34, 35, 133
convergence, adaptive, 183–184, 187, 205–207

backwash of from New World to Old, absence of, 329

and collapse of New World empires, 332, 344, 384

depopulation from, 18, 322, 324–326, 333, 336

domesticated animals, role in evolution of, 327, 331

effects on society, 326, 334, 335

genetic lineage loss and, 204–205, 336–337

language loss and, 139, 195, 335–336

susceptibility of New World populations to, 322, 328, 330–332

vectors for, 329

influenza, 18, 322, 327

information

exchange between dispersed groups, 210–211, 221, 233, 235, 302–303

storage, human capacity for, 222, 227–228, 231

Ingstad, Helge, 319

inheritance

mitochondrial, Y chromosome and autosomal DNA, illustration of, 158f

process described, 155–156

Innuitian ice sheet, 32

insolation, 3, 13, 26–28, 30f–31, 49, 54, 59, 351, 352

instant flight reflex, 331

intelligence testing, as replacement for craniometry, 171

interglacials, 26

International Commission on Stratigraphy, definition of Pleistocene, 29, 350

International Union for Quaternary Research, 29

iridium, 56, 57

Irving, William, 109

Ishman, Scott, 57

isolate, languages, 139, 140, 143, 147, 362

isolates, projectile points, 250, 253–254, 271, 279, 302, 310

isolation

language change, effect on, 140, 362

genetic change, effect on 156, 164, 174, 180, 204, 206–207, 234, 329 338

of New World and Old World populations, 4, 18, 330, 332

isotatic rebound, 50, 130, 194, 353

isotope analysis, 340

isotope, defined, 7, 28

ivory artifacts, scarcity of, 245

"Jackie Robinson Rule," 109, 122

Jamestown colony (Virginia), 219, 332

Jantz, Richard, 175, 176, 181, 203, 367

jasper, use in stone tool production, 247, 249

Jefferson, Thomas, 44, 63–64, 323, 335

Jelderks, Judge John, 180, 339

Jenner, Edward, 323, 381

jet stream, 3, 39, 52, 60

Jinmium site (Australia), 108, 360

Jodry, Margaret, 299

Johnston, W. A., 241

Jones, George, 286, 314, 315

Jones-Miller site (Colorado), 291, 316, 380

Judge, James, 269

jump kills, 289–291

"junk DNA," 157

Justinian pandemic (plague), 327

Kansan glacial stage, 26

Karafet, Tatiana, 163

katabatic winds, 37, 38

Kelly, Robert, 215, 218, 228, 233, 371

Kemp, Brian, 169, 364

Kennett, James, 55–56

Kennewick remains (Washington), x, 15, 177–180, 337–339

"Caucasoids," 15, 178, 184, 186

legal battle over, xi, 19, 179–180, 205, 337–341, 383

and Native American Grave Protection and Repatriation Act (NAGPRA), 19, 177, 179, 337–341

Kidder, A. V., 82, 86, 89, 90, 124, 171, 325, 337

kills, animal. See also bison, mammoth, mastodon

with adjoining camps, 299

jump, 289–291

large, communal, 289, 297

seasonal trends, 297

of small game, 302

small-scale, 289

targeting cow-calf herds, 297

kill sites, 13, 277, 289

Clovis, 259–261, 263–264, 269–271, 273

Shimek, Bohumil, 80

Short Tandem Repeats (STRs). *See* mutations

showing off, 218, 243, 273–274. *See also* costly signaling

shrines, 316

Siberia
mammals, 34, 273
peopling of, 19, 21–22, 132, 151, 162, 170, 189, 193, 198, 212

sickle sheen, 279

Signor-Lipps effect, 264

Silberbauer, George, 215, 231, 273

Simms, Steven, 92

Simpson, Ruth "Dee," 98

Simpson projectile points, 285

simulation of colonization process, 196–197

Single Nucleotide Polymorphism (SNP). *See* mutations

Sinodonts, 149, 151, 154, 155*t*, 191

Sino-Tibetan language family, 146

skeletal remains, 15, 19, 170–176, 338, 339, 340.
See also Kennewick,
Buhl skeleton, 275, 316
Late Pleistocene and Early Holocene from North America, 172*t*–173*t*
ownership of (*see* Native American Graves Protection and Repatriation Act (NAGPRA))
"Red Lady" of Paviland, 65–66
Taber child, 102–103, 107

skinning knives, 297

"Skraelings," 204, 319, 320, 380

slavery, 333, 336

Sloan projectile points, 310, 317

Sloan site (Missouri), 317–318

Slobodin, Sergei, 188, 189

smallpox (*Variola major*), 195, 322, 331, 381
Antoine pandemic, 327
Aztec victims of, 343*f*, 344
earliest records of, 327
Native Americans, epidemics, 18, 195, 322–324, 326, 328, 329, 336
vaccine, 323, 381

Smilodon fatalis (saber-toothed cat), 48, 259

Smith, Bruce, 78

Smith, Captain John, 323

snow lines, 43

social disruption. *See* infectious disease, effects on society

social isolation, role in language differentiation, 140

social networks, role in adaptation, 219, 235, 237, 253, 286

social rank, 317

solar radiation. *See* insolation

Solutrean culture, 190
supposed colonization of America, 15, 185–188, 191, 368

Somme Valley sites (France), 66, 67

sound correspondences. *See* language(s)

source populations. *See* genetic(s)

specialized hunting, 92, 218, 255

Spirit Cave (Nevada), 315

springs, 268, 269, 275, 311, 314

Stafford, Michael, 317

Stafford, Thomas, 254, 265

staging areas, migration, 254

Stalker, Archie, 102–103

stalking, 215

stampede, as hunting technique, 261, 288, 291, 298, 379

Stanfield-Worley rockshelter (Alabama), 312

Stanford, Dennis, 126, 186–188, 368, 380

Steadman, David, 258

Steele, Gentry, 174, 176, 279

Steensen, Niels. *See* Steno, Nicolaus

Steinbeck, John, 281

Steno, Nicholaus, 100, 106, 111, 284

Steno's law of stratigraphic superposition, 101–102. *See also* stratigraphy

Stewart, T. Dale, 329

Stiger, Mark, 300

stock market, 1929 collapse of, 281

stone
acquisition of, 131, 134, 225–226, 247–250, 252, 285–286, 309
caches, 236, 250–252, 271
conveyance zones, 286
engravings on, 187, 246–247, 304*f*
heat-treated stone, 249
outcrops, 225, 226*f*, 248, 250, 285
tool technology, 20, 21, 70, 88, 97, 116, 193, 245, 249, 315
type of, as indicator of mobility, 236, 247–249